Linear Algebra and Matrix Theory

Linear Algebra and Matrix Theory

second edition

Evar D. Nering
Professor of Mathematics
Arizona State University

JOHN WILEY & SONS

New York • Chichester • Brisbane • Toronto

Library of Congress Catalog Card Number: 76-91646

SBN 471 63178 7

Printed in the United States of America

12 13 14 15 16 17 18 19 20

Preface to first edition

The underlying spirit of this treatment of the theory of matrices is that of a concept and its representation. For example, the abstract concept of an integer is the same for all cultures and presumably should be the same to a being from another planet. But the various symbols we write down and carelessly refer to as "numbers" are really only representations of the abstract numbers. These representations should be called "numerals" and we should not confuse a numeral with the number it represents. Numerals of different types are the inventions of various cultures and individuals, and the superiority of one system of numerals over another lies in the ease with which they can be manipulated and the insight they give us into the nature of the numbers they represent.

We happen to use numerals to represent things other than numbers. For example, we put numerals (not numbers) on the backs of football players to represent and identify them. This does not attribute to the football players any of the properties of the corresponding numbers, and the usual operations of arithmetic have no meaning in this context. No one would think of adding the halfback, 20, to the fullback, 40, to obtain the guard, 60.

Matrices are used to represent various concepts with a wide variety of different properties. To cover these possibilities a number of different manipulations with matrices are introduced. In each situation the appropriate manipulations that should be performed on a matrix or a set of matrices depend critically on the concepts represented. The student who learns the formalisms of matrix "arithmetic" without learning the underlying concepts is in serious danger of performing operations which have no meaning for the problem at hand.

The formal manipulations with matrices are relatively easy to learn, and few students have any difficulty performing the operations accurately, if somewhat slowly. If a course in matrix theory, however, places too much emphasis on these formal steps, the student acquires an ill-founded self-assurance. If he makes an error exactly analogous to adding halfbacks to fullbacks to obtain guards, he usually considers this to be a minor error (since each step was performed accurately) and does not appreciate how serious a blunder he has made.

In even the simplest problems matrices can appear as representing several different types of concepts. For example, it is typical to have a problem in which some matrices represent vectors, some represent linear transformations, and others represent changes of bases. This alone should make it clear that an understanding of the things represented is essential to a meaningful manipulation of the representing symbols. In courses in vector analysis and differential geometry many students have difficulty with the concepts of covariant vectors and contravariant vectors. The troubles stem almost entirely from a confusion between the representing symbols and the things represented. As long as a student thinks of n-tuples (the representing symbols) as being identical with vectors (the things represented) he must think there are two different "kinds" of vectors all jumbled together and he sees no way to distinguish between them. There are, in fact, not two "kinds" of vectors. There are two entirely distinct vector spaces; their representations happen to look alike, but they have to be manipulated differently.

Although the major emphasis in this treatment of matrix theory is on concepts and proofs that some may consider abstract, full attention is given to practical computational procedures. In different problems different patterns of steps must be used, because the information desired is not the same in all. Fortunately, although the patterns change, there are only a few different types of steps. The computational techniques chosen here are not only simple and effective, but require a variety of steps that is particularly small. Because these same steps occur often, ample practice is provided and the student should be able to obtain more than adequate confidence in his skill.

A single pattern of discussion recurs regularly throughout the book. First a concept is introduced. A coordinate system is chosen so that the concept can be represented by n-tuples or matrices. It is shown how the representation of the concept must be changed if the coordinate system is chosen in a different way. Finally, an effective computational procedure is described by which a particular coordinate system can be chosen so that the representing n-tuples or matrices are either simple or yield significant information. In this way computational skill is supported by an understanding of the reasons for the various procedures and why they differ. Lack of this understanding is the most serious single source of difficulty for many students of matrix theory.

The material contained in the first five chapters is intended for use in a one-semester, or one-quarter, course in the theory of matrices. The topics have been carefully selected and very few omissions should be made. The sections recommended for omission are designated with asterisks. The part of Section 4 of Chapter III following the proof of the Hamilton-Cayley Theorem can also be omitted without harm. No other omission in the first three chapters is recommended. For a one-quarter course the following additional omissions are recommended: Chapter IV, Sections 4 and 5; Chapter V, Sections 3, 4, 5, and 9.

Chapter V contains two parallel developments leading to roughly analogous results. One, through Sections 1, 6, 7, 8, 10, 11, and 12, is rather formal and is based on the properties of matrices; the other, through Sections 1, 3, 4, 5, 9, and 11, is more theoretical and is based on the properties of linear functionals. This latter material is, in turn, based on the material in Sections 1–5 of Chapter IV. For this reason these omissions should not be made in a random manner. Omissions can be made to accommodate a one-quarter course, or for other purposes, by carrying one line of development to its completion and curtailing the other.

The exercises are an important part of this book. The computational exercises in particular should be worked. They are intended to illustrate the theory, and their cumulative effect is to provide assurance and understanding. Numbers have been chosen so that arithmetic complications are minimized. A student who understands what he is doing should not find them lengthy or tedious. The theoretical problems are more difficult. Frequently they are arranged in sequence so that one exercise leads naturally into the next. Sometimes an exercise has been inserted mainly to provide a suggestive context for the next exercise. In other places large steps have been broken down into a sequence of smaller steps, and these steps arranged in a sequence of exercises. For this reason a student may find it easier to work ten exercises in a row than to work five of them by taking every other one. These exercises are numerous and the student should not let himself become bogged down by them. It is more important to keep on with the pace of development and to obtain a picture of the subject as a whole than it is to work every exercise.

The last chapter on selected applications of linear algebra contains a lot of material in rather compact form. For a student who has been through the first five chapters it should not be difficult, but it is not easy reading for someone whose previous experience with linear algebra is in a substantially different form. Many with experience in the applications of mathematics will be unfamiliar with the emphasis on abstract and general methods. I have had enough experience in full-time industrial work and as a consultant to know that these abstract ideas are fully as practical as any concrete methods, and anyone who takes the trouble to familiarize himself with these ideas will find that this is true. In all engineering problems and most scientific problems it is necessary to deal with particular cases and with particular numbers. This is usually, however, only the last phase of the work. Initially, the problem must be dealt with in some generality until some rather important decisions can be made (and this is by far the more interesting and creative part of the work). At this stage methods which lead to understanding are to be preferred to methods which obscure understanding in unnecessary formalisms.

This book is frankly a textbook and not a treatise, and I have not attempted to give credits and references at each point. The material is a mosaic that draws on many sources. I must, however, acknowledge my tremendous debt to Professor Emil Artin, who was my principal source of mathematical education and inspiration from my junior year through my Ph.D. The viewpoint presented here, that matrices are mere representations of things more basic, is as close to his viewpoint as it is possible for an admiring student to come in adopting the ideas of his teacher. During my student days the book presenting this material in a form most in harmony with this viewpoint was Paul Halmos' elegant treatment in *Finite Dimensional Vector Spaces*, the first edition. I was deeply impressed by this book, and its influence on the organization of my text is evident.

EVAR D. NERING

Tempe, Arizona
January, 1963

Preface to second edition

This edition differs from the first primarily in the addition of new material. Although there are significant changes, essentially the first edition remains intact and embedded in this book. In effect, the material carried over from the first edition does not depend logically on the new material. Therefore, this first edition material can be used independently for a one-semester or one-quarter course. For such a one-semester course, the first edition usually required a number of omissions as indicated in its preface. This omitted material, together with the added material in the second edition, is suitable for the second semester of a two-semester course.

The concept of duality receives considerably expanded treatment in this second edition. Because of the aesthetic beauty of duality, it has long been a favorite topic in abstract mathematics. I am convinced that a thorough understanding of this concept also should be a standard tool of the applied mathematician and others who wish to apply mathematics. Several sections of the chapter concerning applications indicate how duality can be used. For example, in Section 3 of Chapter V, the inner product can be used to avoid introducing the concept of duality. This procedure is often followed in elementary treatments of a variety of subjects because it permits doing some things with a minimum of mathematical preparation. However, the cost in loss of clarity is a heavy price to pay to avoid linear functionals. Using the inner product to represent linear functionals in the vector space overlays two different structures on the same space. This confuses concepts that are similar but essentially different. The lack of understanding which usually accompanies this shortcut makes facing a new context an unsure undertaking. I think that the use of the inner product to allow the cheap and early introduction of some manipulative techniques should be avoided. It is far better to face the necessity of introducing linear functionals at the earliest opportunity.

I have made a number of changes aimed at clarification and greater precision. I am not an advocate of rigor for rigor's sake since it usually adds nothing to understanding and is almost always dull. However, rigor is not the same as precision, and algebra is a mathematical subject capable of both precision and beauty. However, tradition has allowed several situations to arise in which different words are used as synonyms, and all are applied indiscriminately to concepts that are not quite identical. For this reason, I have chosen to use "eigenvalue" and "characteristic value" to denote non-identical concepts; these terms are not synonyms in this text. Similarly, I have drawn a distinction between dual transformations and adjoint transformations.

Many people were kind enough to give me constructive comments and observations resulting from their use of the first edition. All these comments were seriously considered, and many resulted in the changes made in this edition. In addition, Chandler Davis (Toronto), John H. Smith (Boston College), John V. Ryff (University of Washington), and Peter R. Christopher (Worcester Polytechnic) went through the first edition or a preliminary version of the second edition in detail. Their advice and observations were particularly valuable to me. To each and every one who helped me with this second edition, I want to express my debt and appreciation.

Evar D. Nering

Tempe, Arizona
September, 1969

Contents

xi

Introduction

Many of the most important applications of mathematics involve what are known as linear methods. The idea of what is meant by a linear method applied to a linear problem or a linear system is so important that it deserves attention in its own right. We try to describe intuitively what is meant by a linear system and then give some idea of the reasons for the importance of the concept.

As an example, consider an electrical network. When the network receives an input, an output from the network results. As is customary, we can consider combining two inputs by adding them and then putting their sum through the system. This sum input will also produce an output. If the output of the sum is the sum of the outputs the system is said to be additive. We can also modify an input by changing its magnitude, by multiplying the input by a constant factor. If the resulting output is also multiplied by the same factor the system is said to be homogeneous. If the system is both additive and homogeneous it is said to be linear.

The simplification in the analysis of a system that results from the knowledge, or the assumption, that the system is linear is enormous. If we know the outputs for a collection of different inputs, we know the outputs for all inputs that can be obtained by combining these inputs in various ways. Suppose, for example, that we are considering all inputs that are periodic functions of time with a given period. The theory of Fourier series tells us that, under reasonable restrictions, these periodic functions can be represented as sums of simple sine functions. Thus in analyzing the response of a linear system to a periodic input it is sufficient to determine the response when the input is a simple sine function.

So many of the problems that we encounter are assumed to be linear problems and so many of the mathematical techniques developed are

1

inherently linear that a catalog of the possibilities would be a lengthy undertaking. Potential theory, the theory of heat, and the theory of small vibrations of mechanical systems are examples of linear theories. In fact, it is not easy to find brilliantly successful applications of mathematics to non-linear problems. In many applications the system is assumed to be linear even though it is not. For example, the differential equation for a simple pendulum is not linear since the restoring force is proportional to the sine of the displacement angle. We usually replace the sine of the angle by the angle in radians to obtain a linear differential equation. For small angles this is a good approximation, but the real justification is that linear methods are available and easily applied.

In mathematics itself the operations of differentiation and integration are linear. The linear differential equations studied in elementary courses are linear in the sense intended here. In this case the unknown function is the input and the differential operator is the system. Any physical problem that is describable by a linear differential equation, or system of linear differential equations, is also linear.

Matrix theory, vector analysis, Fourier series, Fourier transforms, and Laplace transforms are examples of mathematical techniques which are particularly suited for handling linear problems. In order for the linear theory to apply to the linear problem it is necessary that what we have called "inputs and outputs" and "linear systems" be representable within the theory. In Chapter I we introduce the concept of vector space. The laws of combination which will be defined for vector spaces are intended to make precise the meaning of our vague statement, "combining these inputs in various ways." Generally, one vector space will be used for the set of inputs and one vector space will be used for the set of outputs. We also need something to represent the "linear system," and for this purpose we introduce the concept of linear transformation in Chapter II.

The next step is to introduce a practical method for performing the needed calculations with vectors and linear transformations. We restrict ourselves to the case in which the vector spaces are finite dimensional. Here it is appropriate to represent vectors by n-tuples and to represent linear transformations by matrices. These representations are also introduced in Chapters I and II.

Where the vector spaces are infinite dimensional other representations are required. In some cases the vectors may be represented by infinite sequences, or Fourier series, or Fourier transforms, or Laplace transforms. For example, it is now common practice in electrical engineering to represent inputs and outputs by Laplace transforms and to represent linear systems by still other Laplace transforms called transfer functions.

The point is that the concepts of vector spaces and linear transformations are common to all linear methods while matrix theory applies to only those linear problems that are finite dimensional. Thus it is of practical value to discuss vector spaces and linear transformations as much as possible before introducing the formalisms of n-tuples and matrices. And, generally, proofs that can be given without recourse to n-tuples and matrices will be shorter, simpler, and clearer.

The correspondences that can be set up between vectors and the n-tuples which represent them, and between linear transformations and the matrices which represent them, are not unique. Therefore, we have to study the totality of all possible ways to represent vectors and linear transformations and the relations between these different representations. Each possible correspondence is called a coordinization of the vector space, and the process of changing from one correspondence to another is called a change of coordinates.

Any property of the vector space or linear transformation which is independent of any particular coordinatization is called an invariant or geometric property. We are primarily interested in those properties of vectors and linear transformations which are invariant, and, if we use a coordinatization to establish such a property, we are faced with the problem of showing that the conclusion does not depend on the particular coordinatization being used. This is an additional reason for preferring proofs which do not make use of any coordinatization.

On the other hand, if a property is known to be invariant, we are free to choose any coordinate system we wish. In such a case it is desirable and advantageous to select a coordinate system in which the problem we wish to handle is particularly simple, or in which the properties we wish to establish are clearly revealed. Chapter III and V are devoted to methods for selecting these advantageous coordinate systems.

In Chapter IV we introduce ideas which allow us to define the concept of distance in our vector spaces. This accounts for the principal differences between the discussions in Chapters III and V. In Chapter III there is no restriction on the coordinate systems which are permitted. In Chapter V the only coordinate systems permitted are "Cartesian"; that is, those in which the theorem of Pythagoras holds. This additional restriction in permissible coordinate systems means that it is more difficult to find advantageous coordinate systems.

In addition to allowing us to introduce the concept of distance the material in Chapter IV is of interest in its own right. There we study linear forms, bilinear forms, and quadratic forms. They have application to a number of important problems in physics, chemistry, and engineering. Here too, coordinate systems are introduced to allow specific calculations, but proofs given without reference to any coordinate systems are preferred.

Historically, the term "linear algebra" was originally applied to the study of linear equations, bilinear and quadratic forms, and matrices, and their changes under a change of variables. With the more recent studies of Hilbert spaces and other infinite dimensional vector spaces this approach has proved to be inadequate. New techniques have been developed which depend less on the choice or introduction of a coordinate system and not at all upon the use of matrices. Fortunately, in most cases these techniques are simpler than the older formalisms, and they are invariably clearer and more intuitive.

These newer techniques have long been known to the working mathematician, but until very recently a curious inertia has kept them out of books on linear algebra at the introductory level.

These newer techniques are admittedly more abstract than the older formalisms, but they are not more difficult. Also, we should not identify the word "concrete" with the word "useful." Linear algebra in this more abstract form is just as useful as in the more concrete form, and in most cases it is easier to see how it should be applied. A problem must be understood, formulated in mathematical terms, and analyzed before any meaningful computation is possible. If numerical results are required, a computational procedure must be devised to give the results with sufficient accuracy and reasonable efficiency. All the steps to the point where numerical results are considered are best carried out symbolically. Even though the notation and terminology of matrix theory is well suited for computation, it is not necessarily the best notation for the preliminary work.

It is a curious fact that if we look at the work of an engineer applying matrix theory we will seldom see any matrices at all. There will be symbols standing for matrices, and these symbols will be carried through many steps in reasoning and manipulation. Only occasionally or at the end will any matrix be written out in full. This is so because the computational aspects of matrices are burdensome and unnecessary during the early phases of work on a problem. All we need is an algebra of rules for manipulating with them. During this phase of the work it is better to use some concept closer to the concept in the field of application and introduce matrices at the point where practical calculations are needed.

An additional advantage in our studying linear algebra in its invariant form is that there are important applications of linear algebra where the underlying vector spaces are infinite dimensional. In these cases matrix theory must be supplanted by other techniques. A case in point is quantum mechanics which requires the use of Hilbert spaces. The exposition of linear algebra given in this book provides an easy introduction to the study of such spaces.

In addition to our concern with the beauty and logic of linear algebra in this form we are equally concerned with its utility. Although some hints of the applicability of linear algebra are given along with its development, Chapter VI is devoted to a discussion of some of the more interesting and representative applications.

I | Vector spaces

In this chapter we introduce the concepts of a vector space and a basis for that vector space. We assume that there is at least one basis with a finite number of elements, and this assumption enables us to prove that the vector space has a vast variety of different bases but that they all have the same number of elements. This common number is called the dimension of the vector space.

For each choice of a basis there is a one-to-one correspondence between the elements of the vector space and a set of objects we shall call *n*-tuples. A different choice for a basis will lead to a different correspondence between the vectors and the *n*-tuples. We regard the vectors as the fundamental objects under consideration and the *n*-tuples a representations of the vectors. Thus, how a particular vector is represented depends on the choice of the basis, and these representations are non-invariant. We call the *n*-tuple the coordinates of the vector it represents; each basis determines a coordinate system.

We then introduce the concept of subspace of a vector space and develop the algebra of subspaces. Under the assumption that the vector space is finite dimensional, we prove that each subspace has a basis and that for each basis of the subspace there is a basis of the vector space which includes the basis of the subspace as a subset.

1 | Definitions

To deal with the concepts that are introduced we adopt some notational conventions that are commonly used. We usually use sans-serif italic letter to denote sets.

$\alpha \in S$ means α is an *element* of the set S.

$\alpha \notin S$ means α is *not an element* of the set S.

$S \subset T$ means S is a *subset* of the set T.

$S \cap T$ denotes the *intersection* of the sets S and T, the set of elements in both S and T.

$S \cup T$ denotes the *union* of the sets S and T, the set of elements in S or T.

$T - S$ denotes the set of elements in T but not in S. In case T is the set of all objects under consideration, we shall call $T - S$ the *complement* of S and denote it by CS.

$S_\mu : \mu \in M$ denotes a collection of sets indexed so that one set S_μ is specified for each element $\mu \in M$. M is called the *index set*.

$\cap_{\mu \in M} S_\mu$ denotes the intersection of all sets $S_\mu : \mu \in M$.

$\cup_{\mu \in M} S_\mu$ denotes the union of all sets $S_\mu : \mu \in M$.

$\emptyset$ denotes the set with no elements, the *empty set*.

A set will often be specified by listing the elements in the set or by giving a property which characterizes the elements of the set. In such cases we use braces: $\{\alpha, \beta\}$ is the set containing just the elements α and β, $\{\alpha \mid P\}$ is the set of all α with property P, $\{\alpha_\mu \mid \mu \in M\}$ denotes the set of all α_μ corresponding to μ in the index set M. We have such frequent use for the set of all integers or a subset of the set of all integers as an index set that we adopt a special convention for these cases. $\{\alpha_i\}$ denotes a set of elements indexed by a subset of the set of integers. Usually the same index set is used over and over. In such cases it is not necessary to repeat the specifications of the index set and often designation of the index set will be omitted. Where clarity requires it, the index set will be specified. We are careful to distinguish between the set $\{\alpha_i\}$ and an element α_i of that set.

Definition. By a *field* F we mean a non-empty set of elements with two laws of combination, which we call addition and multiplication, satisfying the following conditions:

*F*1. To every pair of elements a, $b \in F$ there is associated a unique element, called their sum, which we denote by $a + b$.

*F*2. Addition is associative; $(a + b) + c = a + (b + c)$.

*F*3. There exists an element, which we denote by 0, such that $a + 0 = a$ for all $a \in F$.

*F*4. For each $a \in F$ there exists an element, which we denote by $-a$, such that $a + (-a) = 0$. Following usual practice we write $b + (-a) = b - a$.

*F*5. Addition is commutative; $a + b = b + a$.

*F*6. To every pair of element a, $b \in F$ there is associated a unique element, called their product, which we denote by ab, or $a \cdot b$.

*F*7. Multiplication is associative; $(ab)c = a(bc)$.

*F*8. There exists an element different from 0, which we denote by 1, such that $a \cdot 1 = a$ for all $a \in F$.

*F*9. For each $a \in F$, $a \neq 0$, there exists an element which we denote by a^{-1}, such that $a \cdot a^{-1} = 1$.

*F*10. Multiplication is commutative: $ab = ba$.

*F*11. Multiplication is distributive with respect to addition:

$$(a + b)c = ac + bc.$$

The elements of *F* are called *scalars*, and will generally be denoted by lower case Latin italic letters.

The rational numbers, real numbers, and complex numbers are familiar and important examples of fields, but they do not exhaust the possibilities. As a less familiar example, consider the set $\{0, 1\}$ where addition is defined by the rules: $0 + 0 = 1 + 1 = 0, 0 + 1 = 1$; and multiplication is defined by the rules: $0 \cdot 0 = 0 \cdot 1 = 0, 1 \cdot 1 = 1$. This field has but two elements, and there are other fields with finitely many elements.

We do not develop the various properties of abstract fields and we are not concerned with any specific field other than the rational numbers, the real numbers, and the complex numbers. We find it convenient and desirable at the moment to leave the exact nature of the field of scalars unspecified because much of the theory of vector spaces and matrices is valid for arbitrary fields.

The student unacquainted with the theory of abstract fields will not be handicapped for it will be sufficient to think of *F* as being one of the familiar fields. All that matters is that we can perform the operations of addition and subtraction, multiplication and division, in the usual way. Later we have to restrict *F* to either the field of real numbers or the field of complex numbers in order to obtain certain classical results; but we postpone that moment as long as we can. At another point we have to make a very mild assumption, that is, $1 + 1 \neq 0$, a condition that happens to be false in the example given above. The student interested mainly in the properties of matrices with real or complex coefficients should consider this to be no restriction.

Definition. A *vector space V* over *F* is a non-empty set of elements, called *vectors*, with two laws of combination, called *vector addition* (or *addition*) and *scalar multiplication*, satisfying the following conditions:

*A*1. To every pair of vectors α, $\beta \in V$ there is associated a unique vector in *V* called their *sum*, which we denote by $\alpha + \beta$.

*A*2. Addition is associative; $(\alpha + \beta) + \gamma = \alpha + (\beta + \gamma)$.

*A*3. There exists a vector, which we denote by 0, such that $\alpha + 0 = \alpha$ for all $\alpha \in V$.

*A*4. For each $\alpha \in V$ there exists an element, which we denote by $-\alpha$, such that $\alpha + (-\alpha) = 0$.

*A*5. Addition is commutative; $\alpha + \beta = \beta + \alpha$.

*B*1. To every scalar $a \in F$ and vector $\alpha \in V$, there is associated a unique vector, called the *product* of a and α, which we denote by $a\alpha$.

*B*2. Scalar multiplication is associative: $a(b\alpha) = (ab)\alpha$.

*B*3. Scalar multiplication is distributive with respect to vector addition; $a(\alpha + \beta) = a\alpha + a\beta$.

*B*4. Scalar multiplication is distributive with respect to scalar addition; $(a + b)\alpha = a\alpha + b\alpha$.

*B*5. $1 \cdot \alpha = \alpha$ (where $1 \in F$).

We generally use lower case Greek letters to denote vectors. An exception is the zero vector of *A*3. From a logical point of view we should not use the same symbol "0 ' for both the zero scalar and the zero vector, but this practice is rooted in a long tradition and it is not as confusing as it may seem at first.

The vector space axioms concerning addition alone have already appeared in the definition of a field. A set of elements satisfying the first four axioms is called a *group*. If the set of elements also satisfies *A*5 it is called a *commutative* group or *abelian* group. Thus both fields and vector spaces are abelian groups under addition. The theory of groups is well developed and our subsequent discussion would be greatly simplified if we were to assume a prior knowledge of the theory of groups. We do not assume a prior knowledge of the theory of groups; therefore, we have to develop some of their elementary properties as we go along, although we do not stop to point out that what was proved is properly a part of group theory. Except for specific applications in Chapter VI we do no more than use the term "group" to denote a set of elements satisfying these conditions.

First, we give some examples of vector spaces. Any notation other than "*F*" for a field and "*V*" for a vector space is used consistently in the same way throughout the rest of the book, and these examples serve as definitions for these notations:

(1) Let F be any field and let $V = P$ be the set of all polynomials in an indeterminate x with coefficients in F. Vector addition is defined to be the ordinary addition of polynomials, and multiplication is defined to be the ordinary multiplication of a polynomial by an element of F.

(2) For any positive integer n, let P_n be the set of all polynomials in x with coefficients in F of degree $\leq n - 1$, together with the zero polynomial. The operations are defined as in Example (1).

(3) Let $F = R$, the field of real numbers, and take V to be the set of all real-valued functions of a real variable. If f and g are functions we define vector addition and scalar multiplication by the rules

$$(f + g)(x) = f(x) + g(x),$$
$$(af)(x) = a[f(x)].$$

$$(1.1)$$

(4) Let $F = R$, and let V be the set of continuous real-valued functions of a real variable. The operations are defined as in Example (3). The point of this example is that it requires a theorem to show that $A1$ and $B1$ are satisfied.

(5) Let $F = R$, and let V be the set of real-valued functions defined on the interval $[0, 1]$ and integrable over that interval. The operations are defined as in Example (3). Again, the main point is to show that $A1$ and $B1$ are satisfied.

(6) Let $F = R$, and let V be the set of all real-valued functions of a real variable differentiable at least m times (m a positive integer). The operations are defined as in Example (3).

(7) Let $F = R$, and let V be the set of all real-valued functions differentiable at least twice and satisfying the differential equation $\dfrac{d^2y}{dx^2} + y = 0$.

(8) Let $F = R$, and let $V = R^n$ be the set of all real ordered n-tuples, $\alpha = (a_1, a_2, \ldots, a_n)$ with $a_i \in F$. Vector addition and scalar multiplication are defined by the rules

$$(a_1, \ldots, a_n) + (b_1, \ldots, b_n) = (a_1 + b_1, \ldots, a_n + b_n),$$
$$a(a_1, \ldots, a_n) = (aa_1, \ldots, aa_n). \tag{1.2}$$

We call this vector space the *n-dimensional real coordinate space* or the *real affine n-space*. (The name "Euclidean *n*-space" is sometimes used, but that term should be reserved for an affine *n*-space in which distance is defined.)

(9) Let F^n be the set of all n-tuples of elements of F. Vector addition and scalar multiplication are defined by the rules (1.2). We call this vector space an *n-dimensional coordinate space*.

An immediate consequence of the axioms defining a vector space is that the zero vector, whose existence is asserted in $A3$, and the negative vector, whose existence is asserted in $A4$, are unique. Specifically, suppose 0 satisfies $A3$ for *all* vectors in V and that for *some* $\alpha \in V$ there is a $0'$ satisfying the condition $\alpha + 0 = \alpha + 0' = \alpha$. Then $0' = 0' + 0 = 0' + (\alpha + (-\alpha)) = (0' + \alpha) + (-\alpha) = (\alpha + 0') + (-\alpha) = \alpha + (-\alpha) = 0$. Notice that we have proved not merely that the zero vector satisfying $A3$ for *all* α is unique; we have proved that a vector satisfying the condition of $A3$ for *some* α must be the zero vector, which is a much stronger statement.

Also, suppose that to a given α there were two negatives, $(-\alpha)$ and $(-\alpha)'$, satisfying the conditions of $A4$. Then $(-\alpha)' = (-\alpha)' + 0 = (-\alpha)' + \alpha + (-\alpha) = (-\alpha) + \alpha + (-\alpha)' = (-\alpha) + 0 = (-\alpha)$. Both these demonstrations used the commutative law, $A5$. Use of this axiom could have been avoided, but the necessary argument would then have been somewhat longer.

Uniqueness enables us to prove that $0\alpha = 0$. (Here is an example of the seemingly ambiguous use of the symbol "0." The "0" on the left side is a scalar while that on the right is a vector. However, no other interpretation could be given the symbols and it proves convenient to conform to the convention rather than introduce some other symbol for the zero vector.) For each $\alpha \in V$, $\alpha = 1 \cdot \alpha = (1 + 0)\alpha = 1 \cdot \alpha + 0 \cdot \alpha = \alpha + 0 \cdot \alpha$. Thus $0 \cdot \alpha = 0$. In a similar manner we can show that $(-1)\alpha = -\alpha$. $\alpha + (-1)\alpha = (1 - 1)\alpha = 0 \cdot \alpha = 0$. Since the negative vector is unique we see that $(-1)\alpha = -\alpha$. It also follows similarly that $a \cdot 0 = 0$.

EXERCISES

1 to 4. What theorems must be proved in each of the Examples (4), (5), (6), and (7) to verify *A*1? To verify *B*1? (These axioms are usually the ones which require most specific verification. For example, if we establish that the vector space described in Example (3) satisfies all the axioms of a vector space, then *A*1 and *B*1 are the only ones that must be verified for Examples (4), (5), (6), and (7). Why?)

5. Let P^+ be the set of polynomials with real coefficients and positive constant term. Is P^+ a vector space? Why?

6. Show that if $a\alpha = 0$ and $a \neq 0$, then $\alpha = 0$. (*Hint:* Use axiom *F*9 for fields.)

7. Show that if $a\alpha = 0$ and $\alpha \neq 0$, then $a = 0$.

8. Show that the ξ such that $\alpha + \xi = \beta$ is (uniquely) $\xi = \beta + (-\alpha)$.

9. Let $\alpha = (2, -5, 0, 1)$ and $\beta = (-3, 3, 1, -1)$ be vectors in the coordinate space R^4. Determine

    (*a*) $\alpha + \beta$.
    (*b*) $\alpha - \beta$.
    (*c*) 3α.
    (*d*) $2\alpha + 3\beta$.

10. Show that any field can be considered to be a vector space over itself.

11. Show that the real numbers can be considered to be a vector space over the rational numbers.

12. Show that the complex numbers can be considered to be a vector space over the real numbers.

13. Prove the uniqueness of the zero vector and the uniqueness of the negative of each vector without using the commutative law, *A*5.

2 | Linear Independence and Linear Dependence

Because of the associative law for vector addition, we can omit the parentheses from expressions like $a_1\alpha_1 + (a_2\alpha_2 + a_3\alpha_3) = (a_1\alpha_1 + a_2\alpha_2) + a_3\alpha_3$ and write them in the simpler form $a_1\alpha_1 + a_2\alpha_2 + a_3\alpha_3 = \sum_{i=1}^{3} a_i\alpha_i$. It is clear that this convention can be extended to a sum of any number of such terms provided that only a finite number of coefficients are different from zero. Thus, whenever we write an expression like $\sum_i a_i \alpha_i$ (in which we do not specify the range of summation), it will be assumed, tacitly if not explicitly, that the expression contains only a finite number of non-zero coefficients.

If $\beta = \sum_i a_i \alpha_i$, we say that β is a *linear combination* of the α_i. We also say that β is *linearly dependent* on the α_i if β can be expressed as a linear combination of the α_i. An expression of the form $\sum_i a_i \alpha_i = 0$ is called a *linear relation* among the α_i. A relation with all $a_i = 0$ is called a *trivial linear relation*; a relation in which at least one coefficient is non-zero is called a *non-trivial linear relation*.

Definition. A set of vectors is said to be *linearly dependent* if there exists a non-trivial linear relation among them. Otherwise, the set is said to be *linearly independent*.

It should be noted that any set of vectors that includes the zero vector is linearly dependent. A set consisting of exactly one non-zero vector is linearly independent. For if $a\alpha = 0$ with $a \neq 0$, then $\alpha = 1 \cdot \alpha = (a^{-1} \cdot a)\alpha = a^{-1}(a\alpha) = a^{-1} \cdot 0 = 0$. Notice also that the empty set is linearly independent.

It is clear that the concept of linear independence of a set would be meaningless if a vector from a set could occur arbitrarily often in a possible relation. If a set of vectors is given, however, by itemizing the vectors in the set it is a definite inconvenience to insist that all the vectors listed be distinct. The burden of counting the number of times a vector can appear in a relation is transferred to the index set. For each index in the index set, we require that a linear relation contain but one term corresponding to that index. Similarly, when we specify a set by itemizing the vectors in the set, we require that one and only one vector be listed for each index in the index set. But we allow the possibility that several indices may be used to identify the same vector. Thus the set $\{\alpha_1, \alpha_2\}$, where $\alpha_1 = \alpha_2$ is linearly dependent, and any set with any vector listed at least twice is linearly dependent. To be precise, the concept of linear independence is a property of *indexed sets* and not a property of sets. In the example given above, the relation $\alpha_1 - \alpha_2 = 0$ involves two terms in the indexed set $\{\alpha_i \mid i \in \{1, 2\}\}$ while the set $\{\alpha_1, \alpha_2\}$ actually contains only one vector. We should refer to the linear dependence of an *indexed* set rather than the linear dependence of a set. The conventional terminology, which we are adopting, is inaccurate. This usage, however, is firmly rooted in tradition and, once understood, it is a convenience and not a source of difficulty. We speak of the linear dependence of a set, but the concept always refers to an indexed set. For a linearly independent indexed set, no vector can be listed twice; so in this case the inaccuracy of referring to a set rather than an indexed set is unimportant.

The concept of linear dependence and independence is used in essentially two ways. (1) If a set $\{\alpha_i\}$ of vectors is known to be linearly dependent, there exists a non-trivial linear relation of the form $\sum_i a_i \alpha_i = 0$. (This relation is not unique, but that is usually incidental.) There is at least one non-zero coefficient; let a_k be non-zero. Then $\alpha_k = \sum_{i \neq k} (-a_k^{-1} a_i)\alpha_i$; that is one of the vectors of the set $\{\alpha_i\}$ is a linear combination of the others. (2) If a set $\{\alpha_i\}$ of vectors is known to be linearly independent and a linear relation $\sum_i a_i \alpha_i = 0$ is obtained, we can conclude that all $a_i = 0$. This seemingly trivial observation is surprisingly useful.

In Example (1) the zero vector is the polynomial with all coefficients equal to zero. Thus the set of monomials $\{1, x, x^2, \ldots\}$ is a linearly independent set. The set $\{1, x, x^2, x^2 + x + 1\}$ is linearly dependent since $1 + x + x^2 - (x^2 + x + 1) = 0$. In P_n of Example (2), any $n + 1$ polynomials form a linearly dependent set.

In R^3 consider the vectors $\{\alpha = (1, 1, 0),\ \beta = (1, 0, 1),\ \gamma = (0, 1, 1),\ \delta = (1, 1, 1)\}$. These four vectors are linearly dependent since $\alpha + \beta + \gamma - 2\delta = 0$, yet any three of these four vectors are linearly independent.

Theorem 2.1. *If α is linearly dependent on $\{\beta_i\}$ and each β_i is linearly dependent on $\{\gamma_j\}$, then α is linearly dependent on $\{\gamma_j\}$.*

PROOF. From $\alpha = \sum_i b_i \beta_i$ and $\beta_i = \sum_j c_{ij} \gamma_j$ it follows that $\alpha = \sum_i b_i (\sum_j c_{ij}\gamma_j) = \sum_j (\sum_i b_i c_{ij})\gamma_j$. $\square$

Theorem 2.2. *A set of non-zero vectors $\{\alpha_1, \alpha_2, \ldots\}$ is linearly dependent if and only if some α_k is a linear combination of the α_j with $j < k$.*

PROOF. Suppose the vectors $\{\alpha_1, \alpha_2, \ldots\}$ are linearly dependent. Then there is a non-trivial linear relation among them; $\sum_i a_i \alpha_i = 0$. Since a positive finite number of coefficients are non-zero, there is a last non-zero coefficient a_k. Furthermore, $k \geq 2$ since $\alpha_1 \neq 0$. Thus $\alpha_k = -a_k^{-1} \sum_{i=1}^{k-1} a_i \alpha_i = \sum_{i=1}^{k-1} (-a_k^{-1} a_i)\alpha_i$.

The converse is obvious. $\square$

For any subset A of V the set of all linear combinations of vectors in A is called the set *spanned* by A, and we denote it by $\langle A \rangle$. We also say that A *spans* $\langle A \rangle$. It is a part of this definition that $A \subset \langle A \rangle$. We also agree that the empty set $\emptyset$ spans the set consisting of the zero vector alone. It is readily apparent that if $A \subset B$, then $\langle A \rangle \subset \langle B \rangle$.

In this notation Theorem 2.1 is equivalent to the statement: If $A \subset \langle B \rangle$ and $B \subset \langle C \rangle$, then $A \subset \langle C \rangle$.

Theorem 2.3. *The set $\{\alpha_i\}$ of non-zero vectors is linearly independent if and only if for each k, $\alpha_k \notin \langle \alpha_1, \ldots, \alpha_{k-1} \rangle$.* (To follow our definitions exactly, the set spanned by $\{\alpha_1, \ldots, \alpha_{k-1}\}$ should be denoted by $\langle \{\alpha_1, \ldots, \alpha_{k-1}\} \rangle$.

We shall use the symbol $\langle \alpha_1, \ldots, \alpha_{k-1} \rangle$ instead since it is simpler and there is no danger of ambiguity.)

PROOF. This is merely Theorem 2.2 in contrapositive form and stated in new notation. □

Theorem 2.4. *If B and C are any subsets such that* $B \subset \langle C \rangle$, *then* $\langle B \rangle \subset \langle C \rangle$.
PROOF. Set $A = \langle B \rangle$ in Theorem 2.1. Then $B \subset \langle C \rangle$ implies that $\langle B \rangle = A \subset \langle C \rangle$. □

Theorem 2.5. *If* $\alpha_k \in A$ *is dependent on the other vectors in* A, *then* $\langle A \rangle = \langle A - \{\alpha_k\} \rangle$.
PROOF. The assumption that α_k is dependent on $A - \{\alpha_k\}$ means that $A \subset \langle A - \{\alpha_k\} \rangle$. It then follows from Theorem 2.4 that $\langle A \rangle \subset \langle A - \{\alpha_k\} \rangle$. The equality follows from the fact that the inclusion in the other direction is evident. □

Theorem 2.6. *For any set* C, $\langle\langle C \rangle\rangle = \langle C \rangle$.
PROOF. Setting $B = \langle C \rangle$ in Theorem 2.4 we obtain $\langle\langle C \rangle\rangle = \langle B \rangle \subset \langle C \rangle$. Again, the inclusion in the other direction is obvious. □

Theorem 2.7. *If a finite set* $A = \{\alpha_1, \ldots, \alpha_n\}$ *spans* V, *then every linearly independent set contains at most* n *elements.*
PROOF. Let $B = \{\beta_1, \beta_2, \ldots\}$ be a linearly independent set. We shall successively replace the α_i by the β_j, obtaining at each step a new n-element set that spans V. Thus, suppose that $A_k = \{\beta_1, \ldots, \beta_k, \alpha_{k+1}, \ldots, \alpha_n\}$ is an n-element set that spans V. (Our starting point, the hypothesis of the theorem, is the case $k = 0$.) Since A_k spans V, β_{k+1} is dependent on A_k. Thus the set $\{\beta_1, \ldots, \beta_k, \beta_{k+1}, \alpha_{k+1}, \ldots, \alpha_n\}$ is linearly dependent. In any non-trivial relation that exists the non-zero coefficients cannot be confined to the β_j, for they are linearly independent. Thus one of the $\alpha_i (i > k)$ is dependent on the others, and after reindexing $\{\alpha_{k+1}, \ldots, \alpha_n\}$ if necessary we may assume that it is α_{k+1}. By Theorem 2.5 the set $A_{k+1} = \{\beta_1, \ldots, \beta_{k+1}, \alpha_{k+2}, \ldots, \alpha_n\}$ also spans V.

If there were more than n elements in B, we would in this manner arrive at the spanning set $A_n = \{\beta_1, \ldots, \beta_n\}$. But then the dependence of β_{n+1} on A_n would contradict the assumed linear independence of B. Thus B contains at most n elements. □

Theorem 2.7 is stated in slightly different forms in various books. The essential feature of the proof is the step-by-step replacement of the vectors in one set by the vectors in the other. The theorem is known as the Steinitz replacement theorem.

EXERCISES

1. In the vector space P of Example (1) let $p_1(x) = x^2 + x + 1$, $p_2(x) = x^2 - x - 2$, $p_3(x) = x^2 + x - 1$, $p_4(x) = x - 1$. Determine whether or not the set $\{p_1(x), p_2(x), p_3(x), p_4(x)\}$ is linearly independent. If the set is linearly dependent, express one element as a linear combination of the others.

2. Determine $\langle\{p_1(x), p_2(x), p_3(x), p_4(x)\}\rangle$, where the $p_i(x)$ are the same polynomials as those defined in Exercise 1. (The set required is infinite, so that we cannot list all its elements. What is required is a description; for example, "all polynomials of a certain degree ór less," "all polynomials with certain kinds of coefficients," etc.)

3. A linearly independent set is said to be *maximal* if it is contained in no larger linearly independent set. In this definition the emphasis is on the concept of set inclusion and not on the number of elements in a set. In particular, the definition allows the possibility that two different maximal linearly independent sets might have different numbers of elements. Find all the maximal linearly independent subsets of the set given in Exercise 1. How many elements are in each of them?

4. Show that no finite set spans P; that is, show that there is no maximal finite linearly independent subset of P. Why are these two statements equivalent?

5. In Example (2) for $n = 4$, find a spanning set for P_4. Find a minimal spanning set. Use Theorem 2.7 to show that no other spanning set has fewer elements.

6. In Example (1) or (2) show that $\{1, x + 1, x^2 + x + 1, x^3 + x^2 + x + 1, x^4 + x^3 + x^2 + x + 1\}$ is a linearly independent set.

7. In Example (1) show that the set of all polynomials divisible by $x - 1$ cannot span P.

8. Determine which of the following set in R^4 are linearly independent over R.
   (a) $\{(1, 1, 0, 1), \quad (1, -1, 1, 1), \quad (2, 2, 1, 2), \quad (0, 1, 0, 0)\}$.
   (b) $\{(1, 0, 0, 1), \quad (0, 1, 1, 0), \quad (1, 0, 1, 0), \quad (0, 1, 0, 1)\}$.
   (c) $\{(1, 0, 0, 1), \quad (0, 1, 0, 1), \quad (0, 0, 1, 1), \quad (1, 1, 1, 1)\}$.

9. Show that $\{e_1 = (1, 0, 0, \ldots, 0), e_2 = (0, 1, 0, \ldots, 0), \ldots, e_n = (0, 0, 0, \ldots, 1)\}$ is linearly independent in F^n over F.

10. In Exercise 11 of Section 1 it was shown that we may consider the real numbers to be a vector space over the rational numbers. Show that $\{1, \sqrt{2}\}$ is a linearly independent set over the rationals. (This is equivalent to showing that $\sqrt{2}$ is irrational.) Using this result show that $\{1, \sqrt{2}, \sqrt{3}\}$ is linearly independent.

11. Show that if one vector of a set is the zero vector, then the set is linearly dependent.

12. Show that if an indexed set of vectors has one vector listed twice, the set is linearly dependent.

13. Show that if a subset of S is linearly dependent, then S is linearly dependent.

14. Show that if a set S is linearly independent, then every subset of S is linearly independent.

15. Show that if the set $A = \{\alpha_1, \ldots, \alpha_n\}$ is linearly independent and $\{\alpha_1, \ldots, \alpha_n, \beta\}$ is linearly dependent, then β is dependent on A.

16. Show that, if each of the vectors $\{\beta_0, \beta_1, \ldots, \beta_n\}$ is a linear combination of the vectors $\{\alpha_1, \ldots, \alpha_n\}$, then $\{\beta_0, \beta_1, \ldots, \beta_n\}$ is linearly dependent.

3 | Bases of Vector Spaces

Definition. A linearly independent set spanning a vector space V is called a *basis* or *base* (the plural is *bases*) of V.

If $A = \{\alpha_1, \alpha_2, \ldots\}$ is a basis of V, by definition an $\alpha \in V$ can be written in the form $\alpha = \sum_i a_i\alpha_i$. The interesting thing about a basis, as distinct from other spanning sets, is that the coefficients are uniquely determined by α. For suppose that we also have $\alpha = \sum_i b_i\alpha_i$. Upon subtraction we get the linear relation $\sum_i (a_i - b_i)\alpha_i = 0$. Since $\{\alpha_i\}$ is a linearly independent set, $a_i - b_i = 0$ and $a_i = b_i$ for each i. A related fact is that a basis is a particularly efficient spanning set, as we shall see.

In Example (1) the vectors $\{\alpha_i = x^i \mid i = 0, 1, \ldots\}$ form a basis. We have already observed that this set is linearly independent, and it clearly spans the space of all polynomials. The space P_n has a basis with a finite number of elements; $\{1, x, x^2, \ldots, x^{n-1}\}$.

The vector spaces in Examples (3), (4), (5), (6), and (7) do not have bases with a finite number of elements.

In Example (8) every R^n has a finite basis consisting of $\{\alpha_i \mid \alpha_i = (\delta_{1i}, \delta_{2i}, \ldots, \delta_{ni})\}$. (Here δ_{ij} is the useful symbol known as the Kronecker delta. By definition $\delta_{ij} = 0$ if $i \neq j$ and $\delta_{ii} = 1$.)

Theorem 3.1. *If a vector space has one basis with a finite number of elements, then all other bases are finite and have the same number of elements.*

PROOF. Let A be a basis with a finite number n of elements, and let B be any other basis. Since A spans V and B is linearly independent, by Theorem 2.7 the number m of element in B must be at most n. This shows that B is finite and $m \leq n$. But then the roles of A and B can be interchanged to obtain the inequality in the other order so that $m = n$. $\square$

A vector space with a finite basis is called a *finite dimensional* vector space, and the number of elements in a basis is called the *dimension* of the space. Theorem 3.1 says that the dimension of a finite dimensional vector space is well defined. The vector space with just one element, the zero vector, has one linearly independent subset, the empty set. The empty set is also a spanning set and is therefore a basis of $\{0\}$. Thus $\{0\}$ has dimension zero. There are very interesting vector spaces with infinite bases; for example, P of Example (1). Moreover, many of the theorems and proofs we give are also valid for infinite dimensional vector spaces. It is not our intention, however, to deal with infinite dimensional vector spaces as such, and whenever we speak of the dimension of a vector space without specifying whether it is finite or infinite dimensional we mean that the dimension is finite.

Among the examples we have discussed so far, each P_n and each R^n is n-dimensional. We have already given at least one basis for each. There are many others. The bases we have given happen to be conventional and convenient choices.

Theorem 3.2. *Any* $n + 1$ *vectors in an n-dimensional vector space are linearly dependent.*

PROOF. Their independence would contradict Theorem 2.7. □

We have already seen that the four vectors $\{\alpha = (1, 1, 0)$, $\beta = (1, 0, 1)$, $\gamma = (0, 1, 1)$, $\delta = (1, 1, 1)\}$ form a linearly dependent set in R^3. Since R^3 is 3-dimensional we see that this must be expected for any set containing at least four vectors from R^3. The next theorem shows that each subset of three is a basis.

Theorem 3.3. *A set of n vectors in an n-dimensional vector space V is a basis if and only if it is linearly independent.*

PROOF. The "only if" is part of the definition of a basis. Let $A = \{\alpha_1, \ldots, \alpha_n\}$ be a linearly independent set and let α be any vector in V. Since $\{\alpha_1, \ldots, \alpha_n, \alpha\}$ contains $n + 1$ elements it must be linearly dependent. Any non-trivial relation that exists must contain α with a non-zero coefficient, for if that coefficient were zero the relation would amount to a relation in A. Thus α is dependent on A. Hence A spans V and is a basis. □

Theorem 3.4. *A set of n vectors in an n-dimensional vector space V is a basis if and only if it spans V.*

PROOF. The "only if" is part of the definition of a basis. If n vectors did span V and were linearly dependent, then (by Theorem 2.5) a proper subset would also span V, contrary to Theorem 2.7. □

We see that a basis is a maximal linearly independent set and a minimal spanning set. This idea is made explicit in the next two theorems.

Theorem 3.5. *In a finite dimensional vector space, every spanning set contains a basis.*

PROOF. Let B be a set spanning V. If $V = \{0\}$, then $\emptyset \subset B$ is a basis of $\{0\}$. If $V \neq \{0\}$, then B must contain at least one non-zero vector α_1. We now search for another vector in B which is not dependent on $\{\alpha_1\}$. We call this vector α_2 and search for another vector in B which is not dependent on the linearly independent set $\{\alpha_1, \alpha_2\}$. We continue in this way as long as we can, but the process must terminate as we cannot find more than n

linearly independent vectors in B. Thus suppose we have obtained the set $A = \{\alpha_1, \ldots, \alpha_m\}$ with the property that every vector in B is linearly dependent on A. Then because of Theorem 2.1 the set A must also span V and it is a basis. $\square$

To drop the assumption that the vector space is n-dimensional would change the complexion of Theorem 3.5 entirely. As it stands the theorem is interesting but minor, and not difficult to prove. Without this assumption the theorem would assert that *every* vector space has a basis since every vector space is spanned by itself. Discussion of such a theorem is beyond the aims of this treatment of the subject of vector spaces.

Theorem 3.6. *In a finite dimensional vector space any linearly independent set of vectors can be extended to a basis.*

PROOF. Let $A = \{\alpha_1, \ldots, \alpha_n\}$ be a basis of V, and let $B = \{\beta_1, \ldots, \beta_m\}$ be a linearly independent set $(m \leq n)$. The set $\{\beta_1, \ldots, \beta_m, \alpha_1, \ldots, \alpha_n\}$ spans V. If this set is linearly dependent (and it surely is if $m > 0$) then some element is a linear combination of the preceding elements (Theorem 2.2). This element cannot be one of the β_i's for then B would be linearly dependent. But then this α_i can be removed to obtain a smaller set spanning V (Theorem 2.5). We continue in this way, discarding elements as long as we have a linearly dependent spanning set. At no stage do we discard one of the β_i's. Since our spanning set is finite this process must terminate with a basis containing B as a subset. $\square$

Theorem 3.6 is one of the most frequently used theorems in the book. It is often used in the following way. A non-zero vector with a certain desired property is selected. Since the vector is non-zero, the set consisting of that vector alone is a linearly independent set. An application of Theorem 3.6 shows that there is a basis containing that vector. This is usually the first step of a proof by induction in which a basis is obtained for which all the vectors in the basis have the desired property.

Let $A = \{\alpha_1, \ldots, \alpha_n\}$ be an arbitrary basis of V, a vector space of dimension n over the field F. Let α be any vector in V. Since A is a spanning set α can be represented as a linear combination of the form $\alpha = \sum_{i=1}^{n} a_i\alpha_i$. Since A is linearly independent this representation is unique, that is, the coefficients a_i are uniquely determined by α (for the given basis A). On the other hand, for each n-tuple $(a_1, \ldots, a_n)$ there is a vector in V of the form $\sum_{i=1}^{n} a_i\alpha_i$. Thus there is a one-to-one correspondence between the vectors in V and the n-tuples $(a_1, \ldots, a_n) \in F^n$.

If $\alpha = \sum_{i=1}^{n} a_i\alpha_i$, the scalar a_i is called the *i-th coordinate* of α, and $a_i\alpha_i$ is called the *i-th component* of α. Generally, coordinates and components depend on the choice of the entire basis and cannot be determined from individual vectors in the basis. Because of the rather simple correspondence between coordinates and components there is a tendency to confuse them and to use both terms for both concepts. Since the intended meaning is usually clear from context, this is seldom a source of difficulty.

If $\alpha = \sum_{i=1}^{n} a_i \alpha_i$ corresponds to the n-tuple $(a_1, \ldots, a_n)$ and $\beta = \sum_{i=1}^{n} b_i \alpha_i$ corresponds to the n-tuple $(b_1, \ldots, b_n)$, then $\alpha + \beta = \sum_{i=1}^{n} (a_i + b_i)\alpha_i$ corresponds to the n-tuple $(a_1 + b_1, \ldots, a_n + b_n)$. Also, $a\alpha = \sum_{i=1}^{n} aa_i \alpha_i$ corresponds to the n-tuple $(aa_1, \ldots, aa_n)$. Thus the definitions of vector addition and scalar multiplication among n-tuples defined in Example (9) correspond exactly to the corresponding operations in V among the vectors which they represent. When two sets of objects can be put into a one-to-one correspondence which preserves all significant relations among their elements, we say the two sets are *isomorphic*; that is, they have the same form. Using this terminology, we can say that every vector space of dimension n over a given field F is isomorphic to the n-dimensional coordinate space F^n. Two sets which are isomorphic differ in details which are not related to their internal structure. They are essentially the same. Furthermore, since two sets isomorphic to a third are isomorphic to each other we see that all n-dimensional vector spaces over the same field of scalars are isomorphic.

The set of n-tuples together with the rules for addition and scalar multiplication forms a vector space in its own right. However, when a basis is chosen in an abstract vector space V the correspondence described above establishes an isomorphism between V and F^n. In this context we consider F^n to be a *representation* of V. Because of the existence of this isomorphism a study of vector spaces could be confined to a study of coordinate spaces. However, the exact nature of the correspondence between V and F^n depends upon the choice of a basis in V. If another basis were chosen in V a correspondence between the $\alpha \in V$ and the n-tuples would exist as before, but the correspondence would be quite different. We choose to regard the vector space V and the vectors in V as the basic concepts and their representation by n-tuples as a tool for computation and convenience. There are two important benefits from this viewpoint. Since we are free to choose the basis we can try to choose a coordinatization for which the computations are particularly simple or for which some fact that we wish to demonstrate is particularly evident. In fact, the choice of a basis and the consequences of a change in basis is the central theme of matrix theory. In addition, this distinction between a vector and its representation removes the confusion that always occurs when we define a vector as an n-tuple and then use another n-tuple to represent it.

Only the most elementary types of calculations can be carried out in the abstract. Elaborate or complicated calculations usually require the introduction of a representing coordinate space. In particular, this will be required extensively in the exercises in this text. But the introduction of coordinates can result in confusions that are difficult to clarify without extensive verbal description or awkward notation. Since we wish to avoid cumbersome notation and keep descriptive material at a minimum in the exercises, it is helpful to spend some time clarifying conventional notations and circumlocutions that will appear in the exercises.

The introduction of a coordinate representation for V involves the selection of a basis $\{\alpha_1, \ldots, \alpha_n\}$ for V. With this choice α_1 is represented by $(1, 0, \ldots, 0)$, α_2 is represented by $(0, 1, 0, \ldots, 0)$, etc. While it may be necessary to find a basis with certain desired properties the basis that is introduced at first is arbitrary and serves only to express whatever problem we face in a form suitable for computation. Accordingly, it is customary to suppress specific reference to the basis given initially. In this context it is customary to speak of "the vector $(a_1, a_2, \ldots, a_n)$" rather than "the vector α whose representation with respect to the given basis $\{\alpha_1, \ldots, \alpha_n\}$ is $(a_1, a_2, \ldots, a_n)$." Such short-cuts may be disgracefully inexact, but they are so common that we must learn how to interpret them.

For example, let V be a two-dimensional vector space over R. Let $A = \{\alpha_1, \alpha_2\}$ be the selected basis. If $\beta_1 = \alpha_1 + \alpha_2$ and $\beta_2 = -\alpha_1 + \alpha_2$, then $B = \{\beta_1, \beta_2\}$ is also a basis of V. With the convention discussed above we would identify α_1 with $(1, 0)$, α_2 with $(0, 1)$, β_1 with $(1, 1)$, and β_2 with $(-1, 1)$. Thus, we would refer to the basis $B = \{(1, 1), (-1, 1)\}$. Since $\alpha_1 = \frac{1}{2}\beta_1 - \frac{1}{2}\beta_2$, α_1 has the representation $(\frac{1}{2}, -\frac{1}{2})$ with respect to the basis B. If we are not careful we can end up by saying that "$(1, 0)$ is represented by $(\frac{1}{2}, -\frac{1}{2})$."

EXERCISES

To show that a given set is a basis by direct appeal to the definition means that we must show the set is linearly independent and that it spans V. In any given situation, however, the task is very much simpler. Since V is n-dimensional a proposed basis must have n elements. Whether this is the case can be told at a glance. In view of Theorems 3.3 and 3.4 if a set has n elements, to show that it is a basis it suffices to show either that it spans V or that it is linearly independent.

1. In R^3 show that $\{(1, 1, 0), (1, 0, 1), (0, 1, 1)\}$ is a basis by showing that it is linearly independent.

2. Show that $\{(1, 1, 0), (1, 0, 1), (0, 1, 1)\}$ is a basis by showing that $\langle(1, 1, 0), (1, 0, 1), (0, 1, 1)\rangle$ contains $(1, 0, 0)$, $(0, 1, 0)$ and $(0, 0, 1)$. Why does this suffice?

3. In R^4 let $A = \{(1, 1, 0, 0), (0, 0, 1, 1), (1, 0, 1, 0), (0, 1, 0, -1)\}$ be a basis (is it?) and let $B = \{(1, 2, -1, 1), (0, 1, 2, -1)\}$ be a linearly independent set (is it?). Extend B to a basis of R^4. (There are many ways to extend B to a basis. It is intended here that the student carry out the steps of the proof of Theorem 3.6 for this particular case.)

4. Find a basis of R^4 containing the vector $(1, 2, 3, 4)$. (This is another even simpler application of the proof of Theorem 3.6. This, however, is one of the most important applications of this theorem, to find a basis containing a particular vector.)

5. Show that a maximal linearly independent set is a basis.

6. Show that a minimal spanning set is a basis.

4 | Subspaces

Definition. A *subspace* W of a vector space V is a non-empty subset of V which is itself a vector space with respect to the operations of addition and scalar multiplication defined in V. In particular, the subspace must be a vector space over the same field F.

The first problem that must be settled is the problem of determining the conditions under which a subset W is in fact a subspace. It should be clear that axioms $A2$, $A5$, $B2$, $B3$, $B4$, and $B5$ need not be checked as they are valid in any subset of V. The most innocuous conditions seem to be $A1$ and $B1$, but it is precisely these conditions that must be checked. If $B1$ holds for a non-empty subset W, there is an $\alpha \in W$ so that $0\alpha = 0 \in W$. Also, for each $\alpha \in W$, $(-1)\alpha = -\alpha \in W$. Thus $A3$ and $A4$ follow from $B1$ in any non-empty subset of a vector space and it is sufficient to check that W is non-empty and closed under addition and scalar multiplication.

The two closure conditions can be combined into one statement: if α, $\beta \in W$ and a, $b \in F$, then $a\alpha + b\beta \in W$. This may seem to be a small change, but it is a very convenient form of the conditions. It is also equivalent to the statement that all linear combinations of elements in W are also in W; that is, $\langle W \rangle = W$. It follows directly from this statement that for any subset A, $\langle A \rangle$ is a subspace. Thus, instead of speaking of the subset spanned by A, we speak of the subspace spanned by A.

Every vector space V has V and the zero space $\{0\}$ as subspaces. As a rule we are interested in subspaces other than these and to distinguish them we call the subspaces other than V and $\{0\}$ *proper* subspaces. In addition, if W is a subspace we designate subspaces of W other than W and $\{0\}$ as proper subspaces of W.

In Examples (1) and (2) we can take a fixed finite set $\{x_1, x_2, \ldots, x_m\}$ of elements of F and define W to be the set of all polynomials such that $p(x_1) = p(x_2) = \cdots = p(x_m) = 0$. To show that W is a subspace it is sufficient to show that the sum of two polynomials which vanish at the x_i also vanishes at the x_i, and the product of a scalar and a polynomial vanishing at the x_i also vanishes at the x_i. What is the situation in P_n if $m > n$? Similar subspaces can be defined in examples (3), (4), (5), (6), and (7).

The space P_m is a subspace of P, and also a subspace of P_n for $m \leq n$.

In R^n, for each m, $0 \leq m \leq n$, the set of all $\alpha = (a_1, a_2, \ldots, a_n)$ such that $a_1 = a_2 = \cdots = a_m = 0$ is a subspace of R^n. This subspace is proper if $0 < m < n$.

Notice that the set of all n-tuples of rational numbers is a subset of R^n and it is a vector space over the rational numbers, but it is not a subspace of R^n since it is not a vector space over the real numbers. Why?

Theorem 4.1. *The intersection of any collection of subspaces is a subspace.*

PROOF. Let $W_\mu : \mu \in M$ be an indexed collection of subspaces of V. $\cap_{\mu \in M} W_\mu$ is not empty since it contains 0. Let α, $\beta \in \cap_{\mu \in M} W_\mu$ and $a, b \in F$. Then α, $\beta \in W_\mu$ for each $\mu \in M$. Since W_μ is a subspace $a\alpha + b\beta \in W_\mu$ for each $\mu \in M$, and hence $a\alpha + b\beta \in \cap_{\mu \in M} W_\mu$. Thus $\cap_{\mu \in M} W_\mu$ is a subspace. $\square$

Let A be any subset of V, not necessarily a subspace. There exist subspaces $W_\mu \subset V$ which contain A; in fact, V is one of them. The intersection $\cap_{A \subset W_\mu} W_\mu$ of all such subspaces is a subspace containing A. It is the smallest subspace containing A.

Theorem 4.2. *For any $A \subset V$, $\cap_{A \subset W_\mu} W_\mu = \langle A \rangle$; that is, the smallest subspace containing A is exactly the subspace spanned by A.*

PROOF. Since $\cap_{A \subset W_\mu} W_\mu$ is a subspace containing A, it contains all linear combinations of elements of A. Thus $\langle A \rangle \subset \cap_{A \subset W_\mu} W_\mu$. On the other hand $\langle A \rangle$ is a subspace containing A, that is, $\langle A \rangle$ is one of the W_μ and hence $\cap_{A \subset W_\mu} W_\mu \subset \langle A \rangle$. Thus $\cap_{A \subset W_\mu} W_\mu = \langle A \rangle$. $\square$

$W_1 + W_2$ is defined to be the set of all vectors of the form $\alpha_1 + \alpha_2$ where $\alpha_1 \in W_1$ and $\alpha_2 \in W_2$.

Theorem 4.3. *If W_1 and W_2 are subspaces of V, then $W_1 + W_2$ is a subspace of V.*

PROOF. If $\alpha = \alpha_1 + \alpha_2 \in W_1 + W_2$, $\beta = \beta_1 + \beta_2 \in W_1 + W_2$, and a, $b \in F$, then $a\alpha + b\beta = a(\alpha_1 + \alpha_2) + b(\beta_1 + \beta_2) = (a\alpha_1 + b\beta_1) + (a\alpha_2 + b\beta_2) \in W_1 + W_2$. Thus $W_1 + W_2$ is a subspace. $\square$

Theorem 4.4. *$W_1 + W_2$ is the smallest subspace containing both W_1 and W_2; that is, $W_1 + W_2 = \langle W_1 \cup W_2 \rangle$. If A_1 spans W_1 and A_2 spans W_2, then $A_1 \cup A_2$ spans $W_1 + W_2$.*

PROOF. Since $0 \in W_1$, $W_2 \subset W_1 + W_2$. Similarly, $W_1 \subset W_1 + W_2$. Since $W_1 + W_2$ is a subspace containing $W_1 \cup W_2$, $\langle W_1 \cup W_2 \rangle \subset W_1 + W_2$. For any $\alpha \in W_1 + W_2$, α can be written in the form $\alpha = \alpha_1 + \alpha_2$ where $\alpha_1 \in W_1$ and $\alpha_2 \in W_2$. Then $\alpha_1 \in W_1 \subset \langle W_1 \cup W_2 \rangle$ and $\alpha_2 \in W_2 \subset \langle W_1 \cup W_2 \rangle$. Since $\langle W_1 \cup W_2 \rangle$ is a subspace, $\alpha = \alpha_1 + \alpha_2 \in \langle W_1 \cup W_2 \rangle$. Thus $W_1 + W_2 = \langle W_1 \cup W_2 \rangle$.

The second part of the theorem now follows directly. $W_1 = \langle A_1 \rangle \subset \langle A_1 \cup A_2 \rangle$ and $W_2 = \langle A_2 \rangle \subset \langle A_1 \cup A_2 \rangle$ so that $W_1 \cup W_2 \subset \langle A_1 \cup A_2 \rangle \subset \langle W_1 \cup W_2 \rangle$, and hence $\langle W_1 \cup W_2 \rangle = \langle A_1 \cup A_2 \rangle$. $\square$

Theorem 4.5. *A subspace W of an n-dimensional vector space V is a finite dimensional vector space of dimension $m \leq n$.*

PROOF. If $W = \{0\}$, then W is 0-dimensional. Otherwise, there is a non-zero vector $\alpha_1 \in W$. If $\langle \alpha_1 \rangle = W$, W is 1-dimensional. Otherwise, there is an $\alpha_2 \notin \langle \alpha_1 \rangle$ in W. We continue in this fashion as long as possible. Suppose we have obtained the linearly independent set $\{\alpha_1, \ldots, \alpha_k\}$ and that it does not span W. Then there exists an $\alpha_{k+1} \in W$, $\alpha_{k+1} \notin \langle \alpha_1, \ldots, \alpha_k \rangle$. In a linear relation of the form $\sum_{i=1}^{k+1} a_i \alpha_i = 0$ we could not have $a_{k+1} \neq 0$ for then $\alpha_{k+1} \in \langle \alpha_1, \ldots, \alpha_k \rangle$. But then the relation reduces to the form $\sum_{i=1}^{k} a_i \alpha_i = 0$. Since $\{\alpha_1, \ldots, \alpha_k\}$ is linearly independent, all $a_i = 0$. Thus $\{\alpha_1, \ldots, \alpha_k, \alpha_{k+1}\}$ is linearly independent. In general, any linearly independent set in W that does not span W can be expanded into a larger linearly independent set in W. This process cannot go on indefinitely for in that event we would obtain more than n linearly independent vectors in V. Thus there exists an m such that $\langle \alpha_1, \ldots, \alpha_m \rangle = W$. It is clear that $m \leq n$. $\square$

Theorem 4.6. *Given any subspace W of dimension m in an n-dimensional vector space V, there exists a basis $\{\alpha_1, \ldots, \alpha_m, \alpha_{m+1}, \ldots, \alpha_n\}$ of V such that $\{\alpha_1, \ldots, \alpha_m\}$ is a basis of W.*

PROOF. By the previous theorem we see that W has a basis $\{\alpha_1, \ldots, \alpha_m\}$. This set is also linearly independent when considered in V, and hence by Theorem 3.6 it can be extended to a basis of V. $\square$

Theorem 4.7. *If two subspaces U and W of a vector space V have the same finite dimension and $U \subset W$, then $U = W$.*

PROOF. By the previous theorem there exists a basis of U which can be extended to a basis of W. But since dim $U =$ dim W, the basis of W can have no more elements than does the basis of U. This means a basis of U is also a basis of W; that is, $U = W$. $\square$

Theorem 4.8. *If W_1 and W_2 are any two subspaces of a finite dimensional vector space V, then dim $(W_1 + W_2) =$ dim $W_1 +$ dim $W_2 -$ dim $(W_1 \cap W_2)$.*

PROOF. Let $\{\alpha_1, \ldots, \alpha_r\}$ be a basis of $W_1 \cap W_2$. This basis can be extended to a basis $\{\alpha_1, \ldots, \alpha_r, \beta_1, \ldots, \beta_s\}$ of W_1 and also to a basis $\{\alpha_1, \ldots, \alpha_r, \gamma_1, \ldots, \gamma_t\}$ of W_2. It is clear that $\{\alpha_1, \ldots, \alpha_r, \beta_1, \ldots, \beta_s, \gamma_1, \ldots, \gamma_t\}$ spans $W_1 + W_2$; we wish to show that this set is linearly independent. Suppose $\sum_i a_i \alpha_i + \sum_j b_j \beta_j + \sum_k c_k \gamma_k = 0$ is a linear relation. Then $\sum_i a_i \alpha_i + \sum_j b_j \beta_j = -\sum_k c_k \gamma_k$. The left side is in W_1 and the right side is in W_2, and hence both are in $W_1 \cap W_2$. Each side is then expressible as a linear combination of the $\{\alpha_i\}$. Since any representation of an element as a linear combination of the $\{\alpha_1, \ldots, \alpha_r, \beta_1, \ldots, \beta_s\}$ is unique, this means that

$b_j = 0$ for all j. By a symmetric argument we see that all $c_k = 0$. Finally, this means that $\sum_i a_i \alpha_i = 0$ from which it follows that all $a_i = 0$. This shows that the spanning set $\{\alpha_1, \ldots, \alpha_r, \beta_1, \ldots, \beta_s, \gamma_1, \ldots, \gamma_t\}$ is linearly independent and a basis of $W_1 + W_2$. Thus dim $(W_1 + W_2) = r + s + t = (r + s) + (r + t) - r = \dim W_1 + \dim W_2 - \dim (W_1 \cap W_2)$. □

As an example, consider in R^3 the subspaces $W_1 = \langle (1, 0, 2), (1, 2, 2) \rangle$ and $W_2 = \langle (1, 1, 0), (0, 1, 1) \rangle$. Both subspaces are of dimension 2. Since $W_1 \subset W_1 + W_2 \subset R^3$ we see that $2 \leq \dim (W_1 + W_2) \leq 3$. Because of Theorem 4.8 this implies that $1 \leq \dim (W_1 \cap W_2) \leq 2$. In more familiar terms, W_1 and W_2 are planes in a 3-dimensional space. Since both planes contain the origin, they do intersect. Their intersection is either a line or, in case they coincide, a plane. The first problem is to find a basis for $W_1 \cap W_2$. Any $\alpha \in W_1 \cap W_2$ must be expressible in the forms $\alpha = a(1, 0, 2) + b(1, 2, 2) = c(1, 1, 0) + d(0, 1, 1)$. This leads to the three equations:

$$a + b = c$$
$$2b = c + d$$
$$2a + 2b = \quad d.$$

These equations have the solutions $b = -3a$, $c = -2a$, $d = -4a$. Thus $\alpha = a(1, 0, 2) - 3a(1, 2, 2) = a(-2, -6, -4)$. As a check we also have $\alpha = -2a(1, 1, 0) - 4a(0, 1, 1) = a(-2, -6, -4)$. We have determined that $\{(1, 3, 2)\}$ is a basis of $W_1 \cap W_2$. Also $\{(1, 3, 2), (1, 0, 2)\}$ is a basis of W_1 and $\{(1, 3, 2), (1, 1, 0)\}$ is a basis of W_2.

We are all familiar with the theorem from solid geometry to the effect that two non-parallel planes intersect in a line, and the example above is an illustration of that theorem. In spaces of dimension higher than 3, however, it is possible for two subspaces of dimension 2 to have but one point in common. For example, in R^4 the subspaces $W_1 = \langle (1, 0, 0, 0), (0, 1, 0, 0) \rangle$ and $W_2 = \langle (0, 0, 1, 0), (0, 0, 0, 1) \rangle$ are each 2-dimensional and $W_1 \cap W_2 = \{0\}$, $W_1 + W_2 = R^4$.

Those cases in which dim $(W_1 \cap W_2) = 0$ deserve special mention. If $W_1 \cap W_2 = \{0\}$ we say that the sum $W_1 + W_2$ is *direct*: $W_1 + W_2$ is a *direct sum* of W_1 and W_2. To indicate that a sum is direct we use the notation, $W_1 \oplus W_2$. For $\alpha \in W_1 \oplus W_2$ there exist $\alpha_1 \in W_1$ and $\alpha_2 \in W_2$ such that $\alpha = \alpha_1 + \alpha_2$. This much is true for any sum of two subspaces. If the sum is direct, however, α_1 and α_2 are uniquely determined by α. For if $\alpha = \alpha_1 + \alpha_2 = \alpha_1' + \alpha_2'$, then $\alpha_1 - \alpha_1' = \alpha_2' - \alpha_2$. Since the left side is in W_1 and the right side is in W_2, both are in $W_1 \cap W_2$. But this means $\alpha_1 - \alpha_1' = 0$ and $\alpha_2 - \alpha_2' = 0$; that is, the decomposition of α into a sum of an element in W_1 plus an element in W_2 is unique. If V is the direct sum of W_1 and W_2, we say that W_1 and W_2 are *complementary* and that W_2 is a *complementary subspace* of W_1, or a *complement* of W_1.

The notion of a direct sum can be extended to a sum of any finite number of subspaces. The sum $W_1 + \cdots + W_k$ is said to be *direct* if for each i, $W_i \cap (\sum_{j \neq i} W_j) = \{0\}$. If the sum of several subspaces is direct, we use the notation $W_1 \oplus W_2 \oplus \cdots \oplus W_k$. In this case, too, $\alpha \in W_1 \oplus \cdots \oplus W_k$ can be expressed uniquely in the form $\alpha = \sum_i \alpha_i$, $\alpha_i \in W_i$.

Theorem 4.9. *If W is a subspace of V there exists a subspace W' such that $V = W \oplus W'$.*

PROOF. Let $\{\alpha_1, \ldots, \alpha_m\}$ be a basis of W. Extend this linearly independent set to a basis $\{\alpha_1, \ldots, \alpha_m, \alpha_{m+1}, \ldots, \alpha_n\}$ of V. Let W' be the subspace spanned by $\{\alpha_{m+1}, \ldots, \alpha_n\}$. Clearly, $W \cap W' = \{0\}$ and the sum $V = W + W'$ is direct. $\square$

Thus every subspace of a finite dimensional vector space has a complementary subspace. The complement is not unique, however. If for W there exists a subspace W' such that $V = W \oplus W'$, we say that W is a *direct summand* of V.

Theorem 4.10. *For a sum of several subspaces of a finite dimensional vector space to be direct it is necessary and sufficient that* $\dim (W_1 + \cdots + W_k) = \dim W_1 + \cdots + \dim W_k$.

PROOF. This is an immediate consequence of Theorem 4.8 and the principle of mathematical induction. $\square$

EXERCISES

1. Let P be the space of all polynomials with real coefficients. Determine which of the following subsets of P are subspaces.

   (a) $\{p(x) \mid p(1) = 0\}$.
   (b) $\{p(x) \mid \text{constant term of } p(x) = 0\}$.
   (c) $\{p(x) \mid \text{degree of } p(x) = 3\}$.
   (d) $\{p(x) \mid \text{degree of } p(x) \leq 3\}$.

(Strictly speaking, the zero polynomial does not have a degree associated with it. It is sometimes convenient to agree that the zero polynomial has degree less than any integer, positive or negative. With this convention the zero polynomial is included in the set described above, and it is not necessary to add a separate comment to include it.)

   (e) $\{p(x) \mid \text{degree of } p(x) \text{ is even}\} \cup \{0\}$.

2. Determine which of the following subsets of R^n are subspaces.

   (a) $\{(x_1, x_2, \ldots, x_n) \mid x_1 = 0\}$.
   (b) $\{(x_1, x_2, \ldots, x_n) \mid x_1 \geq 0\}$.
   (c) $\{(x_1, x_2, \ldots, x_n) \mid x_1 + 2x_2 = 0\}$.
   (d) $\{(x_1, x_2, \ldots, x_n) \mid x_1 + 2x_2 = 1\}$.
   (e) $\{(x_1, x_2, \ldots, x_n) \mid x_1 + 2x_2 \geq 0\}$.
   (f) $\{(x_1, x_2, \ldots, x_n) \mid m_i < x_i < M_i: i = 1, 2, \ldots, n$ where the m_i and M_i are constants$\}$.
   (g) $\{(x_1, x_2, \ldots, x_n) \mid x_1 = x_2 = \cdots = x_n\}$.

3. What is the essential difference between the condition used to define the subset in (c) of Exercise 2 and the condition used in (d)? Is the lack of a non-zero constant term important in (c)?

4. What is the essential difference between the condition used to define the subset in (c) of Exercise 2 and the condition used in (e)? What, in general, are the differences between the conditions in (a), (c), and (g) and those in (b), (e), and (f)?

5. Show that $\{(1, 1, 0, 0), (1, 0, 1, 1)\}$ and $\{(2, -1, 3, 3), (0, 1, -1, -1)\}$ span the same subspace of R^4.

6. Let W be the subspace of R^5 spanned by $\{(1, 1, 1, 1, 1), (1, 0, 1, 0, 1), (0, 1, 1, 1, 0), (2, 0, 0, 1, 1), (2, 1, 1, 2, 1), (1, -1, -1, -2, 2), (1, 2, 3, 4, -1)\}$. Find a basis for W and the dimension of W.

7. Show that $\{(1, -1, 2, -3), (1, 1, 2, 0), (3, -1, 6, -6)\}$ and $\{(1, 0, 1, 0), (0, 2, 0, 3)\}$ do not span the same subspace.

8. Let $W = \langle(1, 2, 3, 6), (4, -1, 3, 6), (5, 1, 6, 12)\rangle$ and $W_2 = \langle(1, -1, 1, 1), (2, -1, 4, 5)\rangle$ be subspaces of R^4. Find bases for $W_1 \cap W_2$ and $W_1 + W_2$. Extend the basis of $W_1 \cap W_2$ to a basis of W_1, and extend the basis of $W_1 \cap W_2$ to a basis of W_2. From these bases obtain a basis of $W_1 + W_2$.

9. Let P be the space of all polynomials with real coefficients, and let $W_1 = \{p(x) \mid p(1) = 0\}$ and $W_2 = \{p(x) \mid p(2) = 0\}$. Determine $W_1 \cap W_2$ and $W_1 + W_2$. (These spaces are infinite dimensional and the student is not expected to find bases for these subspaces. What is expected is a simple criterion or description of these subspaces.)

10. We have already seen (Section 1, Exercise 11) that the real numbers form a vector space over the rationals. Show that $\{1, \sqrt{2}\}$ and $\{1 - \sqrt{2}, 1 + \sqrt{2}\}$ span the same subspace.

11. Show that if W_1 and W_2 are subspaces, then $W_1 \cup W_2$ is not a subspace unless one is a subspace of the other.

12. Show that the set of all vectors $(x_1, x_2, x_3, x_4) \in R^4$ satisfying the equations

$$3x_1 - 2x_2 - x_3 - 4x_4 = 0$$
$$x_1 + x_2 - 2x_3 - 3x_4 = 0$$

is a subspace of R^4. Find a basis for this subspace. (*Hint:* Solve the equations for x_1 and x_2 in terms of x_3 and x_4. Then specify various values for x_3 and x_4 to obtain as many linearly independent vectors as are needed.)

13. Let S, T, and T^* be three subspaces of V (of finite dimension) for which (a) $S \cap T = S \cap T^*$, (b) $S + T = S + T^*$, (c) $T \subset T^*$. Show that $T = T^*$.

14. Show by example that it is possible to have $S \oplus T = S \oplus T^*$ without having $T = T^*$.

15. If $V = W_1 \oplus W_2$ and W is any subspace of V such that $W_1 \subset W$, show that $W = (W \cap W_1) \oplus (W \cap W_2)$. Show by an example that the condition $W_1 \subset W$ (or $W_2 \subset W$) is necessary.

chapter II | Linear transformations and matrices

In this chapter we define linear transformations and various operations: addition of two linear transformations, multiplication of two linear transformations, and multiplication of a linear transformation by a scalar. Linear transformations are functions of vectors in one vector space U with values which are vectors in the same or another vector space V which preserve linear combinations. They can be represented by matrices in the same sense that vectors can be represented by n-tuples. This representation requires that operations of addition, multiplication, and scalar multiplication of matrices be defined to correspond to these operations with linear transformations. Thus we establish an algebra of matrices by means of the conceptually simpler algebra of linear transformations.

The matrix representing a linear transformation of U into V depends on the choice of a basis in U and a basis in V. Our first problem, a recurrent problem whenever matrices are used to represent anything, is to see how a change in the choice of bases determines a corresponding change in the matrix representing the linear transformation. Two matrices which represent the same linear transformation with respect to different sets of bases must have some properties in common. This leads to the idea of equivalence relations among matrices. The exact nature of this equivalence relation depends on the bases which are permitted.

In this chapter no restriction is placed on the bases which are permitted and we obtain the widest kind of equivalence. In Chapter III we identify U and V and require that the same basis be used in both. This yields a more restricted kind of equivalence, and a study of this equivalence is both interesting and fruitful. In Chapter V we make further restrictions in the permissible bases and obtain an even more restricted equivalence.

When no restriction is placed on the bases which are permitted, the equivalence is so broad that it is relatively uninteresting. Very useful results are obtained, however, when we are permitted to change basis only in the image space V. In every set of mutually equivalent matrices we select one, representative of all of them, which we call a normal form, in this case the Hermite normal form. The Hermite normal form is one of our most important and effective computational tools, far exceeding in utility its application to the study of this particular equivalence relation.

The pattern we have described is worth conscious notice since it is recurrent and the principal underlying theme in this exposition of matrix theory. We define a concept, find a representation suitable for effective computation, change bases to see how this change affects the representation, and then seek a normal form in each class of equivalent representations.

1 | Linear Transformations

Let U and V be vector spaces over the same field of scalars F.

Definition.   A *linear transformation* σ of U into V is a single-valued mapping of U into V which associates to each element $\alpha \in U$ a unique element $\sigma(\alpha) \in V$ such that for all α, $\beta \in U$ and all a, $b \in F$ we have

$$\sigma(a\alpha + b\beta) = a\sigma(\alpha) + b\sigma(\beta). \tag{1.1}$$

We call $\sigma(\alpha)$ the *image* of α under the linear transformation σ. If $\bar{\alpha} \in V$, then any vector $\alpha \in U$ such that $\sigma(\alpha) = \bar{\alpha}$ is called an *inverse image* of $\bar{\alpha}$. The set of *all* $\alpha \in U$ such that $\sigma(\alpha) = \bar{\alpha}$ is called the *complete inverse image* of $\bar{\alpha}$, and it is denoted by $\sigma^{-1}(\bar{\alpha})$. Generally, $\sigma^{-1}(\bar{\alpha})$ need not be a single element as there may be more than one $\alpha \in U$ such that $\sigma(\alpha) = \bar{\alpha}$.

By taking particular choices for a and b we see that for a linear transformation $\sigma(\alpha + \beta) = \sigma(\alpha) + \sigma(\beta)$ and $\sigma(a\alpha) = a\sigma(\alpha)$. Loosely speaking, the image of the sum is the sum of the images and the image of the product is the product of the images. This descriptive language has to be interpreted generously since the operations before and after applying the linear transformation may take place in different vector spaces. Furthermore, the remark about scalar multiplication is inexact since we do not apply the linear transformation to scalars; the linear transformation is defined only for vectors in U. Even so, the linear transformation does preserve the structural operations in a vector space and this is the reason for its importance. Generally, in algebra a structure-preserving mapping is called a *homomorphism*. To describe the special role of the elements of F in the condition, $\sigma(a\alpha) = a\sigma(\alpha)$, we say that a linear transformation is a homomorphism *over* F, or an *F-homomorphism*.

If for $\alpha \neq \beta$ it necessarily follows that $\sigma(\alpha) \neq \sigma(\beta)$, the homomorphism σ is said to be *one-to-one* and it is called a *monomorphism*. If A is any subset of

U, $\sigma(A)$ will denote the set of all images of elements of A; $\sigma(A) = \{\bar{\alpha} \mid \bar{\alpha} = \sigma(\alpha)$ for some $\alpha \in A\}$. $\sigma(A)$ is called the *image* of A. $\sigma(U)$ is often denoted by $\text{Im}(\sigma)$ and is called the *image* of σ. If $\text{Im}(\sigma) = V$ we shall say that the homomorphism is a mapping *onto* V and it is called an *epimorphism*.

We call the set U, on which the linear transformation σ is defined, the *domain* of σ. We call V, the set in which the images of σ are defined, the *codomain* of σ. Strictly speaking, a linear transformation must specify the domain and codomain as well as the mapping. For example, consider the linear transformation that maps every vector of U onto the zero vector of V. This mapping is called the *zero mapping*. If W is any subspace of V, there is also a zero mapping of U into W, and this mapping has the same effect on the elements of U as the zero mapping of U into V. However, they are different linear transformations since they have different codomains. This may seem like an unnecessarily fine distinction. Actually, for most of this book we could get along without this degree of precision. But the more deeply we go into linear algebra the more such precision is needed. In this book we need this much care when we discuss dual spaces and dual transformations in Chapter IV.

A homomorphism that is both an epimorphism and a monomorphism is called an *isomorphism*. If $\bar{\alpha} \in V$, the fact that σ is an epimorphism says that here is an $\alpha \in U$ such that $\sigma(\alpha) = \bar{\alpha}$. The fact that σ is a monomorphism says that this α is unique. Thus, for an isomorphism, we can define an inverse mapping σ^{-1} that maps $\bar{\alpha}$ onto α.

Theorem 1.1. *The inverse σ^{-1} of an isomorphism is also an isomorphism.*

PROOF. Since σ^{-1} is obviously one-to-one and onto, it is necessary only to show that it is linear. If $\bar{\alpha} = \sigma(\alpha)$ and $\bar{\beta} = \sigma(\beta)$, then $\sigma(a\alpha + b\beta) = a\bar{\alpha} + b\bar{\beta}$ so that $\sigma^{-1}(a\bar{\alpha} + b\bar{\beta}) = a\alpha + b\beta = a\sigma^{-1}(\bar{\alpha}) + b\sigma^{-1}(\bar{\beta})$. $\square$

For the inverse isomorphism $\sigma^{-1}(\bar{\alpha})$ is an element of U. This conflicts with the previously given definition of $\sigma^{-1}(\bar{\alpha})$ as a complete inverse image in which $\sigma^{-1}(\bar{\alpha})$ is a subset of U. However, the symbol σ^{-1}, standing alone, will always be used to denote an isomorphism, and in this case there is no difficulty caused by the fact that $\sigma^{-1}(\bar{\alpha})$ might denote either an element or a one-element set.

Let us give some examples of linear transformations. Let $U = V = P$, the space of polynomials in x with coefficients in R. For $\alpha = \sum_{i=0}^{n} a_i x^i$, define $\sigma(\alpha) = \dfrac{d\alpha}{dx} = \sum_{i=0}^{n} ia^{i-1}$. In calculus one of the very first things proved about the derivative is that it is linear, $\dfrac{d(\alpha + \beta)}{dx} = \dfrac{d\alpha}{dx} + \dfrac{d\beta}{dx}$ and $\dfrac{d(a\alpha)}{dx} = a\dfrac{d\alpha}{dx}$. The mapping $\tau(\alpha) = \sum_{i=0}^{n} \dfrac{a_i}{i+1} x^{i+1}$ is also linear. Notice that this is not the indefinite integral since we have specified that the constant of integration shall be zero. Notice that σ is onto but not one-to-one and τ is one-to-one but not onto.

Let $U = R^n$ and $V = R^m$ with $m \leq n$. For each $\alpha = (a_1, \ldots, a_n) \in R^n$ define $\sigma(\alpha) = (a_1, \ldots, a_m) \in R^m$. It is clear that this linear transformation is one-to-one if and only if $m = n$, but it is onto. For each $\beta = (b_1, \ldots, b_m) \in R^m$ define $\tau(\beta) = (b_1, \ldots, b_m, 0, \ldots, 0) \in R^n$. This linear transformation is one-to-one, but it is onto if and only if $m = n$.

Let $U = V$. For a given scalar $a \in F$ the mapping of α onto $a\alpha$ is linear since

$$a(\alpha + \beta) = a\alpha + a\beta = a(\alpha) + a(\beta),$$

and

$$a(b\alpha) = (ab)\alpha = (ba)\alpha = b \cdot a(\alpha).$$

To simplify notation we also denote this linear-transformation by a. Linear transformations of this type are called *scalar transformations*, and there is a one-to-one correspondence between the field of scalars and the set of scalar transformations. In particular, the linear transformation that leaves every vector fixed is denoted by 1. It is called the *identity transformation* or *unit transformation*. If linear transformations in several vector spaces are being discussed at the same time, it may be desirable to identify the space on which the identity transformation is defined. Thus 1_U will denote the identity transformation on U.

When a basis of a finite dimensional vector space V is used to establish a correspondence between vectors in V and n-tuples in F^n, this correspondence is an isomorphism. The required arguments have already been given in Section I-3. Since V and F^n are isomorphic, it is theoretically possible to discuss the properties of V by examining the properties of F^n. However, there is much interest and importance attached to concepts that are independent of the choice of a basis. If a homomorphism or isomorphism can be defined uniquely by intrinsic properties independent of a choice of basis the mapping is said to be *natural* or *canonical*. In particular, any two vector spaces of dimension n over F are isomorphic. Such an isomorphism can be established by setting up an isomorphism between each one and F^n. This isomorphism will be dependent on a choice of a basis in each space. Such an isomorphism, dependent upon the arbitrary choice of bases, is not canonical.

Next, let us define the various operations between linear transformations. For each pair σ, τ of linear transformation of U into V, define $\sigma + \tau$ by the rule

$$(\sigma + \tau)(\alpha) = \sigma(\alpha) + \tau(\alpha) \qquad \text{for all} \qquad \alpha \in V.$$

$\sigma + \tau$ is a linear transformation since

$$(\sigma + \tau)(a\alpha + b\beta) = \sigma(a\alpha + b\beta) + \tau(a\alpha + b\beta) = a\sigma(\alpha) + b\sigma(\beta)$$
$$+ a\tau(\alpha) + b\tau(\beta) = a[\sigma(\alpha) + \tau(\alpha)] + b[\sigma(\beta) + \tau(\beta)]$$
$$= a(\sigma + \tau)(\alpha) + b(\sigma + \tau)(\beta).$$

Observe that addition of linear transformation is commutative; $\sigma + \tau = \tau + \sigma$.

For each linear transformation σ and $a \in F$ define $a\sigma$ by the rule; $(a\sigma)(\alpha) = a[\sigma(\alpha)]$. $a\sigma$ is a linear transformation.

It is not difficult to show that with these two operations the set of all linear transformations of U into V is itself a vector space over F. This is a very important fact and we occasionally refer to it and make use of it. However, we wish to emphasize that we define the sum of two linear transformations if and only if they both have the same domain and the same codomain. It is neither necessary nor sufficient that they have the same image, or that the image of one be a subset of the image of the other. It is simply a question of being clear about the terminology and its meaning. The set of all linear transformations of U into V will be denoted by $\text{Hom}(U, V)$.

There is another, entirely new, operation that we need to define. Let W be a third vector space over F. Let σ be a linear transformation of U into V and τ a linear transformation of V into W. By $\tau\sigma$ we denote the linear transformation of U into W defined by the rule: $(\tau\sigma)(\alpha) = \tau[\sigma(\alpha)]$. Notice that in this context $\sigma\tau$ has no meaning. We refer to this operation as either *iteration* or *multiplication* of linear transformation, and $\tau\sigma$ is called the *product* of τ and σ.

The operations between linear transformations are related by the following rules:

1. Multiplication is associative: $\pi(\tau\sigma) = (\pi\tau)\sigma$. Here π is a linear transformation of W into a fourth vector space X.

2. Multiplication is distributive with respect to addition:

$$(\tau_1 + \tau_2)\sigma = \tau_1\sigma + \tau_2\sigma \qquad \text{and} \qquad \tau(\sigma_1 + \sigma_2) = \tau\sigma_1 + \tau\sigma_2.$$

3. Scalar multiplication commutes with multiplication: $a(\tau\sigma) = \tau(a\sigma)$. These properties are easily proved and are left to the reader.

Notice that if $W \neq U$, then $\tau\sigma$ is defined but $\sigma\tau$ is not. If all linear transformations under consideration are mappings of a vector space U into itself, then these linear transformations can be multiplied in any order. This means that $\tau\sigma$ and $\sigma\tau$ would both be defined, but it would not mean that $\tau\sigma = \sigma\tau$.

The set of linear transformation of a vector space into itself is a vector space, as we have already observed, and now we have defined a product which satisfies the three conditions given above. Such a space is called an *associative algebra*. In our case the algebra consists of linear transformation and it is known as a *linear algebra*. However, the use of terms is always in a state of flux, and today this term is used in a more inclusive sense. When referring to a particular set with an algebraic structure, "linear algebra" still denotes what we have just described. But when referring to an area of study, the term "linear algebra includes virtually every concept in which linear transformations play a role, including linear transformations between different vector spaces (in which the linear transformations cannot always be multiplied), sequences of vector spaces, and even mappings of sets of linear transformations (since they also have the structure of a vector space).

Theorem 1.2. $\mathrm{Im}(\sigma)$ *is a subspace of* V.

PROOF.   If $\bar{\alpha}$ and $\bar{\beta}$ are elements of $\mathrm{Im}(\sigma)$, there exist α, $\beta \in U$ such that $\sigma(\alpha) = \bar{\alpha}$ and $\sigma(\beta) = \bar{\beta}$. For any a, $b \in F$, $\sigma(a\alpha + b\beta) = a\sigma(\alpha) + b\sigma(\beta) = a\bar{\alpha} + b\bar{\beta} \in \mathrm{Im}(\sigma)$. Thus $\mathrm{Im}(\sigma)$ is a subspace of V. $\square$

Corollary 1.3.   *If* U_1 *is a subspace of* U, *then* $\sigma(U_1)$ *is a subspace of* V. $\square$

It follows from this corollary that $\sigma(0) = 0$ where 0 denotes the zero vector of U and the zero vector of V. It is even easier, however, to show it directly. Since $\sigma(0) = \sigma(0 + 0) = \sigma(0) + \sigma(0)$ it follows from the uniqueness of the zero vector that $\sigma(0) = 0$.

For the rest of this book, unless specific comment is made, we assume that all vector spaces under consideration are finite dimensional. Let $\dim U = n$ and $\dim V = m$.

The dimension of the subspace $\mathrm{Im}(\sigma)$ is called the *rank* of the linear transformation σ. The rank of σ is denoted by $\rho(\sigma)$.

Theorem 1.4. $\rho(\tau) \leq \{m, n\}$.

PROOF.   If $\{\alpha_1, \ldots, \alpha_s\}$ is linearly dependent in U, there exists a non-trivial relation of the form $\sum_i a_i\alpha_i = 0$. But then $\sum_i a_i\sigma(\alpha_i) = \sigma(0) = 0$; that is, $\{\sigma(\alpha_1), \ldots, \sigma(\alpha_s)\}$ is linearly dependent in V. A linear transformation preserves linear relations and transforms dependent sets into dependent sets. Thus, there can be no more than n linearly independent elements in $\mathrm{Im}(\sigma)$. In addition, $\mathrm{Im}(\sigma)$ is a subspace of V so that $\dim \mathrm{Im}(\sigma) \leq m$. Thus $\rho(\sigma) = \dim \mathrm{Im}(\sigma) \leq \min \{m, n\}$. $\square$

Theorem 1.5.   *If* W *is a subspace of* V, *the set* $\sigma^{-1}(W)$ *of all* $\alpha \in U$ *such that* $\sigma(\alpha) \in W$ *is a subspace of* U.

PROOF.   If α, $\beta \in \sigma^{-1}(W)$, then $\sigma(a\alpha + b\beta) = a\sigma(\alpha) + b\sigma(\beta) \in W$. Thus $a\alpha + b\beta \in \sigma^{-1}(W)$ and $\sigma^{-1}(W)$ is a subspace. $\square$

The subspace $K(\sigma) = \sigma^{-1}(0)$ is called the *kernel* of the linear transformation σ. The dimension of $K(\sigma)$ is called the *nullity* of σ. The nullity of σ is denoted by $\nu(\sigma)$.

Theorem 1.6. $\rho(\tau) + \nu(\tau) = n$.

PROOF.   Let $\{\alpha_1, \ldots, \alpha_\nu, \beta_1, \ldots, \beta_k\}$ be a basis of U such that $\{\alpha_1, \ldots, \alpha_\nu\}$ is a basis of $K(\sigma)$. For $\alpha = \sum_i a_i\alpha_i + \sum_j b_j\beta_j \in U$ we see that $\sigma(\alpha) = \sum_i a_i\sigma(\alpha_i) + \sum_j b_j\sigma(\beta_j) = \sum_j b_j\sigma(\beta_j)$. Thus $\{\sigma(\beta_1), \ldots, \sigma(\beta_k)\}$ spans $\mathrm{Im}(\sigma)$. On the other hand if $\sum_j c_j\sigma(\beta_j) = 0$, then $\sigma(\sum_j c_j\beta_j) = \sum_j c_j\sigma(\beta_j) = 0$; that is, $\sum_j c_j \beta_j \in K(\sigma)$. In this case there exist coefficients d_i such that $\sum_j c_j \beta_j = \sum_i d_i \alpha_i$. If any of these coefficients were non-zero we would have a non-trivial relation among the elements of $\{\alpha_1, \ldots, \alpha_\nu, \beta_1, \ldots, \beta_k\}$. Hence, all $c_j = 0$ and $\{\sigma(\beta_1), \ldots, \sigma(\beta_k)\}$ is linearly independent. But then it is a basis of $\mathrm{Im}(\sigma)$ so that $k = \rho(\sigma)$. Thus $\rho(\sigma) + \nu(\sigma) = n$. □

Theorem 1.6 has an important geometric interpretation. Suppose that a 3-dimensional vector space R^3 were mapped onto a 2-dimensional vector space R^2. In this case, it is simplest and sufficiently accurate to think of σ as the linear transformation which maps $(a_1, a_2, a_3) \in R^3$ onto $(a_1, a_2) \in R^2$ which we can identify with $(a_1, a_2, 0) \in R^3$. Since $\rho(\sigma) = 2$, $\nu(\sigma) = 1$. Clearly, every point $(0, 0, a_3)$ on the x_3-axis is mapped onto the origin. Thus $K(\sigma)$ is the x_3-axis, the line through the origin in the direction of the projection, and $\{(0, 0, 1) = \alpha_1\}$ is a basis of $K(\sigma)$. It should be evident that any plane through the origin not containing $K(\sigma)$ will be projected onto the $x_1 x_2$-plane and that this mapping is one-to-one and onto. Thus the complementary subspace $\langle \beta_1, \beta_2 \rangle$ can be taken to be any plane through the origin not containing the x_3-axis. This illustrates the wide latitude of choice possible for the complementary subspace $\langle \beta_1, \ldots, \beta_\rho \rangle$.

Theorem 1.7. *A linear transformation σ of U into V is a monomorphism if and only if $\nu(\sigma) = 0$, and it is an epimorphism if and only if $\rho(\sigma) = \dim V$.*

PROOF. $K(\sigma) = \{0\}$ if and only if $\nu(\sigma) = 0$. If σ is a monomorphism, then certainly $K(\sigma) = \{0\}$ and $\nu(\sigma) = 0$. On the other hand, if $\nu(\sigma) = 0$ and $\sigma(\alpha) = \sigma(\beta)$, then $\sigma(\alpha - \beta) = 0$ so that $\alpha - \beta \in K(\sigma) = \{0\}$. Thus, if $\nu(\sigma) = 0$, σ is a monomorphism.

It is but a matter of reading the definitions to see that σ is an epimorphism if and only if $\rho(\sigma) = \dim V$. □

If $\dim U = n < \dim V = m$, then $\rho(\sigma) = n - \nu(\sigma) \leq n < m$ so that σ cannot be an epimorphism. If $n > m$, then $\nu(\sigma) = n - \rho(\sigma) \geq n - m > 0$, so that σ cannot be a monomorphism. Any linear transformation from a vector space into a vector space of higher dimension must fail to be an epimorphism. Any linear transformation from a vector space into a vector space of lower dimension must fail to be a monomorphism.

Theorem 1.8. *Let U and V have the same finite dimension n. A linear transformation σ of U into V is an isomorphism if and only if it is an epimorphism. σ is an isomorphism if and only if it is a monomorphism.*

PROOF. It is part of the definition of an isomorphism that it is both an epimorphism and a monomorphism. Suppose σ is an epimorphism. $\rho(\sigma) = n$ and $\nu(\sigma) = 0$ by Theorem 1.6. Hence, σ is a monomorphism. Conversely if σ is a monomorphism, then $\nu(\sigma) = 0$ and, by Theorem 1.6, $\rho(\sigma) = n$. Hence, σ is an epimorphism. □

Thus a linear transformation σ of U into V is an isomorphism if two of the following three conditions are satisfied: (1) dim $U =$ dim V, (2) σ is an epimorphism, (3) σ is a monomorphism.

Theorem 1.9. $\rho(\sigma) = \rho(\tau\sigma) + $ dim $\{\text{Im}(\sigma) \cap K(\tau)\}$.

PROOF. Let τ' be a new linear transformation defined on $\text{Im}(\sigma)$ mapping $\text{Im}(\sigma)$ into W so that for all $\alpha \in \text{Im}(\sigma)$, $\tau'(\alpha) = \tau(\alpha)$. Then $K(\tau') = \text{Im}(\sigma) \cap K(\tau)$ and $\rho(\tau') = $ dim $\tau[\text{Im}(\sigma)] = $ dim $\tau\sigma(U) = \rho(\tau\sigma)$. Then Theorem 1.6 takes the form

$$\rho(\tau') + \nu(\tau') = \text{dim Im}(\sigma),$$

or

$$\rho(\tau\sigma) + \text{dim } \{\text{Im}(\sigma) \cap K(\tau)\} = \rho(\sigma). \ \square$$

Corollary 1.10. $\rho(\tau\sigma) = $ dim $\{\text{Im}(\sigma) + K(\tau)\} - \nu(\tau)$.

PROOF. This follows from Theorem 1.9 by application of Theorem 4.8 of Chapter I. $\square$

Corollary 1.11. If $K(\tau) \subset \text{Im}(\sigma)$, then $\rho(\sigma) = \rho(\tau\sigma) + \nu(\tau)$. $\square$

Theorem 1.12. *The rank of a product of linear transformations is less than or equal to the rank of either factor:* $\rho(\tau\sigma) \leq $ min $\{\rho(\tau), \rho(\sigma)\}$.

PROOF. The rank of $\tau\sigma$ is the dimension of $\tau[\sigma(U)] \subset \tau(V)$. Thus considering dim $\sigma(U)$ as the "n" and dim $\tau(V)$ as the "m" of Theorem 1.3 we see that dim $\tau\sigma(U) = \rho(\tau\sigma) \leq $ min $\{\text{dim } \sigma(U), \text{dim } \tau(V)\} = $ min $\{\rho(\sigma), \rho(\tau)\}$. $\square$

Theorem 1.13. *If σ is an epimorphism, then $\rho(\tau\sigma) = \rho(\tau)$. If τ is a monomorphism, then $\rho(\tau\sigma) = \rho(\sigma)$.*

PROOF. If σ is an epimorphism, then $K(\tau) \subset \text{Im}(\sigma) = V$ and Corollary 1.11 applies. Thus $\rho(\tau\sigma) = \rho(\sigma) - \nu(\tau) = m - \nu(\tau) = \rho(\tau)$. If τ is a monomorphism, then $K(\tau) = \{0\} \subset \text{Im}(\sigma)$ and Corollary 1.11 applies. Thus $\rho(\tau\sigma) = \rho(\sigma) - \nu(\tau) = \rho(\sigma)$. $\square$

Corollary 1.14. *The rank of a linear transformation is not changed by multiplication by an isomorphism (on either side).* $\square$

Theorem 1.15. *σ is an epimorphism if and only if $\tau\sigma = 0$ implies $\tau = 0$. τ is a monomorphism if and only if $\tau\sigma = 0$ implies $\sigma = 0$.*

PROOF. Suppose σ is an epimorphism. Assume $\tau\sigma$ is defined and $\tau\sigma = 0$. If $\tau \neq 0$, there is a $\beta \in V$ such that $\tau(\beta) \neq 0$. Since σ is an epimorphism, there is an $\alpha \in U$ such that $\sigma(\alpha) = \beta$. Then $\tau\sigma(\alpha) = \tau(\beta) \neq 0$. This is a contradiction and hence $\tau = 0$. Now, suppose $\tau\sigma = 0$ implies $\tau = 0$. If σ is not an epimorphism then $\text{Im}(\sigma)$ is a subspace of V but $\text{Im}(\sigma) \neq V$. Let $\{\beta_1, \ldots, \beta_r\}$ be a basis of $\text{Im}(\sigma)$, and extend this independent set to a basis $\{\beta_1, \ldots, \beta_r, \ldots, \beta_m\}$ of V. Define $\tau(\beta_i) = \beta_i$ for $i > r$ and $\tau(\beta_i) = 0$ for

$i \leq r$. Then $\tau\sigma = 0$ and $\tau \neq 0$. This is a contradiction and, hence, σ is an epimorphism.

Now, assume $\tau\sigma$ is defined and $\tau\sigma = 0$. Suppose τ is a monomorphism. If $\sigma \neq 0$, there is an $\alpha \in U$ such that $\sigma(\alpha) \neq 0$. Since τ is a monomorphism, $\tau\sigma(\alpha) \neq 0$. This is a contradiction and, hence, $\sigma = 0$. Now assume $\tau\sigma = 0$ implies $\sigma = 0$. If τ is not a monomorphism there is an $\alpha \in U$ such that $\alpha \neq 0$ and $\tau(\alpha) = 0$. Let $\{\alpha_1, \ldots, \alpha_n\}$ be any basis of U. Define $\sigma(\alpha_i) = \alpha$ for each i. Then $\tau\sigma(\alpha_i) = \tau(\alpha) = 0$ for all i and $\tau\sigma = 0$. This is a contradiction and, hence, τ is a monomorphism. $\square$

Corollary 1.16. *σ is an epimorphism if and only if $\tau_1\sigma = \tau_2\sigma$ implies $\tau_1 = \tau_2$. τ is a monomorphism if and only if $\tau\sigma_1 = \tau\sigma_2$ implies $\sigma_1 = \sigma_2$.*

The statement that $\tau_1\sigma = \tau_2\sigma$ implies $\tau_1 = \tau_2$ is called a *right-cancellation*, and the statement that $\tau\sigma_1 = \tau\sigma_2$ implies $\sigma_1 = \sigma_2$ is called a *left-cancellation*. Thus, an epimorphism is a linear transformation that can be cancelled on the right, and a monomorphism is a linear transformation that can be cancelled on the left.

Theorem 1.17. *Let $A = \{\alpha_1, \ldots, \alpha_n\}$ be any basis of U. Let $B = \{\beta_1, \ldots, \beta_n\}$ be any n vectors in V (not necessarily linearly independent). There exists a uniquely determined linear transformation σ of U into V such that $\sigma(\alpha_i) = \beta_i$ for $i = 1, 2, \ldots, n$.*

PROOF.   Since A is a basis of U, any vector $\alpha \in U$ can be expressed uniquely in the form $\alpha = \sum_{i=1}^{n} a_i\alpha_i$. If σ is to be linear we must have

$$\sigma(\alpha) = \sum_{i=1}^{n} a_i\sigma(\alpha)_i = \sum_{i=1}^{n} a_i\beta_i \in U.$$

It is a simple matter to verify that the mapping so defined is linear. $\square$

Corollary 1.18. *Let $C = \{\gamma_1, \ldots, \gamma_r\}$ be any linearly independent set in U, where U is finite dimensional. Let $D = \{\delta_1, \ldots, \delta_r\}$ be any r vectors in V. There exists a linear transformation σ of U into V such that $\sigma(\gamma_i) = \delta_i$ for $i = 1, \ldots, r$.*

PROOF.   Extend C to a basis of U. Define $\sigma(\gamma_i) = \delta_i$ for $i = 1, \ldots, r$, and define the values of σ on the other elements of the basis arbitrarily. This will yield a linear transformation σ with the desired properties. $\square$

It should be clear that, if C is not already a basis, there are many ways to define σ. It is worth pointing out that the independence of the set C is crucial to proving the existence of the linear transformation with the desired properties. Otherwise, a linear relation among the elements of C would impose a corresponding linear relation among the elements of D, which would mean that D could not be arbitrary.

Theorem 1.17 establishes, for one thing, that linear transformations really do exist. Moreover, they exist in abundance. The real utility of this theorem and its corollary is that it enables us to establish the existence of a linear transformation with some desirable property with great convenience. All we have to do is to define this function on an independent set.

Definition. A linear transformation π of V into itself with the property that $\pi^2 = \pi$ is called a *projection*.

Theorem 1.19. *If π is a projection of V into itself, then $V = \mathrm{Im}(\pi) \oplus K(\pi)$ and π acts like the identity on $\mathrm{Im}(\pi)$.*

PROOF. For $\alpha \in V$, let $\alpha_1 = \pi(\alpha)$. Then $\pi(\alpha_1) = \pi^2(\alpha) = \pi(\alpha) = \alpha_1$. This shows that π acts like the identity on $\mathrm{Im}(\pi)$. Let $\alpha_2 = \alpha - \alpha_1$. Then $\pi(\alpha_2) = \pi(\alpha) - \pi(\alpha_1) = \alpha_1 - \alpha_1 = 0$. Thus $\alpha = \alpha_1 + \alpha_2$ where $\alpha_1 \in \mathrm{Im}(\pi)$ and $\alpha_2 \in K(\pi)$. Clearly, $\mathrm{Im}(\pi) \cap K(\pi) = \{0\}$. $\square$

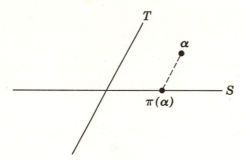

Fig. 1

If $S = \mathrm{Im}(\pi)$ and $T = K(\pi)$, we say that π is a projection of V onto S along T. In the case where V is the real plane, Fig. 1 indicates the interpretation of these words. α is projected onto a point of S in a direction parallel to T.

EXERCISES

1. Show that $\sigma((x_1, x_2)) = (x_2, x_1)$ defines a linear transformation of R^2 into itself.

2. Let $\sigma_1((x_1, x_2)) = (x_2, -x_1)$ and $\sigma_2((x_1, x_2)) = (x_1, -x_2)$. Determine $\sigma_1 + \sigma_2$, $\sigma_1\sigma_2$ and $\sigma_2\sigma_1$.

3. Let $U = V = R^n$ and let $\sigma((x_1, x_2, \ldots, x_n)) = (x_1, x_2, \ldots, x_k, 0, \ldots, 0)$ where $k < n$. Describe $\mathrm{Im}(\sigma)$ and $K(\sigma)$.

4. Let $\sigma((x_1, x_2, x_3, x_4)) = (3x_1 - 2x_2 - x_3 - 4x_4, x_1 + x_2 - 2x_3 - 3x_4)$. Show that σ is a linear transformation. Determine the kernel of σ.

5. Let $\sigma((x_1, x_2, x_3)) = (2x_1 + x_2 + 3x_3, 3x_1 - x_2 + x_3, -4x_1 + 3x_2 + x_3)$. Find

a basis of $\sigma(U)$. (*Hint:* Take particular values of the x_i to find a spanning set for $\sigma(U)$.) Find a basis of $K(\sigma)$.

6. Let D denote the operator of differentiation,

$$D(y) = \frac{dy}{dx}, \ D^2(y) = D[D(y)] = \frac{d^2y}{dx^2}, \ \text{etc.}$$

Show that D^n is a linear transformation, and also that $p(D)$ is a linear transformation if $p(D)$ is a polynomial in D with constant coefficients. (Here we must assume that the space of functions on which D is defined contains only functions differentiable at least as often as the degree of $p(D)$.)

7. Let $U = V$ and let σ and τ be linear transformations of U into itself. In this case $\sigma\tau$ and $\tau\sigma$ are both defined. Construct an example to show that it is not always true that $\sigma\tau = \tau\sigma$.

8. Let $U = V = P$, the space of polynomials in x with coefficients in R. For $\alpha = \sum_{i=0}^{n} a_i x^i$ let

$$\sigma(\alpha) = \sum_{i=0}^{n} i a_i x^{i-1}$$

and

$$\tau(\alpha) = \sum_{i=0}^{n} \frac{a_i}{i+1} x^{i+1}.$$

Show that $\sigma\tau = 1$, but that $\tau\sigma \neq 1$.

9. Show that if two scalar transformations coincide on U then the defining scalars are equal.

10. Let σ be a linear transformation of U into V and let $A = \{\alpha_1, \dots, \alpha_n\}$ be a basis of U. Show that if the values $\{\sigma(\alpha_1), \dots, \sigma(\alpha_n)\}$ are known, then the value of $\sigma(\alpha)$ can be computed for each $\alpha \in U$.

11. Let U and V be vector spaces of dimensions n and m, respectively, over the same field F. We have already commented that the set of all linear transformations of U into V forms a vector space. Give the details of the proof of this assertion. Let $A = \{\alpha_1, \dots, \alpha_n\}$ be a basis of U and $B = \{\beta_1, \dots, \beta_m\}$ be a basis of V. Let σ_{ij} be the linear transformation of U into V such that

$$\sigma_{ij}(\alpha_k) = \begin{cases} 0 & \text{if } k \neq j, \\ \beta_i & \text{if } k = j. \end{cases}$$

Show that $\{\sigma_{ij} \mid i = 1, \dots, m; j = 1, \dots, n\}$ is a basis of this vector space.

For the following sequence of problems let dim $U = n$ and dim $V = m$. Let σ be a linear transformation of U into V and τ a linear transformation of V into W.

12. Show that $\rho(\sigma) \leq \rho(\tau\sigma) + \nu(\tau)$. (*Hint:* Let $V' = \sigma(U)$ and apply Theorem 1.6 to τ defined on V'.)

13. Show that max $\{0, \rho(\sigma) + \rho(\tau) - m\} \leq \rho(\tau\sigma) \leq \min \{\rho(\tau), \rho(\sigma)\}$.

14. Show that max $\{n - m + v(\tau), v(\sigma)\} \leq v(\tau\sigma) \leq$ min $\{n, v(\sigma) + v(\tau)\}$. (For $m = n$ this inequality is known as Sylvester's law of nullity.)

15. Show that if $v(\tau) = 0$, then $\rho(\tau\sigma) = \rho(\sigma)$.

16. It is not generally true that $v(\sigma) = 0$ implies $\rho(\tau\sigma) = \rho(\tau)$. Construct an example to illustrate this fact. (*Hint:* Let m be very large.)

17. Show that if $m = n$ and $v(\sigma) = 0$, then $\rho(\tau\sigma) = \rho(\tau)$.

18. Show that if σ_1 and σ_2 are linear transformations of U into V, then

$$\rho(\sigma_1 + \sigma_2) \leq \text{min } \{m, n, \rho(\sigma_1) + \rho(\sigma_2)\}.$$

19. Show that $|\rho(\sigma_1) - \rho(\sigma_2)| \leq \rho(\sigma_1 + \sigma_2)$.

20. If S is any subspace of V there is a subspace T such that $V = S \oplus T$. Then every $x \in V$ can be represented uniquely in the form $\alpha = \alpha_1 + \alpha_2$ where $\alpha_1 \in S$ and $\alpha_2 \in T$. Show that the mapping π which maps α onto α_1 is a linear transformation. Show that T is the kernel of π. Show that $\pi^2 = \pi$. The mapping π is called a *projection of* V onto S along T.

21. (Continuation) Let π be a projection. Show that $1 - \pi$ is also a projection. What is the kernel of $1 - \pi$? Onto what subspace is $1 - \pi$ a projection? Show that $\pi(1 - \pi) = 0$.

2 | Matrices

Definition. A *matrix* over a field F is a rectangular array of scalars. The array will be written in the form

$$\begin{bmatrix} a_{11} & a_{12} & \cdots & a_{1n} \\ a_{21} & a_{22} & \cdots & a_{2n} \\ \cdot & \cdot & & \cdot \\ \cdot & \cdot & & \cdot \\ \cdot & \cdot & & \cdot \\ a_{m1} & a_{m2} & \cdots & a_{mn} \end{bmatrix} \qquad (2.1)$$

whenever we wish to display all the elements in the array or show the form of the array. A matrix with m rows and n columns is called an $m \times n$ matrix. An $n \times n$ matrix is said to be of *order n*.

We often abbreviate a matrix written in the form above to $[a_{ij}]$ where the first index denotes the number of the row and the second index denotes the number of the column. The particular letter appearing in each index position is immaterial; it is the position that is important. With this convention a_{ij} is a scalar and $[a_{ij}]$ is a matrix. Whereas the elements a_{ij} and a_{kl} need not be equal, we consider the matrices $[a_{ij}]$ and $[a_{kl}]$ to be identical since both $[a_{ij}]$ and $[a_{kl}]$ stand for the entire matrix. As a further convenience we often use upper case Latin italic letters to denote matrices; $A = [a_{ij}]$. Whenever we use lower case Latin italic letters to denote the scalars appearing in the matrix, we use the corresponding upper case Latin italic letter to denote the matrix. The matrix in which all scalars are zero is denoted by 0 (the third use of this symbol!). The a_{ij} appearing in the array $[a_{ij}]$ are called the *elements* of $[a_{ij}]$. Two matrices are equal if and only if they have exactly the same elements. The *main diagonal* of the matrix $[a_{ij}]$ is the set of elements $\{a_{11}, \ldots, a_{tt}\}$ where $t = \min \{m, n\}$. A *diagonal matrix* is a square matrix in which the elements not in the main diagonal are zero.

Matrices can be used to represent a variety of different mathematical concepts. The way matrices are manipulated depends on the objects which they represent. Considering the wide variety of situations in which matrices have found application, there is a remarkable similarity in the operations performed on matrices in these situations. There are differences too, however, and to understand these differences we must understand the object represented and what information can be expected by manipulating with the matrices. We first investigate the properties of matrices as representations of linear transformations. Not only do the matrices provide us with a convenient means of doing whatever computation is necessary with linear transformations, but the theory of vector spaces and linear transformations also proves to be a powerful tool in developing the properties of matrices.

Let U be a vector space of dimension n and V a vector space of dimension m, both over the same field F. Let $A = \{\alpha_1, \ldots, \alpha_n\}$ be an arbitrary but fixed basis of U, and let $B = \{\beta_1, \ldots, \beta_m\}$ be an arbitrary but fixed basis of V. Let σ be a linear transformation of U into V. Since $\sigma(\alpha_j) \in V$, $\sigma(\alpha_j)$ can be expressed uniquely as a linear combination of the elements of B;

$$\sigma(\alpha_j) = \sum_{i=1}^{m} a_{ij}\beta_i. \qquad (2.2)$$

We define the *matrix representing σ with respect to the bases A and B* to be the matrix $A = [a_{ij}]$.

The correspondence between linear transformations and matrices is actually one-to-one and onto. Given the linear transformation σ, the a_{ij} exist because B spans V, and they are unique because B is linearly independent. On the other hand, let $A = [a_{ij}]$ be any $m \times n$ matrix. We can define $\sigma(\alpha_j) = \sum_{i=1}^{m} a_{ij}\beta_i$ for each $\alpha_j \in A$, and then we can extend the proposed linear transformation to all of U by the condition that it be linear. Thus, if $\xi = \sum_{j=1}^{n} x_j\alpha_j$, we define

$$\sigma(\xi) = \sum_{j=1}^{n} x_j\sigma(\alpha_j)$$

$$= \sum_{j=1}^{n} x_j \left(\sum_{i=1}^{m} a_{ij}\beta_i \right)$$

$$= \sum_{i=1}^{m} \left(\sum_{j=1}^{n} a_{ij}x_j \right)\beta_i. \qquad (2.3)$$

σ can be extended to all of U because A spans U, and the result is well defined (unique) because A is linearly independent.

Here are some examples of linear transformations and the matrices which represent them. Consider the real plane $R^2 = U = V$. Let $A = B = \{(1, 0), (0, 1)\}$. A $90°$ rotation counterclockwise would send $(1, 0)$ onto $(0, 1)$ and it would send $(0, 1)$ onto $(-1, 0)$. Since $\sigma((1, 0)) = 0 \cdot (1, 0) + 1 \cdot (0, 1)$ and $\sigma((0, 1)) = (-1) \cdot (1, 0) + 0 \cdot (0, 1)$, σ is represented by the matrix

$$\begin{bmatrix} 0 & -1 \\ 1 & 0 \end{bmatrix}.$$

The elements appearing in a column are the coordinates of each image of a basis vector under a transformation.

In general, a rotation counterclockwise through an angle of θ will send $(1, 0)$ onto $(\cos \theta, \sin \theta)$ and $(0, 1)$ onto $(-\sin \theta, \cos \theta)$. Thus this rotation is represented by

$$\begin{bmatrix} \cos \theta & -\sin \theta \\ \sin \theta & \cos \theta \end{bmatrix}. \tag{2.4}$$

Suppose now that τ is another linear transformation of U into V represented by the matrix $B = [b_{ij}]$. Then for the sum $\sigma + \tau$ we have

$$(\sigma + \tau)(\alpha_j) = \sigma(\alpha_j) + \tau(\alpha_j) = \sum_{i=1}^{m} a_{ij}\beta_i + \sum_{i=1}^{m} b_{ij}\beta_i$$

$$= \sum_{i=1}^{m} (a_{ij} + b_{ij})\beta_i. \tag{2.5}$$

Thus $\sigma + \tau$ is represented by the matrix $[a_{ij} + b_{ij}]$. Accordingly, we define the *sum* of two matrices to be that matrix obtained by the addition of the corresponding elements in the two arrays; $A + B = [a_{ij} + b_{ij}]$ is the matrix corresponding to $\sigma + \tau$. The sum of two matrices is defined if and only if the two matrices have the same number of rows and the same number of columns.

If a is any scalar, for the linear transformation $a\sigma$ we have

$$(a\sigma)(\alpha_j) = a \sum_{i=1}^{m} a_{ij}\beta_i = \sum_{i=1}^{m} (aa_{ij})\beta_i. \tag{2.6}$$

Thus $a\sigma$ is represented by the matrix $[aa_{ij}]$. We therefore define *scalar multiplication* by the rule $aA = [aa_{ij}]$.

Let W be a third vector space of dimension r over the field F, and let $C = \{\gamma_1, \ldots, \gamma_r\}$ be an arbitrary but fixed basis of W. If the linear transformation σ of U into V is represented by the $m \times n$ matrix $A = [a_{ij}]$ and the linear transformation τ of V into W is represented by the $r \times m$ matrix $B = [b_{ki}]$, what matrix represents the linear transformation $\tau\sigma$ of U into W?

$$(\tau\sigma)(\alpha_j) = \tau(\sigma(\alpha_j)) = \tau\left(\sum_{i=1}^{m} a_{ij}\beta_i\right)$$

$$= \sum_{i=1}^{m} a_{ij}\tau(\beta_i)$$

$$= \sum_{i=1}^{m} a_{ij}\left(\sum_{k=1}^{r} b_{ki}\gamma_k\right)$$

$$= \sum_{k=1}^{r} \left(\sum_{i=1}^{m} b_{ki}a_{ij}\right)\gamma_k. \tag{2.7}$$

Thus, if we define $c_{kj} = \sum_{i=1}^{m} b_{ki}a_{ij}$, then $C = [c_{kj}]$ is the matrix representing the product transformation $\tau\sigma$. Accordingly, we call C the *matrix product* of B and A, in that order: $C = BA$.

For computational purposes it is customary to write the arrays of B and A side by side. The element c_{kj} of the product is then obtained by multiplying the corresponding elements of row k of B and column j of A and adding. We can trace the elements of row k of B with a finger of the left hand while at the same time tracing the elements of column j of A with a finger of the right hand. At each step we compute the product of the corresponding elements and accumulate the sum as we go along. Using this simple rule we can, with practice, become quite proficient, even to the point of doing "without hands."

Check the process in the following examples:

$$\begin{bmatrix} 1 & 4 & -1 & 2 \\ 0 & 2 & 1 & 3 \\ -2 & 1 & -2 & 2 \end{bmatrix} \begin{bmatrix} 1 & -1 \\ 0 & 2 \\ 2 & 1 \\ 3 & -2 \end{bmatrix} = \begin{bmatrix} 5 & 2 \\ 11 & -1 \\ 0 & -2 \end{bmatrix}.$$

All definitions and properties we have established for linear transformations can be carried over immediately for matrices. For example, we have:

1. $0 \cdot A = 0$. (The "0" on the left is a scalar, the "0" on the right is a matrix with the same number of rows and columns as A.)

2. $1 \cdot A = A$.

3. $A(B + C) = AB + AC$.

4. $(A + B)C = AC + BC$.

5. $A(BC) = (AB)C$.

Of course, in each of the above statements we must assume the operations proposed are well defined. For example, in 3, B and C must be the same size and A must have the same number of columns as B and C have rows.

The *rank* and *nullity* of a matrix A are the rank and nullity of the associated linear transformation, respectively.

Theorem 2.1. *For an* $m \times n$ *matrix* A, *the rank of* A *plus the nullity of* A *is equal to* n. *The rank of a product* BA *is less than or equal to the rank of either factor.*

These statements have been established for linear transformations and therefore hold for their corresponding matrices. $\square$

The rank of σ is the dimension of the subspace $\text{Im}(\sigma)$ of V. Since $\text{Im}(\sigma)$ is spanned by $\{\sigma(\alpha_1), \ldots, \sigma(\alpha_n)\}$, $\rho(\sigma)$ is the number of elements in a maximal linearly independent subset of $\{\sigma(\alpha_1), \ldots, \sigma(\alpha_n)\}$. Expressed in terms of coordinates, $\sigma(\alpha_j) = \sum_{i=1}^{m} a_{ij}\beta_i$ is represented by the m-tuple $(a_{1j}, a_{2j}, \ldots, a_{mj})$, which is the m-tuple in column j of the matrix $[a_{ij}]$. Thus $\rho(\sigma) = \rho(A)$ is also equal to the maximum number of linearly independent columns of A. This is usually called the column rank of a matrix A, and the maximum number of linearly independent rows of A is called the row rank of A. We, however, show before long that the number of linearly independent rows in a matrix is equal to the number of linearly independent columns. Until that time we consider "rank" and "column rank" as synonymous.

Returning to Equation (2.3), we see that, if $\xi \in U$ is represented by $(x_1, \ldots, x_n)$ and the linear transformation σ of U into V is represented by the matrix $A = [a_{ij}]$, then $\sigma(\xi) \in V$ is represented by $(y_1, \ldots, y_m)$ where

$$y_i = \sum_{j=1}^{n} a_{ij}x_j \qquad (i = 1, \ldots, m). \qquad (2.8)$$

In view of the definition of matrix multiplication given by Equation (2.7) we can interpret Equations (2.8) as a matrix product of the form

$$Y = AX \qquad (2.9)$$

where

$$Y = \begin{bmatrix} y_1 \\ \cdot \\ \cdot \\ \cdot \\ y_m \end{bmatrix} \quad \text{and} \quad X = \begin{bmatrix} x_1 \\ \cdot \\ \cdot \\ \cdot \\ x_n \end{bmatrix}.$$

This single matric equation contains the m equations in (2.8).

We have already used the n-tuple $(x_1, \ldots, x_n)$ to represent the vector $\xi = \sum_{i=1}^{n} x_i\alpha_i$. Because of the usefulness of equation (2.9) we also find it convenient to represent ξ by the one-column matrix X. In fact, since it is somewhat wasteful of space and otherwise awkward to display one-column matrices we use the n-tuple $(x_1, \ldots, x_n)$ to represent not only the vector ξ but also the column matrix X. With this convention $[x_1 \cdots x_n]$ is a one-row matrix and $(x_1, \ldots, x_n)$ is a one-column matrix.

Notice that we have now used matrices for two different purposes, (1) to represent linear transformations, and (2) to represent vectors. The single matric equation $Y = AX$ contains some matrices used in each way.

EXERCISES

1. Verify the matrix multiplication in the following examples:

(a)
$$\begin{bmatrix} 3 & 1 & -2 \\ -5 & 2 & 3 \end{bmatrix} \begin{bmatrix} 2 & 1 & -3 \\ -1 & 6 & 1 \\ 1 & 0 & -2 \end{bmatrix} = \begin{bmatrix} 3 & 9 & -4 \\ -9 & 7 & 11 \end{bmatrix}.$$

(b)
$$\begin{bmatrix} 2 & 1 & -3 \\ -1 & 6 & 1 \\ 1 & 0 & -2 \end{bmatrix} \begin{bmatrix} 2 \\ 3 \\ -1 \end{bmatrix} = \begin{bmatrix} 10 \\ 15 \\ 4 \end{bmatrix}.$$

(c)
$$\begin{bmatrix} 3 & 1 & -2 \\ -5 & 2 & 3 \end{bmatrix} \begin{bmatrix} 10 \\ 15 \\ 4 \end{bmatrix} = \begin{bmatrix} 37 \\ -8 \end{bmatrix}.$$

2. Compute
$$\begin{bmatrix} 3 & 9 & -4 \\ -9 & 7 & 11 \end{bmatrix} \begin{bmatrix} 2 \\ 3 \\ -1 \end{bmatrix}.$$

Interpret the answer to this problem in terms of the computations in Exercise 1.

3. Find AB and BA if

$$A = \begin{bmatrix} 1 & 0 & 0 & 1 \\ 0 & 1 & 1 & 0 \\ 1 & 0 & 1 & 0 \\ 0 & 1 & 0 & 1 \end{bmatrix}, \quad B = \begin{bmatrix} 1 & 2 & 3 & 4 \\ 5 & 6 & 7 & 8 \\ -1 & -2 & -3 & -4 \\ -5 & -6 & -7 & -8 \end{bmatrix}.$$

4. Let σ be a linear transformation of R^2 into itself that maps $(1, 0)$ onto $(3, -1)$ and $(0, 1)$ onto $(-1, 2)$. Determine the matrix representing σ with respect to the bases $A = B = \{(1, 0), (0, 1)\}$.

5. Let σ be a linear transformation of R^2 into itself that maps $(1, 1)$ onto $(2, -3)$ and $(1, -1)$ onto $(4, -7)$. Determine the matrix representing σ with respect to the bases $A = B = \{(1, 0), (0, 1)\}$. (*Hint:* We must determine the effect of σ when it is applied to $(1, 0)$ and $(0, 1)$. Use the fact that $(1, 0) = \frac{1}{2}(1, 1) + \frac{1}{2}(1, -1)$ and the linearity of σ.)

6. It happens that the linear transformation defined in Exercise 4 is one-to-one, that is, σ does not map two different vectors onto the same vector. Thus, there is a linear transformation that maps $(3, -1)$ onto $(1, 0)$ and $(-1, 2)$ onto $(0, 1)$. This linear transformation reverses the mapping given by σ. Determine the matrix representing it with respect to the same bases.

7. Let us consider the geometric meaning of linear transformations. A linear transformation of R^2 into itself leaves the origin fixed (why?) and maps straight lines into straight lines. (The word "into" is required here because the image of a straight line may be another straight line or it may be a single point.) Prove that the image of a straight line is a subset of a straight line. (*Hint:* Let σ be represented by the matrix

$$A = \begin{bmatrix} a_{11} & a_{12} \\ a_{21} & a_{22} \end{bmatrix}.$$

Then σ maps (x, y) onto $(a_{11}x + a_{12}y, a_{21}x + a_{22}y)$. Now show that if (x, y) satisfies the equation $ax + by = c$ its image satisfies the equation

$$(aa_{22} - ba_{21})x + (a_{11}b - a_{12}a)y = (a_{11}a_{22} - a_{12}a_{21}c.)$$

8. (Continuation) We say that a straight line is mapped onto itself if every point on the line is mapped onto a point on the line (but not all onto the same point) even though the points on the line may be moved around.

(*a*) A linear transformation maps $(1, 0)$ onto $(-1, 0)$ and $(0, 1)$ onto $(0, -1)$. Show that every line through the origin is mapped onto itself. Show that each such line is mapped onto itself with the sense of direction inverted. This linear transformation is called an *inversion* with respect to the origin. Find the matrix representing this linear transformation with respect to the basis $\{(1, 0), (0, 1)\}$.

(*b*) A linear transformation maps $(1, 1)$ onto $(-1, -1)$ and leaves $(1, -1)$ fixed. Show that every line perpendicular to the line $x_1 + x_2 = 0$ is mapped onto itself with the sense of direction inverted. Show that every point on the line $x_1 + x_2 = 0$ is left fixed. Which lines through the origin are mapped onto themselves? This linear transformation is called a *reflection* about the line $x_1 + x_2 = 0$. Find the matrix representing this linear transformation with respect to the basis $\{(1, 0), (0, 1)\}$. Find the matrix representing this linear transformation with respect to the basis $\{(1, 1), (1, -1)\}$.

(*c*) A liner transformation maps $(1, 1)$ onto $(2, 2)$ and $(1, -1)$ onto $(3, -3)$. Show that the lines through the origin and passing through the points $(1, 1)$ and $(1, -1)$ are mapped onto themselves and that no other lines are mapped onto themselves. Find the matrices representing this linear transformation with respect to the bases $\{(1, 0), (0, 1)\}$ and $\{(1, 1), (1, -1)\}$.

(*d*) A linear transformation leaves $(1, 0)$ fixed and maps $(0, 1)$ onto $(1, 1)$. Show that each line $x_2 = c$ is mapped onto itself and translated within itself a distance equal to c. This linear transformation is called a *shear*. Which lines through the origin are mapped onto themselves? Find the matrix representing this linear transformation with respect to the basis $\{(1, 0), (0, 1)\}$.

(*e*) A linear transformation maps $(1, 0)$ onto $(\frac{5}{13}, \frac{12}{13})$ and $(0, 1)$ onto $(-\frac{12}{13}, \frac{5}{13})$. Show that every line through the origin is rotated counterclockwise through the angle $\theta = \arccos \frac{5}{13}$. This linear transformation is called a *rotation*. Find the matrix representing this linear transformation with respect to the basis $\{(1, 0), (0, 1)\}$.

(*f*) A linear transformation maps $(1, 0)$ onto $(\frac{2}{3}, \frac{2}{3})$ and $(0, 1)$ onto $(\frac{1}{3}, \frac{1}{3})$. Show that each point on the line $2x_1 + x_2 = 3c$ is mapped onto the single point (c, c). The line $x_1 - x_2 = 0$ is left fixed. The only other line through the origin which is mapped into itself is the line $2x_1 + x_2 = 0$. This linear transformation is called a *projection* onto the line $x_1 - x_2 = 0$ parallel to the line $2x_1 + x_2 = 0$. Find the matrices representing this linear transformation with respect to the bases $\{(1, 0), (0, 1)\}$ and $\{(1, 1), (1, -2)\}$.

9. (Continuation) Describe the geometric effect of each of the linear transformations of R^2 into itself represented by the matrices

$$(a) \begin{bmatrix} 0 & 1 \\ 1 & 0 \end{bmatrix} \quad (b) \begin{bmatrix} 0 & 0 \\ 1 & 0 \end{bmatrix} \quad (c) \begin{bmatrix} 1 & 1 \\ 0 & 0 \end{bmatrix}$$

$$(d) \begin{bmatrix} 1 & 0 \\ a & 1 \end{bmatrix} \quad (e) \begin{bmatrix} b & 0 \\ 0 & c \end{bmatrix} \quad (f) \begin{bmatrix} \frac{3}{5} & -\frac{4}{5} \\ \frac{4}{5} & \frac{3}{5} \end{bmatrix}.$$

(*Hint:* In Exercise 7 we have shown that straight lines are mapped into straight lines. We already know that linear transformations map the origin onto the origin. Thus it is relatively easy to determine what happens to straight lines passing through the origin. For example, to see what happens to the x_1-axis it is sufficient to see what happens to the point $(1, 0)$. Among the transformations given appear a rotation, a reflection, two projections, and one shear.)

10. (Continuation) For the linear transformations given in Exercise 9 find all lines through the origin which are mapped onto or into themselves.

11. Let $U = R^2$ and $V = R^3$ and σ be a linear transformation of U into V that maps $(1, 1)$ onto $(0, 1, 2)$ and $(-1, 1)$ onto $(2, 1, 0)$. Determine the matrix that represents σ with respect to the bases $A = \{(1, 0), (0, 1)\}$ in $B = \{(1, 0, 0), (0, 1, 0), (0, 0, 1)\}$ in R^3. (*Hint:* $\frac{1}{2}(1, 1) - \frac{1}{2}(-1, 1) = (1, 0)$.)

12. What is the effect of multiplying an $n \times n$ matrix A by an $n \times n$ diagonal matrix D? What is the difference between AD and DA?

13. Let a and b be two numbers such that $a \neq b$. Find all 2×2 matrices A such that

$$A \begin{bmatrix} a & 0 \\ 0 & b \end{bmatrix} = \begin{bmatrix} a & 0 \\ 0 & b \end{bmatrix} A.$$

14. Show that the matrix $C = [a_i b_j]$ has rank one if not all a_i and not all b_j are zero. (*Hint:* Use Theorem 1.12.)

15. Let a, b, c, and d be given numbers (real or complex) and consider the function

$$f(x) = \frac{ax + b}{cx + d}.$$

Let g be another function of the same form. Show that gf where $gf(x) = g(f(x))$ is a function that can also be written in the same form. Show that each of these functions can be represented by a matrix in such a way that the matrix representing gf is the product of the matrices representing g and f. Show that the inverse function exists if and only if $ad - bc \neq 0$. To what does the function reduce if $ad - bc = 0$?

16. Consider complex numbers of the form $x + yi$ (where x and y are real numbers and $i^2 = -1$) and represent such a complex number by the duple (x, y) in R^2. Let $a + bi$ be a fixed complex number. Consider the function f defined by the rule

$$f(x + yi) = (a + bi)(x + yi) = u + vi.$$

(*a*) Show that this function is a linear transformation of R^2 into itself mapping (x, y) onto (u, v).

(*b*) Find the matrix representing this linear transformation with respect to the basis $\{(1, 0), (0, 1)\}$.

(*c*) Find the matrix which represents the linear transformation obtained by using $c + di$ in place of $a + bi$. Compute the product of these two matrices. Do they commute?

(*d*) Determine the complex number which can be used in place of $a + bi$ to obtain a transformation represented by this matrix product. How is this complex number related to $a + bi$ and $c + di$?

17. Show by example that it is possible for two matrices A and B to have the same rank while A^2 and B^2 have different ranks.

3 | Non-singular Matrices

Let us consider the case where $U = V$, that is, we are considering transformations of V into itself. Generally, a homomorphism of a set into itself is called an *endomorphism*. We consider a fixed basis in V and represent the linear transformation of V into itself with respect to that basis. In this case the matrices are square or $n \times n$ matrices. Since the transformations we are considering map V into itself any finite number of them can be iterated in any order. The commutative law does not hold, however. The same remarks hold for square matrices. They can be multiplied in any order but the commutative law does not hold. For example

$$\begin{bmatrix} 0 & 1 \\ 0 & 0 \end{bmatrix} \begin{bmatrix} 0 & 0 \\ 0 & 1 \end{bmatrix} = \begin{bmatrix} 0 & 1 \\ 0 & 0 \end{bmatrix},$$

$$\begin{bmatrix} 0 & 0 \\ 0 & 1 \end{bmatrix} \begin{bmatrix} 0 & 1 \\ 0 & 0 \end{bmatrix} = \begin{bmatrix} 0 & 0 \\ 0 & 0 \end{bmatrix}.$$

The linear transformation that leaves every element of V fixed is the identity transformation. We denote the identity transformation by 1, the scalar identity. Clearly, the identity transformation is represented by the matrix $I = [\delta_{ij}]$ for any choice of the basis. Notice that $IA = AI = A$ for any $n \times n$ matrix A. I is called the *identity matrix*, or *unit matrix*, of order n. If we wish to point out the dimension of the space we write I_n for the identity matrix of order n. The scalar transformation a is represented by the matrix aI. Matrices of the form aI are called *scalar matrices*.

Definition. A one-to-one linear transformation σ of a vector space onto itself is called an *automorphism*. An automorphism is only a special kind of isomorphism for which the domain and codomain are the same space. If $\sigma(\alpha) = \bar{\alpha}$, the mapping $\sigma^{-1}(\bar{\alpha}) = \alpha$ is called the *inverse transformation* of σ. The rotations represented in Section 2 are examples of automorphisms.

Theorem 3.1. *The inverse σ^{-1} of an automorphism σ is an automorphism.*

Theorem 3.2 *A linear transformation τ of an n-dimensional vector space into itself is an automorphism if and only if it is of rank n; that is, if and only if it is an epimorphism.*

Theorem 3.3. *A linear transformation σ of an n-dimensional vector space into itself is an automorphism if and only if its nullity is 0, that is, if and only if it is a monomorphism.*

PROOF (of Theorems 3.1, 3.2, and 3.3). These properties have already been established for isomorphisms. □

Since it is clear that transformations of rank less than n do not have inverses because they are not onto, we see that automorphisms are the only linear transformations which have inverses. A linear transformation that has an inverse is said to be *non-singular* or *invertible*; otherwise it is said to be *singular*. Let A be the matrix representing the automorphism σ, and let A^{-1} be the matrix representing the inverse transformation σ^{-1}. The matrix $A^{-1}A$ represents the transformation $\sigma^{-1}\sigma$. Since $\sigma^{-1}\sigma$ is the identity transformation, we must have $A^{-1}A = I$. But σ is also the inverse transformation of σ^{-1} so that $\sigma\sigma^{-1} = 1$ and $AA^{-1} = I$. We shall refer to A^{-1} as the *inverse* of A. A matrix that has an inverse is said to be *non-singular* or *invertible*. Only a square matrix can have an inverse.

On the other hand suppose that for the matrix A there exists a matrix B such that $BA = I$. Since I is of rank n, A must also be of rank n and, therefore, A represents an automorphism σ. Furthermore, the linear transformation which B represents is necessarily the inverse transformation σ^{-1} since the product with σ must yield the identity transformation. Thus $B = A^{-1}$. The same kind of argument shows that if C is a matrix such that $AC = I$, then $C = A^{-1}$. Thus we have shown:

Theorem 3.4. *If A and B are square matrices such that $BA = I$, then $AB = I$. If A and B are square matrices such that $AB = I$, then $BA = I$. In either case B is the unique inverse of A.* $\square$

Theorem 3.5. *If A and B are non-singular, then* (1) *AB is non-singular and $(AB)^{-1} = B^{-1}A^{-1}$,* (2) *A^{-1} is non-singular and $(A^{-1})^{-1} = A$,* (3) *for $a \neq 0$, aA is non-singular and $(aA)^{-1} = a^{-1}A^{-1}$.*

PROOF.   In view of the remarks preceding Theorem 3.4 it is sufficient in each case to produce a matrix which will act as a left inverse.

(1) $(B^{-1}A^{-1})(AB) = B^{-1}(A^{-1}A)B = B^{-1}IB = B^{-1}B = I.$

(2) $AA^{-1} = I.$

(3) $(a^{-1}A^{-1})(aA) = (a^{-1}a)(A^{-1}A) = I.$ $\square$

Theorem 3.6. *If A is non-singular, we can solve uniquely the equations $XA = B$ and $AY = B$ for any matrix B of the proper size, but the two solutions need not be equal.*

PROOF.   Solutions exist since $(BA^{-1})A = B(A^{-1}A) = B$ and $A(A^{-1}B) = (AA^{-1})B = B$. The solutions are unique since for any C having the property that $CA = B$ we have $C = CAA^{-1} = BA^{-1}$, and similarly with any solution of $AY = B$. $\square$

As an example illustrating the last statement of the theorem, let

$$A = \begin{bmatrix} 1 & 2 \\ 0 & 1 \end{bmatrix}, \qquad A^{-1} = \begin{bmatrix} 1 & -2 \\ 0 & 1 \end{bmatrix}, \qquad B = \begin{bmatrix} 1 & 0 \\ 2 & 1 \end{bmatrix}.$$

Then

$$X = BA^{-1} = \begin{bmatrix} 1 & -2 \\ 2 & -3 \end{bmatrix}, \qquad \text{and} \qquad Y = A^{-1}B = \begin{bmatrix} -3 & -2 \\ 2 & 1 \end{bmatrix}.$$

We add the remark that for non-singular A, the solution of $XA = B$ exists and is unique if B has n columns, and the solution of $AY = B$ exists and is unique if B has n rows. The proof given for Theorem 3.6 applies without change.

Theorem 3.7.   *The rank of a (not necessarily square) matrix is not changed by multiplication by a non-singular matrix.*

PROOF.   Let A be non-singular and let B be of rank ρ. Then by Theorem 2.1 AB is of rank $r \leq \rho$, and $A^{-1}(AB) = B$ is of rank $\rho \leq r$. Thus $r = \rho$. The proof that BA is of rank ρ is similar. $\square$

Theorem 1.14 states the corresponding property for linear transformations.

The existence or non-existence of the inverse of a square matrix depends on the matrix itself and not on whether it represents a linear transformation of a vector space into itself or a linear transformation of one vector space into another. Thus it is convenient and consistent to extend our usage of the term "non-singular" to include isomorphisms. Accordingly any square matrix with an inverse is *non-singular*.

Let U and V be vector spaces of dimension n over the field F. Let $A = \{\alpha_1, \ldots, \alpha_n\}$ be a basis of U and $B = \{\beta_1, \ldots, \beta_n\}$ be a basis of V. If $\xi = \sum_{i=1}^n x_i \alpha_i$ is any vector in U we can define $\sigma(\xi)$ to be $\sum_{i=1}^n x_i \beta_i$. It is easily seen that σ is an isomorphism and that ξ and $\sigma(\xi)$ are both represented by $(x_1, \ldots, x_n) \in F^n$. Thus any two vector spaces of the same dimension over F are isomorphic. As far as their internal structure is concerned they are indistinguishable. Whatever properties may serve to distinguish them are, by definition, not vector space properties.

EXERCISES

1. Show that the inverse of

$$A = \begin{bmatrix} 1 & 2 & 3 \\ 2 & 3 & 4 \\ 3 & 4 & 6 \end{bmatrix} \quad \text{is} \quad A^{-1} = \begin{bmatrix} -2 & 0 & 1 \\ 0 & 3 & -2 \\ 1 & -2 & 1 \end{bmatrix}.$$

2. Find the square of the matrix

$$A = \tfrac{1}{3} \begin{bmatrix} 1 & 2 & 2 \\ 2 & -2 & 1 \\ 2 & 1 & -2 \end{bmatrix}.$$

What is the inverse of A? (Geometrically, this matrix represents a $180°$ rotation about the line containing the vector $(2, 1, 1)$. The inverse obtained is therefore not surprising.)

3. Compute the image of the vector $(1, -2, 1)$ under the linear transformation represented by the matrix

$$A = \begin{bmatrix} 1 & 2 & 3 \\ 2 & 3 & 4 \\ 0 & 1 & 2 \end{bmatrix}.$$

Show that A cannot have an inverse.

4. Since

$$\begin{bmatrix} x_{11} & x_{12} \\ x_{21} & x_{22} \end{bmatrix} \begin{bmatrix} 3 & -1 \\ -5 & 2 \end{bmatrix} = \begin{bmatrix} 3x_{11} - 5x_{12} & -x_{11} + 2x_{12} \\ 3x_{21} - 5x_{22} & -x_{21} + 2x_{22} \end{bmatrix}$$

we can find the inverse of $\begin{bmatrix} 3 & -1 \\ -5 & 2 \end{bmatrix}$ by solving the equations

$$3x_{11} - 5x_{12} \qquad\qquad = 1$$
$$-x_{11} + 2x_{12} \qquad\qquad = 0$$
$$3x_{21} - 5x_{22} = 0$$
$$-x_{21} + 2x_{22} = 1.$$

Solve these equations and check your answer by showing that this gives the inverse matrix.

We have not as yet developed convenient and effective methods for obtaining the inverse of a given matrix. Such methods are developed later in this chapter and in the following chapter. If we know the geometric meaning of the matrix, however, it is often possible to obtain the inverse with very little work.

5. The matrix $\begin{bmatrix} \frac{3}{5} & -\frac{4}{5} \\ \frac{4}{5} & \frac{3}{5} \end{bmatrix}$ represents a rotation about the origin through the angle $\theta = \arccos \frac{3}{5}$. What rotation would be the inverse of this rotation? What matrix would represent this inverse rotation? Show that this matrix is the inverse of the given matrix.

6. The matrix $\begin{bmatrix} 0 & -1 \\ -1 & 0 \end{bmatrix}$ represents a reflection about the line $x_1 + x_2 = 0$. What operation is the inverse of this reflection? What matrix represents the inverse operation? Show that this matrix is the inverse of the given matrix.

7. The matrix $\begin{bmatrix} 1 & 1 \\ 0 & 1 \end{bmatrix}$ represents a shear. The inverse transformation is also a shear. Which one? What matrix represents the inverse shear? Show that this matrix is the inverse of the given matrix.

8. Show that the transformation that maps (x_1, x_2, x_3) onto $(x_3, -x_1, x_2)$ is an automorphism of F^3. Find the matrix representing this automorphism and its inverse with respect to the basis $\{(1, 0, 0), (0, 1, 0), (0, 0, 1)\}$.

9. Show that an automorphism of a vector space maps every subspace onto a subspace of the same dimension.

10. Find an example to show that there exist non-square matrices A and B such that $AB = I$. Specifically, show that there is an $m \times n$ matrix A and an $n \times m$ matrix B such that AB is the $m \times m$ identity. Show that BA is not the $n \times n$ identity. Prove in general that if $m \neq n$, then AB and BA cannot both be identity matrices.

4 | Change of Basis

We have represented vectors and linear transformations as n-tuples and matrices with respect to arbitrary but fixed bases. A very natural question arises: What changes occur in these representations if other choices for bases are made? The vectors and linear transformations have meaning independent of any particular choice of bases, independent of any coordinate systems, but their representations are entirely dependent on the bases chosen.

Definition. Let $A = \{\alpha_1, \ldots, \alpha_n\}$ and $A' = \{\alpha'_1, \ldots, \alpha'_n\}$ be bases of the vector space U. In a typical "change of basis" situation the representations of various vectors and linear transformations are known in terms of the basis A, and we wish to determine their representations in terms of the basis A'. In this connection, we refer to A as the "old" basis and to A' as the "new" basis. Each α'_j is expressible as a linear combination of the elements of A; that is,

$$\alpha'_j = \sum_{i=1}^{n} p_{ij}\alpha_i. \tag{4.1}$$

The associated matrix $P = [p_{ij}]$ is called the *matrix of transition* from the basis A to the basis A'.

The columns of P are the n-tuples representing the new basis vectors in terms of the old basis. This simple observation is worth remembering as it is usually the key to determining P when a change of basis is made. Since the columns of P are the representations of the basis A' they are linearly independent and P has rank n. Thus P is non-singular.

Now let $\xi = \sum_{i=1}^{n} x_i\alpha_i$ be an arbitrary vector of U and let $\xi = \sum_{i=1}^{n} x'_i\alpha'_i$ be the representation of ξ in terms of the basis A'. Then

$$\xi = \sum_{j=1}^{n} x'_j\alpha'_j = \sum_{j=1}^{n} x'_j \left(\sum_{i=1}^{n} p_{ij}\alpha_i \right)$$

$$= \sum_{i=1}^{n} \left(\sum_{j=1}^{n} p_{ij}x'_j \right)\alpha_i. \tag{4.2}$$

Since the representation of ξ with respect to the basis A is unique we see that $x_i = \sum_{j=1}^{n} p_{ij}x'_j$. Notice that the rows of P are used to express the old coordinates of ξ in terms of the new coordinates. For emphasis and contradistinction, we repeat that the columns of P are used to express the new basis vectors in terms of the old basis vectors.

Let $X = (x_1, \ldots, x_n)$ and $X' = (x'_1, \ldots, x'_n)$ be $n \times 1$ matrices representing the vector ξ with respect to the bases A and A'. Then the set of relations $\{x_i = \sum_{j=1}^{n} p_{ij}x'_j\}$ can be written as the single matric equation

$$X = PX'. \tag{4.3}$$

Now suppose that we have a linear transformation σ of U into V and that $A = [a_{ij}]$ is the matrix representing σ with respect to the bases A in U and $B = \{\beta_1, \ldots, \beta_m\}$ in V. We shall now determine the representation of σ with respect to the bases A' and B.

$$\sigma(\alpha'_j) = \sum_{k=1}^{n} p_{kj}\sigma(\alpha_k) = \sum_{k=1}^{n} p_{kj}\left(\sum_{i=1}^{m} a_{ik}\beta_i\right)$$

$$= \sum_{i=1}^{m} \left(\sum_{k=1}^{n} a_{ik}p_{kj}\right)\beta_i$$

$$= \sum_{i=1}^{m} a'_{ij}\beta_i. \tag{4.4}$$

Since B is a basis, $a'_{ij} = \sum_{k=1}^{n} a_{ik}p_{kj}$ and the matrix $A' = [a'_{ij}]$ representing σ with respect to the bases A' and B is related to A by the matric equation

$$A' = AP. \tag{4.5}$$

This relation can also be demonstrated in a slightly different way. For an arbitrary $\xi = \sum_{j=1}^{n} x_j\alpha_j \in U$ let $\sigma(\xi) = \sum_{i=1}^{m} y_i\beta_i$. Then we have

$$Y = AX = A(PX') = (AP)X'. \tag{4.6}$$

Thus AP is a matrix representing σ with respect to the bases A' and B. Since the matrix representing σ is uniquely determined by the choice of bases we have $A' = AP$.

Now consider the effect of a change of basis in the image space V. Thus let B be replaced by the basis $B' = \{\beta'_1, \ldots, \beta'_m\}$. Let $Q = [q_{ij}]$ be the matrix of transition from B to B', that is, $\beta'_j = \sum_{i=1}^{m} q_{ij}\beta_i$. Then if $A'' = [a''_{ij}]$ represents σ with respect to the bases A and B' we have

$$\sigma(\alpha_j) = \sum_{k=1}^{m} a''_{kj}\beta'_k = \sum_{k=1}^{m} a''_{kj}\left(\sum_{i=1}^{m} q_{ik}\beta_i\right)$$

$$= \sum_{i=1}^{m} \left(\sum_{k=1}^{m} q_{ik}a''_{kj}\right)\beta_i = \sum_{i=1}^{m} a_{ij}\beta_i. \tag{4.7}$$

Since the representation of $\sigma(\alpha_j)$ in terms of the basis B is unique we see that $A = QA''$, or

$$A'' = Q^{-1}A. \tag{4.8}$$

Combining these results, we see that, if both changes of bases are made at once, the new matrix representing σ is $Q^{-1}AP$.

As in the proof of Theorem 1.6 we can choose a new basis $A' = \{\alpha'_1, \ldots, \alpha'_n\}$ of U such that the last $v = n - \rho$ basis elements form a basis of $K(\sigma)$. Since $\{\sigma(\alpha'_1), \ldots, \sigma(\alpha'_\rho)\}$ is a basis of $\sigma(U)$ and is linearly independent in V, it can be extended to a basis B' of V. With respect to the bases A' and B' we have $\sigma(\alpha'_j) = \beta'_j$ for $j \le \rho$ while $\sigma(\alpha'_j) = 0$ for $j > \rho$. Thus the new matrix $Q^{-1}AP$ representing σ is of the form

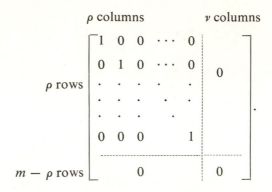

Thus we have

Theorem 4.1. *If A is any $m \times n$ matrix of rank ρ, there exist a non-singular $n \times n$ matrix P and a non-singular $m \times m$ matrix Q such that $A' = Q^{-1}AP$ has the first ρ elements of the main diagonal equal to 1, and all other elements equal to zero.* $\square$

When A and B are unrestricted we can always obtain this relatively simple representation of a linear transformation by a proper choice of bases. More interesting situations occur when A and B are restricted. Suppose, for example, that we take $U = V$ and $A = B$. In this case there is but one basis to change and but one matrix of transition, that is, $P = Q$. In this case it is not possible to obtain a form of the matrix representing σ as simple as that obtained in Theorem 4.1. We say that any two matrices representing the same linear transformation σ of a vector space V into itself are *similar*. This is equivalent to saying that two matrices A and A' are similar if and only if there exists a non-singular matrix of transition P such that $A' = P^{-1}AP$. This case occupies much of our attention in Chapters III and V.

EXERCISES

1. In P_3, the space of polynomials of degree 2 or smaller with coefficients in F, let $A = \{1, x, x^2\}$.

$$A' = \{p_1(x) = x^2 + x + 1, p_2(x) = x^2 - x - 2, p_3(x) = x^2 + x - 1\}$$

is also a basis. Find the matrix of transition from A to A'.

2. In many of the uses of the concepts of this section it is customary to take $A = \{\alpha_i \mid \alpha_i = (\delta_{i1}, \delta_{i2}, \ldots, \delta_{in})\}$ as the old basis in R^n. Thus, in R^2 let $A = \{(1, 0), (0, 1)\}$ and $A' = \{(\frac{1}{2}, \sqrt{3}/2), (-\sqrt{3}/2, \frac{1}{2})\}$. Show that

$$P = \begin{bmatrix} \frac{1}{2} & -\sqrt{3}/2 \\ \sqrt{3}/2 & \frac{1}{2} \end{bmatrix}$$

is the matrix of transition from A to A'.

3. (Continuation) With A' and A as in Exercise 2, find the matrix of transition R from A' to A. (Notice, in particular, that in Exercise 2 the columns of P are the components of the vectors in A' expressed in terms of basis A, whereas in this exercise the columns of R are the components of the vectors in A expressed in terms of the basis A'. Thus these two matrices of transition are determined relative to different bases.) Show that $RP = I$.

4. (Continuation) Consider the linear transformation of σ of R^2 into itself which maps

$$(1, 0) \quad \text{onto} \quad (\tfrac{1}{2}, \sqrt{3}/2)$$
$$(0, 1) \quad \text{onto} \quad (-\sqrt{3}/2, \tfrac{1}{2}).$$

Find the matrix A that represents σ with respect to the basis A.

You should obtain $A = P$. However, A and P do not represent the same thing. To see this, let $\xi = (x_1, x_2)$ be an arbitrary vector in R^2 and compute $\sigma(\xi)$ by means of formula (2.9) and the new coordinates of ξ by means of formula (4.3).

A little reflection will show that the results obtained are entirely reasonable. The matrix A represents a rotation of the real plane counterclockwise through an angle of $\pi/3$. The matrix P represents a rotation of the coordinate axes counterclockwise through an angle of $\pi/3$. In the latter case the motion of the plane relative to the coordinate axes is clockwise through an angle of $\pi/3$.

5. In R^3 let $A = \{(1, 0, 0), (0, 1, 0), (0, 0, 1)\}$ and let $A' = \{(0, 1, 1), (1, 0, 1), (1, 1, 0)\}$. Find the matrix of transition P from A to A' and the matrix of transition P^{-1} from A' to A.

6. Let A, B, and C be three bases of V. Let P be the matrix of transition from A to B and let Q be the matrix of transition from B to C. Is PQ or QP the matrix of transition from A to C? Compare the order of multiplication of matrices of transition and matrices representing linear transformation.

7. Use the results of Exercise 6 to resolve the question raised in the parenthetical remark of Exercise 3, and implicitly assumed in Exercise 5. If P is the matrix of transition from A to A' and Q is the matrix of transition from A' to A, show that $PQ = I$.

5 | Hermite Normal Form

We may also ask how much simplification of the matrix representing a linear transformation σ of U into V can be effected by a change of basis in

V alone. Let $A = \{\alpha_1, \ldots, \alpha_n\}$ be the given basis in U and let $U_k = \langle \alpha_1, \ldots, \alpha_k \rangle$. The subspaces $\sigma(U_k)$ of V form a non-decreasing chain of subspaces with $\sigma(U_{k-1}) \subset \sigma(U_k)$ and $\sigma(U_n) = \sigma(U)$. Since $\sigma(U_k) = \sigma(U_{k-1}) + \langle \sigma(\alpha_k) \rangle$ we see from Theorem 4.8 of Chapter I that dim $\sigma(U_k) \leq$ dim $\sigma(U_{k-1}) + 1$; that is, the dimensions of the $\sigma(U_k)$ do not increase by more than 1 at a time as k increases. Since dim $\sigma(U_n) = \rho$, the rank of σ, an increase of exactly 1 must occur ρ times. For the other times, if any, we must have dim $\sigma(U_k) =$ dim $\sigma(U_{k-1})$ and hence $\sigma(U_k) = \sigma(U_{k-1})$. We have an increase by 1 when $\sigma(\alpha_k) \notin \sigma(U_{k-1})$ and no increase when $\sigma(\alpha_k) \in \sigma(U_{k-1})$.

Let $k_1, k_2, \ldots, k_\rho$ be those indices for which $\sigma(\alpha_{k_i}) \notin \sigma(U_{k_i-1})$. Let $\beta_i' = \sigma(\alpha_{k_i})$. Since $\beta_i' \notin \sigma(U_{k_i-1}) = \langle \beta_1', \ldots, \beta_{i-1}' \rangle$, the set $\{\beta_1', \ldots, \beta_\rho'\}$ is linearly independent (see Theorem 2.3, Chapter I-2). Since $\{\beta_1', \ldots, \beta_\rho'\} \subset \sigma(U)$ and $\sigma(U)$ is of dimension ρ, $\{\beta_1', \ldots, \beta_\rho'\}$ is a basis of $\sigma(U)$. This set can be extended to a basis B' of V. Let us now determine the form of the matrix A' representing σ with respect to the bases A and B'.

Since $\sigma(\alpha_{k_i}) = \beta_i'$, column k_i has a 1 in row i and all other elements of this column are 0's. For $k_i < j < k_{i+1}$, $\sigma(\alpha_j) \in \sigma(U_{k_i})$ so that column j has 0's below row i. In general, there is no restriction on the elements of column j in the first i rows. A' thus has the form

$$
\begin{array}{cc}
\text{column} & \text{column} \\
k_1 & k_2
\end{array}
$$

$$
\begin{bmatrix}
0 & \cdots & 0 & 1 & a'_{1,k_1+1} & \cdots & 0 & a'_{1,k_2+1} & \cdots & \cdot & \cdot \\
0 & \cdots & 0 & 0 & 0 & \cdots & 1 & a'_{2,k_2+1} & \cdots & \cdot & \cdot \\
0 & \cdots & 0 & 0 & 0 & \cdots & 0 & 0 & \cdots & \cdot & \cdot \\
\cdot & & \cdot & \cdot & \cdot & & \cdot & \cdot & & & \\
\cdot & & \cdot & \cdot & \cdot & & \cdot & \cdot & & & \\
\cdot & & \cdot & \cdot & \cdot & & \cdot & \cdot & & & \\
0 & \cdots & 0 & 0 & 0 & \cdots & 0 & 0 & \cdots & \cdot & \cdot
\end{bmatrix}
\tag{5.1}
$$

Once A and σ are given, the k_i and the set $\{\beta_1', \ldots, \beta_\rho'\}$ are uniquely determined. There may be many ways to extend this set to the basis B', but the additional basis vectors do not affect the determination of A' since every element of $\sigma(U)$ can be expressed in terms of $\{\beta_1', \ldots, \beta_\rho'\}$ alone. Thus A' is uniquely determined by A and σ.

Theorem 5.1. *Given any $m \times n$ matrix A of rank ρ, there exists a non-singular $m \times m$ matrix Q such that $A' = Q^{-1}A$ has the following form:*

(1) *There is at least one non-zero element in each of the first ρ rows of A', and the elements in all remaining rows are zero.*

   (2) *The first non-zero element appearing in row i ($i \leq \rho$) is a 1 appearing in column k_i, where $k_1 < k_2 < \cdots < k_\rho$.*

   (3) *In column k_i the only non-zero element is the 1 in row i.*

The form A' is uniquely determined by A.

PROOF. In the applications of this theorem that we wish to make A is usually given alone without reference to any bases A and B, and often without reference to any linear transformation σ. We can, however, introduce any two vector spaces U and V of dimensions n and m over F and let A be any basis of U and B be any basis of V. We can consider A as defining a linear transformation σ of U into V with respect to the bases A and B. The discussion preceding Theorem 5.1 shows that there is at least one non-singular matrix Q such that $Q^{-1}A$ satisfies conditions (1), (2), and (3).

Now suppose there are two non-singular matrices Q_1 and Q_2 such that $Q_1^{-1}A = A_1'$ and $Q_2^{-1}A = A_2'$ both satisfy the conditions of the theorem. We wish to conclude that $A_1' = A_2'$. No matter how the vector spaces U and V are introduced and how the bases A and B are chosen we can regard Q_1 and Q_2 as matrices of transition in V. Thus A_1' represents σ with respect to bases A and B_1' and A_2' represents σ with respect to bases A and B_2'. But condition (3) says that for $i \leq \rho$ the ith basis element in both B_1' and B_2' is $\sigma(\alpha_{k_i})$. Thus the first ρ elements of B_1' and B_2' are identical. Condition (1) says that the remaining basis elements have nothing to do with determining the coefficients in A_1' and A_2'. Thus $A_1' = A_2'$. □   *reduced*

We say that a matrix satisfying the conditions of Theorem 5.1 is in *Hermite normal form*. Often this form is called a *row-echelon form*. And sometimes the term, Hermite normal form, is reserved for a square matrix containing exactly the numbers that appear in the form we obtained in Theorem 5.1 with the change that row i beginning with a 1 in column k_i is moved down to row k_i. Thus each non-zero row begins on the main diagonal and each column with a 1 on the main diagonal is otherwise zero. In this text we have no particular need for this special form while the form described in Theorem 5.1 is one of the most useful tools at our disposal.

The usefulness of the Hermite normal form depends on its form, and the uniqueness of that form will enable us to develop effective and convenient short cuts for determining that form.

Definition. Given the matrix A, the matrix A^T obtained from A by interchanging rows and columns in A is called the *transpose* of A. If $A^T = [a_{ij}']$, the element a_{ij}' appearing in row i column j of A^T is the element a_{ji} appearing in row j column i of A. It is easy to show that $(AB)^T = B^T A^T$. (See Exercise 4.) $\leq . 1$ $\rho(A^T) \leq \rho(A)$ A^T *invertible*

Proposition 5.2. The number of linearly independent rows in a matrix is equal to the number of linearly independent columns.

PROOF. The number of linearly independent columns in a matrix A is its rank ρ. The Hermite normal form $A' = Q^{-1}A$ corresponding to A is also of rank ρ. For A' it is obvious that the number of linearly independent rows in A' is also equal to ρ, that is, the rank of $(A')^T$ is ρ. Since Q^T is non-singular, the rank of $A^T = (QA')^T = (A')^T Q^T$ is also ρ. Thus the number of linearly independent rows in A is ρ. $\square$

EXERCISES

1. Which of the following matrices are in Hermite normal form?

(a)
$$\begin{bmatrix} 0 & 1 & 0 & 0 & 1 \\ 0 & 0 & 1 & 0 & 1 \\ 0 & 0 & 0 & 1 & 0 \\ 0 & 0 & 0 & 0 & 0 \end{bmatrix},$$

(b)
$$\begin{bmatrix} 0 & 0 & 2 & 0 & 4 \\ 0 & 1 & 1 & 0 & 3 \\ 0 & 0 & 0 & 1 & 2 \\ 0 & 0 & 0 & 0 & 0 \end{bmatrix},$$

(c)
$$\begin{bmatrix} 1 & 0 & 0 & 0 & 1 \\ 0 & 1 & 0 & 0 & 1 \\ 0 & 0 & 0 & 1 & 1 \\ 0 & 0 & 0 & 0 & 0 \end{bmatrix},$$

(d)
$$\begin{bmatrix} 0 & 1 & 0 & 1 & 0 & 0 & 1 \\ 0 & 0 & 1 & 0 & 0 & 1 & 0 \\ 0 & 0 & 0 & 0 & 1 & 0 & 0 \\ 0 & 0 & 0 & 0 & 0 & 0 & 0 \end{bmatrix},$$

(e)
$$\begin{bmatrix} 1 & 0 & 1 & 0 & 1 \\ 0 & 1 & 1 & 0 & 0 \\ 0 & 0 & 0 & 1 & 0 \\ 0 & 0 & 0 & 0 & 1 \end{bmatrix}.$$

2. Determine the rank of each of the matrices given in Exercise 1.

3. Let σ and τ be linear transformations mapping R^3 into R^2. Suppose that for a given pair of bases A for R^3 and B for R^2, σ and τ are represented by

$$A = \begin{bmatrix} 1 & 1 & 0 \\ 0 & 0 & 1 \end{bmatrix} \quad \text{and} \quad B = \begin{bmatrix} 1 & 0 & 1 \\ 0 & 1 & 0 \end{bmatrix},$$

respectively. Show that there is no basis B' of R^2 such that B is the matrix representing σ with respect to A and B'.

4. Show that
   (a) $(A + B)^T = A^T + B^T$,
   (b) $(AB)^T = B^T A^T$,
   (c) $(A^{-1})^T = (A^T)^{-1}$.

6 | Elementary Operations and Elementary Matrices

Our purpose in this section is to develop convenient computational methods. We have been concerned with the representations of linear transformations by matrices and the changes these matrices undergo when a basis is changed. We now show that these changes can be effected by elementary operations on the rows and columns of the matrices.

We define three types of *elementary operations* on the rows of a matrix A.

Type I: Multiply a row of A by a non-zero scalar.
Type II: Add a multiple of one row to another row.
Type III: Interchange two rows.

Elementary column operations are defined in an analogous way.

From a logical point of view these operations are redundant. An operation of type III can be accomplished by a combination of operations of types I and II. It would, however, require four such operations to take the place of one operation of type III. Since we wish to develop convenient computational methods, it would not suit our purpose to reduce the number of operations at our disposal. On the other hand, it would not be of much help to extend the list of operations at this point. The student will find that, with practice, he can combine several elementary operations into one step. For example, such a combined operation would be the replacing of a row by a linear combination of rows, provided that the row replaced appeared in the linear combination with a non-zero coefficient. We leave such short cuts to the student.

An elementary operation can also be accomplished by multiplying A on the left by a matrix. Thus, for example, multiplying the second row by the scalar c can be effected by the matrix

$$E_2(c) = \begin{bmatrix} 1 & 0 & 0 & \cdots & 0 \\ 0 & c & 0 & \cdots & 0 \\ 0 & 0 & 1 & \cdots & 0 \\ . & . & . & . & . \\ . & . & . & . & . \\ . & . & . & . & . \\ 0 & 0 & 0 & \cdots & 1 \end{bmatrix}. \tag{6.1}$$

The addition of k times the third row to the first row can be effected by the matrix

$$E_{31}(k) = \begin{bmatrix} 1 & 0 & k & \cdots & 0 \\ 0 & 1 & 0 & \cdots & 0 \\ 0 & 0 & 1 & \cdots & 0 \\ \cdot & \cdot & \cdot & \cdot & \cdot \\ \cdot & \cdot & \cdot & & \cdot \\ \cdot & \cdot & \cdot & & \cdot \\ 0 & 0 & 0 & \cdots & 1 \end{bmatrix}. \qquad (6.2)$$

The interchange of the first and second rows can be effected by the matrix

$$E_{12} = \begin{bmatrix} 0 & 1 & 0 & \cdots & 0 \\ 1 & 0 & 0 & \cdots & 0 \\ 0 & 0 & 1 & \cdots & 0 \\ \cdot & \cdot & \cdot & \cdot & \cdot \\ \cdot & \cdot & \cdot & & \cdot \\ \cdot & \cdot & \cdot & & \cdot \\ 0 & 0 & 0 & \cdots & 1 \end{bmatrix}. \qquad (6.3)$$

These matrices corresponding to the elementary operations are called *elementary matrices*. These matrices are all non-singular and their inverses are also elementary matrices. For example, the inverses of $E_2(c)$, $E_{31}(k)$, and E_{12} are respectively $E_2(c^{-1})$, $E_{31}(-k)$, and E_{12}.

Notice that the elementary matrix representing an elementary operation is the matrix obtained by applying the elementary operation to the unit matrix.

Theorem 6.1. *Any non-singular matrix A can be written as a product of elementary matrices.*

PROOF. At least one element in the first column is non-zero or else A would be singular. Our first goal is to apply elementary operations, if necessary, to obtain a 1 in the upper left-hand corner. If $a_{11} = 0$, we can interchange rows to bring a non-zero element into that position. Thus we may as well suppose that $a_{11} \neq 0$. We can then multiply the first row by a_{11}^{-1}. Thus, to simplify notation, we may as well assume that $a_{11} = 1$. We now add $-a_{i1}$ times the first row to the ith row to make every other element in the first column equal to zero.

The resulting matrix is still non-singular since the elementary operations applied were non-singular. We now wish to obtain a 1 in the position of element a_{22}. At least one element in the second column other than a_{12}

is non-zero for otherwise the first two columns would be dependent. Thus by a possible interchange of rows, not including row 1, and multiplying the second row by a non-zero scalar we can obtain $a_{22} = 1$. We now add $-a_{i2}$ times the second row to the ith row to make every other element in the second column equal to zero. Notice that we also obtain a 0 in the position of a_{12} without affecting the 1 in the upper left-hand corner.

We continue in this way until we obtain the identity matrix. Thus if $E_1, E_2, \ldots, E_r$ are elementary matrices representing the successive elementary operations, we have

$$I = E_r \cdots E_2 E_1 A,$$
or
$$A = E_1^{-1} E_2^{-1} \cdots E_r^{-1}. \ \Box \tag{6.4}$$

In Theorem 5.1 we obtained the Hermite normal form A' from the matrix A by multiplying on the left by the non-singular matrix Q^{-1}. We see now that Q^{-1} is a product of elementary matrices, and therefore that A can be transformed into Hermite normal form by a succession of elementary row operations. It is most efficient to use the elementary row operations directly without obtaining the matrix Q^{-1}.

We could have shown directly that a matrix could be transformed into Hermite normal form by means of elementary row operations. We would then be faced with the necessity of showing that the Hermite normal form obtained is unique and not dependent on the particular sequence of operations used. While this is not particularly difficult, the demonstration is uninteresting and unilluminating and so tedious that it is usually left as an "exercise for the reader." Uniqueness, however, is a part of Theorem 5.1, and we are assured that the Hermite normal form will be independent of the particular sequence of operations chosen. This is important as many possible operations are available at each step of the work, and we are free to choose those that are most convenient.

Basically, the instructions for reducing a matrix to Hermite normal form are contained in the proof of Theorem 6.1. In that theorem, however, we were dealing with a non-singular matrix and thus assured that we could at certain steps obtain a non-zero element on the main diagonal. For a singular matrix, this is not the case. When a non-zero element cannot be obtained with the instructions given we must move our consideration to the next column.

In the following example we perform several operations at each step to conserve space. When several operations are performed at once, some care must be exercised to avoid reducing the rank. This may occur, for example, if we subtract a row from itself in some hidden fashion. In this example we avoid this pitfall, which can occur when several operations of type III are combined, by considering one row as an operator row and adding multiples of it to several others.

Consider the matrix

$$\begin{bmatrix} 4 & 3 & 2 & -1 & 4 \\ 5 & 4 & 3 & -1 & 4 \\ -2 & -2 & -1 & 2 & -3 \\ 11 & 6 & 4 & 1 & 11 \end{bmatrix}$$

as an example.

According to the instructions for performing the elementary row operations we should multiply the first row by $\frac{1}{4}$. To illustrate another possible way to obtain the "1" in the upper left corner, multiply row 1 by -1 and add row 2 to row 1. Multiples of row 1 can now be added to the other rows to obtain

$$\begin{bmatrix} 1 & 1 & 1 & 0 & 0 \\ 0 & -1 & -2 & -1 & 4 \\ 0 & 0 & 1 & 2 & -3 \\ 0 & -5 & -7 & 1 & 11 \end{bmatrix}.$$

Now, multiply row 2 by -1 and add appropriate multiples to the other rows to obtain

$$\begin{bmatrix} 1 & 0 & -1 & -1 & 4 \\ 0 & 1 & 2 & 1 & -4 \\ 0 & 0 & 1 & 2 & -3 \\ 0 & 0 & 3 & 6 & -9 \end{bmatrix}.$$

Finally, we obtain

$$\begin{bmatrix} 1 & 0 & 0 & 1 & 1 \\ 0 & 1 & 0 & -3 & 2 \\ 0 & 0 & 1 & 2 & -3 \\ 0 & 0 & 0 & 0 & 0 \end{bmatrix},$$

which is the Hermite normal form described in Theorem 5.1. If desired, Q^{-1} can be obtained by applying the same sequence of elementary row operations to the unit matrix. However, while the Hermite normal form is necessarily unique, the matrix Q^{-1} need not be unique, as the proof of Theorem 5.1 should show.

Rather than trying to remember the sequence of elementary operations used to reduce A to Hermite normal form, it is more efficient to perform these operations on the unit matrix at the same time we are operating on A. It is suggested that we arrange the work in the following way:

$$
\begin{bmatrix}
4 & 3 & 2 & -1 & 4 & 1 & 0 & 0 & 0 \\
5 & 4 & 3 & -1 & 4 & 0 & 1 & 0 & 0 \\
-2 & -2 & -1 & 2 & -3 & 0 & 0 & 1 & 0 \\
11 & 6 & 4 & 1 & 11 & 0 & 0 & 0 & 1
\end{bmatrix} = [A, I]
$$

$$
\begin{bmatrix}
1 & 1 & 1 & 0 & 0 & -1 & 1 & 0 & 0 \\
0 & -1 & -2 & -1 & 4 & 5 & -4 & 0 & 0 \\
0 & 0 & 1 & 2 & -3 & -2 & 2 & 1 & 0 \\
0 & -5 & -7 & 1 & 11 & 11 & -11 & 0 & 1
\end{bmatrix}
$$

$$
\begin{bmatrix}
1 & 0 & -1 & -1 & 4 & 4 & -3 & 0 & 0 \\
0 & 1 & 2 & 1 & -4 & -5 & 4 & 0 & 0 \\
0 & 0 & 1 & 2 & -3 & -2 & 2 & 1 & 0 \\
0 & 0 & 3 & 6 & -9 & -14 & 9 & 0 & 1
\end{bmatrix}
$$

$$
\begin{bmatrix}
1 & 0 & 0 & 1 & 1 & 2 & -1 & 1 & 0 \\
0 & 1 & 0 & -3 & 2 & -1 & 0 & -2 & 0 \\
0 & 0 & 1 & 2 & -3 & -2 & 2 & 1 & 0 \\
0 & 0 & 0 & 0 & 0 & -8 & 3 & -3 & 1
\end{bmatrix}.
$$

In the end we obtain

$$
Q^{-1} = \begin{bmatrix}
2 & -1 & 1 & 0 \\
-1 & 0 & -2 & 0 \\
-2 & 2 & 1 & 0 \\
-8 & 3 & -3 & 1
\end{bmatrix},
$$

Verify directly that $Q^{-1}A$ is in Hermite normal form.

If A were non-singular, the Hermite normal form obtained would be the identity matrix. In this case Q^{-1} would be the inverse of A. This method of finding the inverse of a matrix is one of the easiest available for hand computation. It is the recommended technique.

EXERCISES

1. Elementary operations provide the easiest methods for determining the rank of a matrix. Proceed as if reducing to Hermite normal form. Actually, it is not necessary to carry out all the steps as the rank is usually evident long before the Hermite normal form is obtained. Find the ranks of the following matrices:

(a) $\begin{bmatrix} 1 & 2 & 3 \\ 4 & 5 & 6 \\ 7 & 8 & 9 \end{bmatrix}$,

(b) $\begin{bmatrix} 0 & 1 & 2 \\ -1 & 0 & 3 \\ -2 & -3 & 0 \end{bmatrix}$,

(c) $\begin{bmatrix} 0 & 1 & 2 \\ 1 & 0 & 3 \\ 2 & 3 & 0 \end{bmatrix}$.

2. Identify the elementary operations represented by the following elementary matrices:

(a) $\begin{bmatrix} 1 & 0 & 0 \\ 0 & 1 & 0 \\ -2 & 0 & 1 \end{bmatrix}$,

(b) $\begin{bmatrix} 0 & 0 & 1 \\ 0 & 1 & 0 \\ 1 & 0 & 0 \end{bmatrix}$,

(c) $\begin{bmatrix} 1 & 0 & 0 \\ 0 & 2 & 0 \\ 0 & 0 & 1 \end{bmatrix}$.

3. Show that the product

$$\begin{bmatrix} -1 & 0 \\ 0 & 1 \end{bmatrix}\begin{bmatrix} 1 & 0 \\ 1 & 1 \end{bmatrix}\begin{bmatrix} 1 & -1 \\ 0 & 1 \end{bmatrix}\begin{bmatrix} 1 & 0 \\ 1 & 1 \end{bmatrix}$$

is an elementary matrix. Identify the elementary operations represented by each matrix in the product.

4. Show by an example that the product of elementary matrices is not necessarily an elementary matrix.

5. Reduce each of the following matrices to Hermite normal form.

(a) $\begin{bmatrix} 2 & 1 & 3 & -2 \\ 2 & -1 & 5 & 2 \\ 1 & 1 & 1 & 1 \end{bmatrix}$,

(b) $\begin{bmatrix} 1 & 2 & 3 & 3 & 10 & 6 \\ 2 & 1 & 0 & 0 & 2 & 3 \\ 2 & 2 & 2 & 1 & 5 & 5 \\ -1 & 1 & 3 & 2 & 5 & 2 \end{bmatrix}$.

6. Use elementary row operations to obtain the inverses of

(a) $\begin{bmatrix} 3 & -1 \\ -5 & 2 \end{bmatrix}$, and

(b) $\begin{bmatrix} 1 & 2 & 3 \\ 2 & 3 & 4 \\ 3 & 4 & 6 \end{bmatrix}$.

7. (a) Show that, by using a sequence of elementary operations of type II only, any two rows of a matrix can be interchanged with one of the two rows multiplied by -1. (In fact, the type II operations involve no scalars other than ± 1.)

(b) Using the results of part (a), show that a type III operation can be obtained by a sequence of type II operations and a single type I operation.

(c) Show that the sign of any row can be changed by a sequence of type II operations and a single type III operation.

8. Show that any matrix A can be reduced to the form described in Theorem 4.1 by a sequence of elementary row operations and a sequence of elementary column operations.

7 | Linear Problems and Linear Equations

For a given linear transformation σ of U into V and a given $\beta \in V$ the problem of finding any or all $\xi \in U$ for which $\sigma(\xi) = \beta$ is called a *linear problem*. Before providing any specific methods for solving such problems, let us see what the set of solutions should look like.

If $\beta \notin \sigma(U)$, then the problem has no solution.

If $\beta \in \sigma(U)$, the problem has at least one solution. Let ξ_0 be one such solution. We call any such ξ_0 a *particular solution*. If ξ is any other solution, then $\sigma(\xi - \xi_0) = \sigma(\xi) - \sigma(\xi_0) = \beta - \beta = 0$ so that $\xi - \xi_0$ is in the kernel of σ. Conversely, if $\xi - \xi_0$ is in the kernel of σ then $\sigma(\xi) = \sigma(\xi_0 + \xi - \xi_0) = \sigma(\xi_0) + \sigma(\xi - \xi_0) = \beta + 0 = \beta$ so that ξ is a solution. Thus the set of all solutions of $\sigma(\xi) = \beta$ is of the form

$$\{\xi_0\} + K(\sigma). \tag{7.1}$$

Since $\{\xi_0\}$ contains just one element, there is a one-to-one correspondence between the elements of $K(\sigma)$ and the elements of $\{\xi_0\} + K(\sigma)$. Thus the size of the set of solutions can be described by giving the dimension of $K(\sigma)$. The set of all solutions of the problem $\sigma(\xi) = \beta$ is not a subspace of U unless $\beta = 0$. Nevertheless, it is convenient to say that the set is of dimension ν, the nullity of σ.

Given the linear problem $\sigma(\xi) = \beta$, the problem $\sigma(\xi) = 0$ is called the *associated homogeneous problem*. The *general solution* is then any particular solution plus the solution of the associated homogeneous problem. The solution of the associated homogeneous problem is the kernel of σ.

Now let σ be represented by the $m \times n$ matrix $A = [a_{ij}]$, β be represented by $B = (b_1, \ldots, b_m)$, and ξ by $X = (x_1, \ldots, x_n)$. Then the linear problem $\sigma(\xi) = \beta$ becomes

$$AX = B \tag{7.2}$$

in matrix form, or

$$\sum_{j=1}^{n} a_{ij}x_j = b_i, \qquad (i = 1, \ldots, m) \tag{7.3}$$

in the form of a system of linear equations.

Given A and B, the *augmented matrix* $[A, B]$ of the system of linear equations is defined to be

$$\overline{A} = \qquad [A, B] = \begin{bmatrix} a_{11} & \cdots & a_{1n} & b_1 \\ \cdot & & \cdot & \cdot \\ \cdot & & \cdot & \cdot \\ \cdot & & \cdot & \cdot \\ a_{m1} & \cdots & a_{mn} & b_m \end{bmatrix}. \tag{7.4}$$

Theorem 7.1. *The system of simultaneous linear equations $AX = B$ has a solution if and only if the rank of A is equal to the rank of the augmented matrix $[A, B]$. Whenever a solution exists, all solutions can be expressed in terms of $\nu = n - \rho$ independent parameters, where ρ is the rank of A.*

PROOF. We have already seen that the linear problem $\sigma(\xi) = \beta$ has a solution if and only if $\beta \in \sigma(U)$. This is the case if and only if β is linearly dependent on $\{\sigma(\alpha_1), \ldots, \sigma(\alpha_n)\}$. But this is equivalent to the condition that B be linearly dependent on the columns of A. Thus adjoining the column of b_i's to form the augmented matrix must not increase the rank. Since the rank of the augmented matrix cannot be less than the rank of A we see that the system has a solution if and only if these two ranks are equal.

Now let Q be a non-singular matrix such that $Q^{-1}A = A'$ is in Hermite normal form. Any solution of $AX = B$ is also a solution of $A'X = Q^{-1}AX = Q^{-1}B = B'$. Conversely, any solution of $A'X = B'$ is also a solution of $AX = QA'X = QB' = B$. Thus the two systems of equations are equivalent.

Now the system $A'X = B'$ is particularly easy to solve since the variable x_{k_i} appears only in the ith equation. Furthermore, non-zero coefficients appear only in the first ρ equations. The condition that $\beta \in \sigma(U)$ also takes on a form that is easily recognizable. The condition that B' be expressible as a linear combination of the columns of A' is simply that the elements of B' below row ρ be zero. The system $A'X = B'$ has the form

$$x_{k_1} + a'_{1,k_1+1}x_{k_1+1} + \cdots 0 + a'_{1,k_2+1}x_{k_2+1} + \cdots = b'_1 \qquad (7.5)$$
$$x_{k_2} + a'_{2,k_2+1}x_{k_2+1} = b'_2$$
$$\cdots$$

Since each x_{k_i} appears in but one equation with unit coefficient, the remaining $n - \rho$ unknowns can be given values arbitrarily and the corresponding values of the x_{k_i} computed. The $n - \rho$ unknowns with indices not the k_i are the $n - \rho$ parameters mentioned in the theorem. □

As an example, consider the system of equations:

$$4x_1 + 3x_2 + 2x_3 - x_4 = 4$$
$$5x_1 + 4x_2 + 3x_3 - x_4 = 4$$
$$-2x_1 - 2x_2 - x_3 + 2x_4 = -3$$
$$11x_1 + 6x_2 + 4x_3 + x_4 = 11.$$

The augmented matrix is

$$\begin{bmatrix} 4 & 3 & 2 & -1 & 4 \\ 5 & 4 & 3 & -1 & 4 \\ -2 & -2 & -1 & 2 & -3 \\ 11 & 6 & 4 & 1 & 11 \end{bmatrix}.$$

This is the matrix we chose for an example in the previous section. There we obtained the Hermite normal form

$$\begin{bmatrix} 1 & 0 & 0 & 1 & 1 \\ 0 & 1 & 0 & -3 & 2 \\ 0 & 0 & 1 & 2 & -3 \\ 0 & 0 & 0 & 0 & 0 \end{bmatrix}.$$

Thus the system of equations $A'X = B'$ corresponding to this augmented matrix is

$$x_1 + x_4 = 1$$
$$x_2 - 3x_4 = 2$$
$$x_3 + 2x_4 = -3.$$

It is clear that this system is very easy to solve. We can take any value whatever for x_4 and compute the corresponding values for x_1, x_2, and x_3. A particular solution, obtained by taking $x_4 = 0$, is $X_0 = (1, 2, -3, 0)$. It is more instructive to write the new system of equations in the form

$$
\begin{aligned}
x_1 &= \quad 1 - \quad x_4 \\
x_2 &= \quad 2 + 3x_4 \\
x_3 &= -3 - 2x_4 \\
x_4 &= \qquad\quad x_4
\end{aligned}
$$

In vector form this becomes

$$(x_1, x_2, x_3, x_4) = (1, 2, -3, 0) + x_4(-1, 3, -2, 1).$$

We can easily verify that $(-1, 3, -2, 1)$ is a solution of the associated homogeneous problem. In fact, $\{(-1, 3, -2, 1)\}$ is a basis for the kernel, and $x_4(-1, 3, -2, 1)$, for an arbitrary x_4, is a general element of the kernel. We have, therefore, expressed the general solution as a particular solution plus the kernel.

The elementary row operations provide us with the recommended technique for solving simultaneous linear equations by hand. This application is the principal reason for introducing elementary row operations rather than column operations.

Theorem 7.2. *The equation $AX = B$ fails to have a solution if and only if there exists a one-row matrix C such that $CA = 0$ and $CB = 1$.*

PROOF. Suppose the equation $AX = B$ has a solution and a C exists such that $CA = 0$ and $CB = 1$. Then we would have $0 = (CA)X = C(AX) = CB = 1$, which is a contradiction.

On the other hand, suppose the equation $AX = B$ has no solution. By Theorem 7.1 this implies that the rank of the augmented matrix $[A, B]$ is greater than the rank of A. Let Q be a non-singular matrix such that $Q^{-1}[A, B]$ is in Hermite normal form. Then if ρ is the rank of A, the $(\rho + 1)$st row of $Q^{-1}[A, B]$ must be all zeros except for a 1 in the last column. If C is the $(\rho + 1)$st row of Q^{-1} this means that

$$C[A, B] = [0 \quad 0 \quad \cdots \quad 0 \quad 1],$$

or

$$CA = 0 \qquad \text{and} \qquad CB = 1. \ \square$$

This theorem is important because it provides a positive condition for a negative conclusion. Theorem 7.1 also provides such a positive condition and it is to be preferred when dealing with a particular system of equations. But Theorem 7.2 provides a more convenient condition when dealing with systems of equations in general.

Although the sytems of linear equations in the exercises that follow are written in expanded form, they are equivalent in form to the matric equation

$AX = B$. From any linear problem in this set, or those that will occur later, it is possible to obtain an extensive list of closely related linear problems that appear to be different. For example, if $AX = B$ is the given linear problem with A an $m \times n$ matrix and Q is any non-singular $m \times m$ matrix, then $A'X = B'$ with $A' = QA$ and $B' = QB$ is a problem with the same set of solutions. If P is a non-singular $n \times n$ matrix, then $A''X'' = B$ where $A'' = AP$ is a problem whose solution X'' is related to the solution X of the original problem by the condition $X'' = P^{-1}X$.

For the purpose of constructing related exercises of the type mentioned, it is desirable to use matrices P and Q that do not introduce tedious numerical calculations. It is very easy to obtain a non-singular matrix P that has only integral elements and such that its inverse also has only integral elements. Start with an identity matrix of the desired order and perform a sequence of elementary operations of types II and III. As long as an operation of type I is avoided, no fractions will be introduced. Furthermore, the inverse operations will be of types II and III so the inverse matrix will also have only integral elements.

For convenience, some matrices with integral elements and inverses with integral elements are listed in an appendix. For some of the exercises that are given later in this book, matrices of transition that satisfy special conditions are also needed. These matrices, known as orthogonal and unitary matrices, usually do not have integral elements. Simple matrices of these types are somewhat harder to obtain. Some matrices of these types are also listed in the appendix.

EXERCISES

1. Show that $\{(1, 1, 1, 0), (2, 1, 0, 1)\}$ spans the subspace of all solutions of the system of linear equations

$$3x_1 - 2x_2 - x_3 - 4x_4 = 0$$
$$x_1 + x_2 - 2x_3 - 3x_4 = 0.$$

2. Find the subspace of all solutions of the system of linear equations

$$x_1 + 2x_2 - 3x_3 + x_4 = 0$$
$$3x_1 - x_2 + 5x_3 - x_4 = 0$$
$$2x_1 + x_2 \qquad\quad x_4 = 0.$$

3. Find all solutions of the following two systems of non-homogeneous linear equations.

(a)
$$x_1 + 3x_2 + 5x_3 - 2x_4 = 11$$
$$3x_1 - 2x_2 - 7x_3 + 5x_4 = 0$$
$$2x_1 + x_2 \qquad + x_4 = 7,$$
(b)
$$x_1 + 3x_2 + 2x_3 + 5x_4 = 10$$
$$3x_1 - 2x_2 - 5x_3 + 4x_4 = -5$$
$$2x_1 + x_2 - x_3 + 5x_4 = 5.$$

4. Find all solutions of the following system of non-homogeneous linear equations

$$2x_1 - x_2 - 3x_3 = 1$$
$$x_1 - x_2 + 2x_3 = -2$$
$$4x_1 - 3x_2 + x_3 = -3$$
$$x_1 \qquad - 5x_3 = 3.$$

5. Find all solutions of the system of equations,

$$7x_1 + 3x_2 + 21x_3 - 13x_4 + x_5 = -14$$
$$10x_1 + 3x_2 + 30x_3 - 16x_4 + x_5 = -23$$
$$7x_1 + 2x_2 + 21x_3 - 11x_4 + x_5 = -16$$
$$9x_1 + 3x_2 + 27x_3 - 15x_4 + x_5 = -20.$$

6. Theorem 7.1 states that a necessary and sufficient condition for the existence of a solution of a system of simultaneous linear equations is that the rank of the augmented matrix be equal to the rank of the coefficient matrix. The most efficient way to determine the rank of each of these matrices is to reduce each to Hermite normal form. The reduction of the augmented matrix to normal form, however, automatically produces the reduced form of the coefficient matrix. How, and where? How is the comparison of the ranks of the coefficient matrix and the augmented matrix evident from the appearance of the reduced form of the augmented matrix?

7. The differential equation $d^2y/dx^2 + 4y = \sin x$ has the general solution $y = C_1 \sin 2x + C_2 \cos 2x + \frac{1}{3} \sin x$. Identify the associated homogeneous problem, the solution of the associated homogeneous problem, and the particular solution.

8 | Other Applications of the Hermite Normal Form

The Hermite normal form and the elementary row operations provide techniques for dealing with problems we have already encountered and handled rather awkwardly.

A Standard Basis for a Subspace

Let $A = \{\alpha_1, \ldots, \alpha_n\}$ be a basis of U and let W be a subspace of U spanned by the set $B = \{\beta_1, \ldots, \beta_r\}$. Since every subspace of U is spanned by a finite set, it is no restriction to assume that B is finite. Let $\beta_i = \sum_{j=1}^n b_{ij}\alpha_j$ so that $(b_{i1}, \ldots, b_{in})$ is the n-tuple representing β_i. Then in the matrix $B = [b_{ij}]$ each row is the representation of a vector in B. Now suppose an elementary row operation is applied to B to obtain B'. Every row of B' is a linear combination of the rows of B and, since an elementary row operation has an inverse, every row of B is a linear combination of the rows of B'. Thus the rows of B and the rows of B' represent sets spanning the same subspace W. We can therefore reduce B to Hermite normal form and obtain a particular set spanning W. Since the non-zero rows of the Hermite normal form are linearly independent, they form a basis of W.

Now let C be another set spanning W. In a similar fashion we can construct a matrix C whose rows represent the vectors in C and reduce this matrix to Hermite normal form. Let C' be the Hermite normal form obtained from C, and let B' be the Hermite normal form obtained from B. We do not assume that B and C have the same number of elements, and therefore B' and C' do not necessarily have the same number of rows. However, in each the number of non-zero rows must be equal to the dimension of W. We claim that the non-zero rows in these two normal forms are identical.

To see this, construct a new matrix with the non-zero rows of C' written beneath the non-zero rows of B' and reduce this matrix to Hermite normal form. Since the rows of C' are dependent on the rows of B', the rows of C' can be removed by elementary operations, leaving the rows of B'. Further reduction is not possible since B' is already in normal form. But by interchanging rows, which are elementary operations, we can obtain a matrix in which the non-zero rows of B' are beneath the non-zero rows of C'. As before, we can remove the rows of B' leaving the non-zero rows of C' as the normal form. Since the Hermite normal form is unique, we see that the non-zero rows of B' and C' are identical. The basis that we obtain from the non-zero rows of the Hermite normal form is the *standard basis* with respect to A for the subspace W.

This gives us an effective method for deciding when two sets span the same subspace. For example, in Chapter I-4, Exercise 5, we were asked to show that $\{(1, 1, 0, 0), (1, 0, 1, 1)\}$ and $\{(2, -1, 3, 3), (0, 1, -1, -1)\}$ span the same space. In either case we obtain $\{(1, 0, 1, 1), (0, 1, -1, -1)\}$ as the standard basis.

The Sum of Two Subspaces

If A_1 is a subset spanning W_1 and A_2 is a subset spanning W_2, then $A_1 \cup A_2$ spans $W_1 + W_2$ (Chapter I, Proposition 4.4). Thus we can find a basis for $W_1 + W_2$ by constructing a large matrix whose rows are the representations of the vectors in $A_1 \cup A_2$ and reducing it to Hermite normal form by elementary row operations.

The Characterization of a Subspace by a Set of Homogeneous Linear Equations

We have already seen that the set of all solutions of a system of homogeneous linear equations is a subspace, the kernel of the linear transformation represented by the matrix of coefficients. The method for solving such a system which we described in Section 7 amounts to passing from a characterization of a subspace as the set of all solutions of a system of equations to its description as the set of all linear combinations of a basis. The question naturally arises: If we are given a spanning set for a subspace W, how can we find a system of simultaneous homogeneous linear equations for which W is exactly the set of solutions?

This is not at all difficult and no new procedures are required. All that is needed is a new look at what we have already done. Consider the homogeneous linear equation $a_1 x_1 + \cdots + a_n x_n = 0$. There is no significant difference between the a_i's and the x_i's in this equation; they appear symmetrically. Let us exploit this symmetry systematically.

If $a_1 x_1 + \cdots + a_n x_n = 0$ and $b_1 x_1 + \cdots + b_n x_n = 0$ are two homogeneous linear equations then $(a_1 + b_1)x_1 + \cdots + (a_n + b_n)x_n = 0$ is a homogeneous linear equation as also is $a a_1 x_1 + \cdots + a a_n x_n = 0$ where $a \in F$. Thus we can consider the set of all homogeneous linear equations in n unknowns as a vector space over F. The equation $a_1 x_1 + \cdots + a_n x_n = 0$ is represented by the n-tuple $(a_1, \ldots, a_n)$.

When we write a matrix to represent a system of equations and reduce that matrix to Hermite normal form we are finding a standard basis for the subspace of the vector space of all homogeneous linear equations in $x_1, \ldots, x_n$ spanned by this system of equations just as we did in the first part of this section for a set of vectors spanning a subspace. The rank of the system of equations is the dimension of the subspace of equations spanned by the given system.

Now let W be a subspace given by a spanning set and solve for the subspace E of all equations satisfied by W. Then solve for the subspace of solutions of the system of equations E. W must be a subspace of the set of all solutions. Let W be of dimension v. By Theorem 7.1 the dimension of E is $n - v$. Then, in turn, the dimension of the set of all solutions of E is $n - (n - v) = v$. Thus W must be exactly the space of all solutions. Thus W and E characterize each other.

If we start with a system of equations and solve it by means of the Hermite normal form, as described in Section 7, we obtain in a natural way a basis for the subspace of solutions. This basis, however, will not be the standard basis. We can obtain full symmetry between the standard system of equations and the standard basis by changing the definition of the standard basis. Instead of applying the elementary row operations by starting with the left-hand column, start with the right-hand column. If the basis obtained in this way is called the standard basis, the equations obtained will be the standard equations, and the solution of the standard equations will be the standard basis. In the following example the computations will be carried out in this way to illustrate this idea. It is not recommended, however, that this be generally done since accuracy with one definite routine is more important.

Let

$$W = \langle (1, 0, -3, 11, -5), (3, 2, 5, -5, 3), (1, 1, 2, -4, 2), (7, 2, 12, 1, 2) \rangle.$$

We now find a standard basis by reducing

$$\begin{bmatrix} 1 & 0 & -3 & 11 & -5 \\ 3 & 2 & 5 & -5 & 3 \\ 1 & 1 & 2 & -4 & 2 \\ 7 & 2 & 12 & 1 & 2 \end{bmatrix}$$

to the form

$$\begin{bmatrix} 2 & 0 & 5 & 0 & 1 \\ 1 & 0 & 2 & 1 & 0 \\ 1 & 1 & 0 & 0 & 0 \\ 0 & 0 & 0 & 0 & 0 \end{bmatrix}.$$

From this we see that the coefficients of our systems of equations satisfy the conditions

$$\begin{aligned} 2a_1 && + 5a_3 && + a_5 &= 0 \\ a_1 && + 2a_3 + a_4 && &= 0 \\ a_1 + a_2 && && &= 0. \end{aligned}$$

The coefficients a_1 and a_3 can be selected arbitrarily and the others computed from them. In particular, we have

$$(a_1, a_2, a_3, a_4, a_5) = a_1(1, -1, 0, -1, -2) + a_3(0, 0, 1, -2, -5).$$

The 5-tuples $(1, -1, 0, -1, -2)$ and $(0, 0, 1, -2, -5)$ represent the two standard linear equations

$$\begin{aligned} x_1 - x_2 && -x_4 - 2x_5 &= 0 \\ && x_3 - 2x_4 - 5x_5 &= 0. \end{aligned}$$

The reader should check that the vectors in W actually satisfy these equations and that the standard basis for W is obtained.

The Intersection of Two Subspaces

Let W_1 and W_2 be subspaces of U of dimensions ν_1 and ν_2, respectively, and let $W_1 \cap W_2$ be of dimension ν. Then $W_1 + W_2$ is of dimension $\nu_1 + \nu_2 - \nu$. Let E_1 and E_2 be the spaces of equations characterizing W_1 and W_2. As we have seen E_1 is of dimension $n - \nu_1$ and E_2 is of dimension $n - \nu_2$. Let the dimension of $E_1 + E_2$ be ρ. Then $E_1 \cap E_2$ is of dimension $(n - \nu_1) + (n - \nu_2) - \rho = 2n - \nu_1 - \nu_2 - \rho$.

Since the vectors in $W_1 \cap W_2$ satisfy the equations in both E_1 and E_2, they satisfy the equations in $E_1 + E_2$. Thus $\nu \leq n - \rho$. On the other hand,

W_1 and W_2 both satisfy the equations in $E_1 \cap E_2$ so that $W_1 + W_2$ satisfies the equations in $E_1 \cap E_2$. Thus $\nu_1 + \nu_2 - \nu \leq n - \{2n - \nu_1 - \nu_2 - \rho\} = \nu_1 + \nu_2 + \rho - n$. A comparison of these two inequalities shows that $\nu = n - \rho$ and hence that $W_1 \cap W_2$ is characterized by $E_1 + E_2$.

Given W_1 and W_2, the easiest way to find $W_1 \cap W_2$ is to determine E_1 and E_2 and then $E_1 + E_2$. From $E_1 + E_2$ we can then find $W_1 \cap W_2$. In effect, this involves solving three systems of equations, and reducing to Hermit normal form three times, but it is still easier than a direct assault on the problem.

As an example consider Exercise 8 of Chapter I-4. Let $W_1 = \langle (1, 2, 3, 6), (4, -1, 3, 6), (5, 1, 6, 12) \rangle$ and $W_2 = \langle (1, -1, 1, 1), (2, -1, 4, 5) \rangle$. Using the Hermite normal form, we find that $E_1 = \langle (-2, -2, 0, 1), (-1, -1, 1, 0) \rangle$ and $E_2 = \langle (-4, -3, 0, 1), (-3, -2, 1, 0) \rangle$. Again, using the Hermite normal form we find that the standard basis for $E_1 + E_2$ is $\{(1, 0, 0, \frac{1}{2}), (0, 1, 0, -1), (0, 0, 1, -\frac{1}{2})\}$. And from this we find quite easily that, $W_1 \cap W_2 = \langle (-\frac{1}{2}, 1, \frac{1}{2}, 1) \rangle$.

Let $B = \{\beta_1, \beta_2, \ldots, \beta_n\}$ be a given finite set of vectors. We wish to solve the problem posed in Theorem 2.2 of Chapter I. How do we show that some β_k is a linear combination of the β_i with $i < k$; or how do we show that no β_k can be so represented?

We are looking for a relation of the form

$$\beta_k = \sum_{i=1}^{k-1} x_{ik}\beta_i. \qquad (8.1)$$

This is not a meaningful numerical problem unless β is a given specific set. This usually means that the β_i are given in terms of some coordinate system, relative to some given basis. But the relation (8.1) is independent of any coordinate system so we are free to choose a different coordinate system if this will make the solution any easier. It turns out that the tools to solve this problem are available.

Let $A = \{\alpha_1, \ldots, \alpha_m\}$ be the given basis and let

$$\beta_j = \sum_{i=1}^{m} a_{ij}\alpha_i, \qquad j = 1, \ldots, n. \qquad (8.2)$$

If $A' = \{\alpha_1', \ldots, \alpha_m'\}$ is the new basis (which we have not specified yet), we would have

$$\beta_j = \sum_{i=1}^{m} a_{ij}'\alpha_i', \qquad j = 1, \ldots, n. \qquad (8.3)$$

What is the relation between $A = [a_{ij}]$ and $A' = [a_{ij}']$? If P is the matrix of transition from the basis A to the basis A', by formula (4.3) we see that

$$A = PA'. \qquad (8.4)$$

Since P is non-singular, it can be represented as a product of elementary matrices. This means A' can be obtained from A by a sequence of elementary row operations.

The solution to (8.1) is now most conveniently obtained if we take A' to be in Hermite normal form. Suppose that A' is in Hermite normal form and use the notation given in Theorem 5.1. Then, for β_{k_i} we would have

$$\beta_{k_i} = \alpha'_i, \tag{8.5}$$

and for j between k_r and k_{r+1} we would have

$$\beta_j = \sum_{i=1}^{r} a'_{ij}\alpha'_i$$

$$= \sum_{i=1}^{r} a'_{ij}\beta_{k_i} \tag{8.6}$$

Since $k_i \leq k_r < j$, this last expression is a relation of the required form. (Actually, every linear relation that exists among the β_i can be obtained from those in (8.6). This assertion will not be used later in the book so we will not take space to prove it. Consider it "an exercise for the reader.")

Since the columns of A and A' represent the vectors in B, the rank of A is equal to the number of vectors in a maximal linearly independent subset of B. Thus, if B is linearly independent the rank of A will be n, this means that the Hermite normal form of A will either show that B is linearly independent or reveal a linear relation in B if it is dependent.

For example, consider the set $\{(1, 0, -3, 11, -5), (3, 2, 5, -5, 3), (1, 1, 2, -4, 2), (7, 2, 12, 1, 2)\}$. The implied context is that a basis $A = \{\alpha_1, \ldots, \alpha_5\}$ is considered to be given and that $\beta_1 = \alpha_1 - 3\alpha_3 + 11\alpha_4 - 5\alpha_5$ etc. According to (8.2) the appropriate matrix is

$$\begin{bmatrix} 1 & 3 & 1 & 7 \\ 0 & 2 & 1 & 2 \\ -3 & 5 & 2 & 12 \\ 11 & -5 & -4 & 1 \\ -5 & 3 & 2 & 2 \end{bmatrix}$$

which reduces to the Hermite normal form

$$\begin{bmatrix} 1 & 0 & 0 & -\frac{3}{4} \\ 0 & 1 & 0 & \frac{23}{4} \\ 0 & 0 & 1 & -\frac{19}{2} \\ 0 & 0 & 0 & 0 \\ 0 & 0 & 0 & 0 \end{bmatrix}.$$

It is easily checked that $-\frac{3}{4}(1, 0, -3, 11, -5) + \frac{23}{4}(3, 2, 5, -5, 3) - \frac{19}{2}(1, 1, 2, -4, 2) = (7, 2, 12, 1, 2)$.

EXERCISES

1. Determine which of the following set in R^4 are linearly independent over R.

    (a) $\{(1, 1, 0, 1), (1, -1, 1, 1), (2, 2, 1, 2), (0, 1, 0, 0)\}$.
    (b) $\{(1, 0, 0, 1), (0, 1, 1, 0), (1, 0, 1, 0), (0, 1, 0, 1)\}$.
    (c) $\{(1, 0, 0, 1), (0, 1, 0, 1), (0, 0, 1, 1), (1, 1, 1, 1)\}$.

This problem is identical to Exercise 8, Chapter I-2.

2. Let W be the subspace of R^5 spanned by $\{(1, 1, 1, 1, 1), (1, 0, 1, 0, 1),$ $(0, 1, 1, 1, 0), (2, 0, 0, 1, 1), (2, 1, 1, 2, 1), (1, -1, -1, -2, 2), (1, 2, 3, 4, -1)\}$. Find a standard basis for W and the dimension of W. This problem is identical to Exercise 6, Chapter I-4.

3. Show that $\{(1, -1, 2, -3), (1, 1, 2, 0), (3, -1, 6, -6)\}$ and $\{(1, 0, 1, 0),$ $(0, 2, 0, 3)\}$ do not span the same subspace. This problem is identical to Exercise 7, Chapter I-4.

4. If $W_1 = \langle(1, 1, 3, -1), (1, 0, -2, 0), (3, 2, 4, -2)\rangle$ and $W_2 = \langle(1, 0, 0, 1),$ $(1, 1, 7, 1)\rangle$ determine the dimension of $W_1 + W_2$.

5. Let $W = \langle(1, -1, -3, 0, 1), (2, 1, 0, -1, 4), (3, 1, -1, 1, 8), (1, 2, 3, 2, 6)\rangle$. Determine the standard basis for W. Find a set of linear equations which characterize W.

6. Let $W_1 = \langle(1, 2, 3, 6), (4, -1, 3, 6), (5, 1, 6, 12)\rangle$ and $W_2 = \langle(1, -1, 1, 1),$ $(2, -1, 4, 5)\rangle$ be subspaces of R^4. Find bases for $W_1 \cap W_2$ and $W_1 + W_2$. Extend the basis of $W_1 \cap W_2$ to a basis of W_1 and extend the basis of $W_1 \cap W_2$ to a basis of W_2. From these bases obtain a basis of $W_1 + W_2$. This problem is identical to Exercise 8, Chapter I-4.

9 | Normal Forms

To understand fully what a normal form is, we must first introduce the concept of an equivalence relation. We say that a *relation* is defined in a set if, for each pair (a, b) of elements in this set, it is decided that "*a* is related to *b*" or "*a* is not related to *b*." If *a* is related to *b*, we write $a \sim b$. An *equivalence relation* in a set S is a relation in S satisfying the following laws:

    Reflexive law: $a \sim a$,
    Symmetric law: If $a \sim b$, then $b \sim a$.
    Transitive law: If $a \sim b$ and $b \sim c$, then $a \sim c$.

If for an equivalence relation we have $a \sim b$, we say that *a* is *equivalent* to *b*.

Examples.   Among rational fractions we can define $a/b \sim c/d$ (for a, b, c, d integers) if and only if $ad = bc$. This is the ordinary definition of equality in rational numbers, and this relation satisfies the three conditions of an equivalence relation.

In geometry we do not ordinarily say that a straight line is parallel to itself. But if we agree to say that a straight line is parallel to itself, the concept of parallelism is an equivalence relation among the straight lines in the plane or in space.

Geometry has many equivalence relations: congruence of triangles, similarity of triangles, the concept of projectivity in projective geometry, etc. In dealing with time we use many equivalence relations: same hour of the day, same day of the week, etc. An equivalence relation is like a generalized equality. Elements which are equivalent share some common or underlying property. As an example of this idea, consider a collection of sets. We say that two sets are equivalent if their elements can be put into a one-to-one correspondence; for example, a set of three battleships and a set of three cigars are equivalent. Any set of three objects shares with any other set of three objects a concept which we have abstracted and called "three." All other qualities which these sets may have are ignored.

It is most natural, therefore, to group mutually equivalent elements together into classes which we call *equivalence classes*. Let us be specific about how this is done. For each $a \in S$, let S_a be the set of all elements in S equivalent to a; that is, $b \in S_a$ if and only if $b \sim a$. We wish to show that the various sets we have thus defined are either disjoint or identical.

Suppose $S_a \cap S_b$ is not empty; that is, there exists a $c \in S_a \cap S_b$ such that $c \sim a$ and $c \sim b$. By symmetry $b \sim c$, and by transitivity $b \sim a$. If d is any element of S_b, $d \sim b$ and hence $d \sim a$. Thus $d \in S_a$ and $S_b \subset S_a$. Since the relation between S_a and S_b is symmetric we also have $S_a \subset S_b$ and hence $S_a = S_b$. Since $a \in S_a$ we have shown, in effect, that a proposed equivalence class can be identified by any element in it. An element selected from an equivalence class will be called a *representative* of that class.

An equivalence relation in a set S defines a partition of that set into equivalence classes in the following sense: (1) Every element of S is in some equivalence class, namely, $a \in S_a$. (2) Two elements are in the same equivalence class if and only if they are equivalent. (3) Non-identical equivalence classes are disjoint. On the other hand, a partition of a set into disjoint subsets can be used to define an equivalence relation; two elements are equivalent if and only if they are in the same subset.

The notions of equivalence relations and equivalence classes are not nearly so novel as they may seem at first. Most students have encountered these ideas before, although sometimes in hidden forms. For example, we may say that two differentiable functions are equivalent if and only if they have the same derivative. In calculus we use the letter "C" in describing the equivalence classes; for example, $\tau^3 + x^2 + 2x + C$ is the set (equivalence class) of all functions whose derivative is $3x^2 + 2x + 2$.

In our study of matrices we have so far encountered four different equivalence relations:

I. The matrices A and B are said to be *left associate* if there exists a non-singular matrix Q such that $B = Q^{-1}A$. Multiplication by Q^{-1} corresponds to performing a sequence of elementary row operations. If A represents a linear transformation σ of U into V with respect to a basis A in U and a basis B in V, the matrix B represents σ with respect to A and a new basis in V.

II. The matrices A and B are said to be *right associate* if there exists a non-singular matrix P such that $B = AP$.

III. The matrices A and B are said to be *associate* if there exist non-singular matrixes P and Q such that $B = Q^{-1}AP$. The term "associate" is not a standard term for this equivalence relation, the term most frequently used being "equivalent." It seems unnecessarily confusing to use the same term for one particular relation and for a whole class of relations. Moreover, this equivalence relation is perhaps the least interesting of the equivalence relations we shall study.

IV. The matrices A and B are said to be *similar* if there exists a non-singular matrix P such that $B = P^{-1}AP$. As we have seen (Section 4) similar matrices are representations of a single linear transformation of a vector space into itself. This is one of the most interesting of the equivalence relations, and Chapter III is devoted to a study of it.

Let us show in detail that the reation we have defined as left associate is an equivalence relation. The matrix Q^{-1} appears in the definition because Q represents the matrix of transition. However, Q^{-1} is just another non-singular matrix, so it is clearly the same thing to say that A and B are left associate if and only if there exists a non-singular matrix Q such that $B = QA$.

(1) $A \sim A$ since $IA = A$.
(2) If $A \sim B$, there is a non-singular matrix Q such that $B = QA$. But then $A = Q^{-1}B$ so that $B \sim A$.
(3) If $A \sim B$ and $B \sim C$, there exist non-singular matrices Q and P such that $B = QA$ and $C = PB$. But then $PQA = PB = C$ and PQ is non-singular so that $A \sim C$.

For a given type of equivalence relation among matrices a *normal form* is a particular matrix chosen from each equivalence class. It is a representative of the entire class of equivalent matrices. In mathematics the terms "normal" and "canonical" are frequently used to mean "standard" in some particular sense. A normal form or canonical form is a standard form selected to represent a class of equivalent elements. A normal form should be selected to have the following two properties: Given any matrix A, (1) it should be possible by fairly direct and convenient methods to find the normal form of the equivalence class containing A, and (2) the method should lead to a unique normal form.

Often the definition of a normal form is compromised with respect to the second of these desirable properties. For example, if the normal form were a matrix with complex numbers in the main diagonal and zeros elsewhere, to make the normal form unique it would be necessary to specify the order of the numbers in the main diagonal. But it is usually sufficient to know the numbers in the main diagonal without regard to their order, so it would be an awkward complication to have to specify their order.

Normal forms have several uses. Perhaps the most important use is that the normal form should yield important or useful information about the concept that the matrix represents. This should be amply illustrated in the case of the concept of left associate and the Hermite normal form. We introduced the Hermite normal form through linear transformations, but we found that it yielded very useful information when the matrix was used to represent linear equations or bases of subspaces.

Given two matrices, we can use the normal form to tell whether they are equivalent. It is often easier to reduce each to normal form and compare the normal forms than it is to transform one into the other. This is the case, for example, in the application described in the first part of Section 8.

Sometimes, knowing the general appearance of the normal form, we can find all the information we need without actually obtaining the normal form. This is the case for the equivalence relation we have called associate. The normal form for this equivalence relation is described in Theorem 4.1. There is just one normal form for each possible value of the rank. The number of different equivalence classes is min $\{m, n\} + 1$. With this notion of equivalence the rank of a matrix is the only property of importance. Any two matrices of the same rank are associate. In practice we can find the rank without actually computing the normal form of Theorem 4.1. And knowing the rank we know the normal form.

We encounter several more equivalence relations among matrices. The type of equivalence introduced will depend entirely on the underlying concepts the matrices are used to represent. It is worth mentioning that for the equivalence relations we introduce there is no necessity to prove, as we did for an example above, that each is an equivalence relation. An underlying concept will be defined without reference to any coordinate system or choice of basis. The matrices representing this concept will transform according to certain rules when the basis is changed. Since a given basis can be retained the relation defined is reflexive. Since a basis changed can be changed back to the original basis, the relation defined is symmetric. A basis changed once and then changed again depends only on the final choice so that the relation is transitive.

For a fixed basis A in U and B in V two different linear transformations σ and τ of U into V are represented by different matrices. If it is possible, however, to choose bases A' in U and B' in V such that the matrix representing τ with respect to A' and B' is the same as the matrix representing σ with respect to A and B, then it is certainly clear that σ and τ share important geometric properties.

For a fixed σ two matrices A and A' representing σ with respect to different bases are related by a matrix equation of the form $A' = Q^{-1}AP$. Since A and A' represent the same linear transformation we feel that they should have some properties in common, those dependent upon σ.

These two points of view are really slightly different views of the same kind of relationship. In the second case, we can consider A and A' as representing two linear transformations with respect to the same basis, instead of the same linear transformation with respect to different bases. For example, in R^2 the matrix $\begin{bmatrix} 1 & 0 \\ 0 & -1 \end{bmatrix}$ represents a reflection about the x_1-axis and $\begin{bmatrix} -1 & 0 \\ 0 & 1 \end{bmatrix}$ represents a reflection about the x_2-axis. When both linear transformations are referred to the same coordinate system they are different. However, for the purpose of discussing properties independent of a coordinate system they are essentially alike. The study of equivalence relations is motivated by such considerations, and the study of normal forms is aimed at determining just what these common properties are that are shared by equivalent linear transformations or equivalent matrices.

To make these ideas precise, let σ and τ be linear transformations of V into itself. We say that σ and τ are *similar* if there exist bases A and B of V such that the matrix representing σ with respect to A is the same as the matrix representing τ with respect to B. If A and B are the matrices representing σ and τ with respect to A and P is the matrix of transition from A to B, then $P^{-1}BP$ is the matrix representing τ with respect to B. Thus σ and τ are similar if $P^{-1}BP = A$.

In a similar way we can define the concepts of left associate, right associate, and associate for linear transformations.

*10 | Quotient Sets, Quotient Spaces

Definition. If S is any set on which an equivalence relation is defined, the collection of equivalence classes is called the *quotient* or *factor set*. Let $\bar{S}$ denote the quotient set. An element of $\bar{S}$ is an equivalence class. If a is an element of S and $\bar{a}$ is the equivalence class containing a, the mapping η that maps a onto $\bar{a}$ is well defined. This mapping is called the *canonical mapping*.

Although the concept of a quotient set might appear new to some, it is certain that almost everyone has encountered the idea before, perhaps in one guise or another. One example occurs in arithmetic. In this setting, let S be the set of all formal fractions of the form a/b where a and b are integers and $b \neq 0$. Two such fractions, a/b and c/d, are equivalent if and only if $ad = bc$. Each equivalence class corresponds to a single rational number. The rules of arithmetic provide methods of computing with rational numbers by performing appropriate operations with formal fractions selected from the corresponding equivalence classes.

Let U be a vector space over F and let K be a subspace of U. We shall call two vectors α, $\beta \in U$ equivalent modulo K if and only if their difference lies in K. Thus $\alpha \sim \beta$ if and only if $\alpha - \beta \in K$. We must first show this defines an equivalence relation. (1) $\alpha \sim \alpha$ because $\alpha - \alpha = 0 \in K$. (2) $\alpha \sim \beta \Rightarrow \alpha - \beta \in K \Rightarrow \beta - \alpha \in K \Rightarrow \beta \sim \alpha$. (3) $\{\alpha \sim \beta \text{ and } \beta \sim \gamma\} \Rightarrow \{\alpha - \beta \in K \text{ and } \beta - \gamma \in K\}$. Since K is a subspace $\alpha - \gamma = (\alpha - \beta) + (\beta - \gamma) \in K$ and, hence, $\alpha \sim \gamma$. Thus "$\sim$" is an equivalence relation.

We wish to define vector addition and scalar multiplication in $\bar{U}$. For $\alpha \in U$, let $\bar{\alpha} \in \bar{U}$ denote the equivalence class containing α. α is called a *representative* of $\bar{\alpha}$. Since $\bar{\alpha}$ may contain other elements besides α, it may happen that $\alpha \neq \alpha'$ and yet $\bar{\alpha} = \bar{\alpha}'$. Let $\bar{\alpha}$ and $\bar{\beta}$ be two elements in $\bar{U}$. Since $\alpha, \beta \in U$, $\alpha + \beta$ is defined. We wish to define $\overline{\alpha + \beta}$ to be the sum of $\bar{\alpha}$ and $\bar{\beta}$. In order for this to be well defined we must end up with the same equivalence class as the sum if different representatives are chosen from $\bar{\alpha}$ and $\bar{\beta}$. Suppose $\bar{\alpha} = \bar{\alpha}'$ and $\bar{\beta} = \bar{\beta}'$. Then $\alpha - \alpha' \in K$, $\beta - \beta' \in K$, and $(\alpha + \beta) - (\alpha' + \beta') \in K$. Thus $\overline{\alpha + \beta} = \overline{\alpha' + \beta'}$ and the sum is well defined. Scalar multiplication is defined similarly. For $a \in F$, $a\bar{\alpha}$ is thus defined to be the equivalence class containing $a\alpha$; that is, $a\bar{\alpha} = \overline{a\alpha}$ These operations in $\bar{U}$ are said to be *induced* by the corresponding operation in U.

Theorem 10.1. *If U is a vector space over F, and K is a subspace of U, the quotient set $\bar{U}$ with vector addition and scalar multiplication defined as above is a vector space over F.*

PROOF. We leave this as an exercise. □

For any $\alpha \in U$, the symbol $\alpha + K$ is used to denote the set of all elements in U that can be written in the form $\alpha + \gamma$ where $\gamma \in K$. (Strictly speaking, we should denote the set by $\{\alpha\} + K$ so that the plus sign combines two objects of the same type. The notation introduced here is traditional and simpler.) The set $\alpha + K$ is called a *coset* of K. If $\beta \in \alpha + K$, then $\beta - \alpha \in K$ and

$\beta \sim \alpha$. Conversely, if $\beta \sim \alpha$, then $\beta - \alpha = \gamma \in K$ so $\beta \in \alpha + K$. Thus $\alpha + K$ is simply the equivalence class $\bar{\alpha}$ containing α. Thus $\alpha + K = \beta + K$ if and only if $\alpha \in \bar{\beta} = \beta + K$ or $\beta \in \bar{\alpha} = \alpha + K$.

The notation $\alpha + K$ to denote $\bar{\alpha}$ is convenient to use in some calculations. For example, $\bar{\alpha} + \bar{\beta} = (\alpha + K) + (\beta + K) = \alpha + \beta + K = \overline{\alpha + \beta}$, and $a\bar{\alpha} = a(\alpha + K) = a\alpha + aK \subset a\alpha + K = \overline{a\alpha}$. Notice that $a\bar{\alpha} = \overline{a\alpha}$ when $\bar{\alpha}$ and $\overline{a\alpha}$ are considered to be elements of $\bar{U}$ and scalar multiplication is the induced operation, but that $a\bar{\alpha}$ and $\overline{a\alpha}$ may not be the same when they are viewed as subsets of U (for example, let $a = 0$). However, since $a\bar{\alpha} \subset \overline{a\alpha}$ the set $a\bar{\alpha}$ determines the desired coset in $\bar{U}$ for the induced operations. Thus we can compute effectively in $\bar{U}$ by doing the corresponding operations with representatives. This is precisely what is done when we compute in residue classes of integers modulo an integer m.

Definition. $\bar{U}$ with the induced operations is called a *factor space* or *quotient space*. In order to designate the role of the subspace K which defines the equivalence relations, $\bar{U}$ is usually denoted by U/K.

In our discussion of solutions of linear problems we actually encountered quotient spaces, but the discussion was worded in such a way as to avoid introducing this more sophisticated concept. Given the linear transformation σ of U into V, let K be the kernel of σ and let $\bar{U} = U/K$ be the corresponding quotient space. If α_1 and α_2 are solutions of the linear problem, $\alpha(\xi) = \beta$, then $\sigma(\alpha_1 - \alpha_2) = 0$ so that α_1 and α_2 are in the same coset of K. Thus for each $\beta \in \text{Im}(\sigma)$ there corresponds precisely one coset of K. In fact the correspondence between U/K and $\text{Im}(\sigma)$ is an isomorphism, a fact which is made more precise in the following theorem.

Theorem 10.2.   (*First homomorphism theorem*). *Let σ be a linear transformation of U into V. Let K be the kernel of σ. Then σ can be written as the product of a canonical mapping η of U onto $\bar{U} = U/K$ and a monomorphism σ_1 of $\bar{U}$ into V.*

PROOF.   The canonical mapping η has already been defined. To define σ_1, for each $\bar{\alpha} \in \bar{U}$ let $\sigma_1(\bar{\alpha}) = \sigma(\alpha)$ where α is any representative of $\bar{\alpha}$. Since $\sigma(\alpha) = \sigma(\alpha')$ for $\alpha \sim \alpha'$, σ_1 is well defined. It is easily seen that σ_1 is a monomorphism since σ must have different values in different cosets. $\square$

The homomorphism theorem is usually stated by saying, "The homomorphic image is isomorphic to the quotient space of U modulo the kernel."

Theorem 10.3.   (*Mapping decomposition theorem*). *Let σ be a linear transformation of U into V. Let K be the kernel of σ and I the image of σ. Then σ can be written as the product $\sigma = \iota\sigma_1\eta$, where η is the canonical mapping of*

U onto $\bar{U} = U/K$, σ_1 is an isomorphism of $\bar{U}$ onto I, and ι is the injection of I into V.

PROOF. Let σ' be the linear transformation of U onto I induced by restricting the codomain of σ to the image of σ. By Theorem 10.2, σ' can be written in the form $\sigma' = \sigma_1 \eta$. $\square$

Theorem 10.4. (*Mapping factor theorem*). *Let S be a subspace of U and let $\bar{U} = U/S$ be the resulting quotient space. Let σ be a linear transformation of U into V, and let K be the kernel of σ. If $S \subset K$, then there exists a linear transformation σ_1 of $\bar{U}$ into V such that $\sigma = \sigma_1 \eta$ where η is the canonical mapping of U onto $\bar{U}$.*

PROOF. For each $\bar{\alpha} \in \bar{U}$, let $\sigma_1(\bar{\alpha}) = \sigma(\alpha)$ where $\alpha \in \bar{\alpha}$. If α' is another representative of $\bar{\alpha}$, then $\alpha - \alpha' \in S \subset K$. Thus $\sigma(\alpha) = \sigma(\alpha')$ and σ_1 is well defined. It is easy to check that σ_1 is linear. Clearly, $\sigma(\alpha) = \sigma_1(\bar{\alpha}) = \sigma_1(\eta(\alpha))$ for all $\alpha \in U$, and $\sigma = \sigma_1 \eta$. $\square$

We say that σ *factors* through $\bar{U}$.

Note that the homomorphism theorem is a special case of the factor theorem in which $K = S$.

Theorem 10.5. (*Induced mapping theorem*). *Let U and V be vector spaces over F, and let τ be a linear transformation of U into V. Let U_0 be a subspace of U and let V_0 be a subspace of V. If $\tau(U_0) \subset V_0$, it is possible to define in a natural way a mapping $\bar{\tau}$ of U/U_0 into V/V_0 such that $\sigma_2 \tau = \bar{\tau} \sigma_1$ where σ_1 is the canonical mapping U onto $\bar{U}$ and σ_2 is the canonical mapping of V onto $\bar{V}$.*

PROOF. Consider $\sigma = \sigma_2 \tau$, which maps U into $\bar{V}$. The kernel of σ is $\tau^{-1}(V_0)$. By assumption, $U_0 \subset \tau^{-1}(V_0)$. Hence, by the mapping factor theorem, there is a linear transformation $\bar{\tau}$ such that $\bar{\tau} \sigma_1 = \sigma_2 \tau$. $\square$

We say that $\bar{\tau}$ is *induced* by τ.

Numerical calculations with quotient spaces can usually be avoided in problems involving finite dimensional vector spaces. If U is a vector space over F and K is a subspace of U, we know from Theorem 4.9 of Chapter I that K is a direct summand. Let $U = K \oplus W$. Then the canonical mapping η maps W isomorphically onto U/K. Thus any calculation involving U/K can be carried out in W.

Although there are many possible choices for the complementary subspace W, the Hermite normal form provides a simple and effective way to select a W and a basis for it. This typically arises in connection with a linear problem. To see this, reexamine the proof of Theorem 5.1. There we let $k_1, k_2, \ldots, k_p$ be those indices for which $\sigma(\alpha_{k_i}) \notin \sigma(U_{k_i-1})$. We showed there that $\{\beta_1', \ldots, \beta_k'\}$ where $\beta_i' = \sigma(\alpha_{k_i})$ formed a basis of $\sigma(U)$. $\{\alpha_{k_1}, \alpha_{k_2}, \ldots, \alpha_{k_p}\}$ is a basis for a suitable W which is complementary to $K(\sigma)$.

Example. Consider the linear transformation σ of R^5 into R^3 represented by the matrix

$$\begin{bmatrix} 1 & 0 & 1 & 0 & 1 \\ 0 & 1 & 0 & 1 & 0 \\ 1 & 0 & 0 & -1 & 0 \end{bmatrix}.$$

It is easy to determine that the kernel K of σ is 2-dimensional with basis $\{(1, -1, -1, 1, 0), (0, 0, -1, 0, 1)\}$. This means that σ has rank 3 and the image of σ is all of R^3. Thus $\overline{R^5} = R^5/K$ is isomorphic to R^3.

Consider the problem of solving the equation $\sigma(\xi) = \beta$, where β is represented by (b_1, b_2, b_3). To solve this problem we reduce the augmented matrix

$$\begin{bmatrix} 1 & 0 & 1 & 0 & 1 & b_1 \\ 0 & 1 & 0 & 1 & 0 & b_2 \\ 1 & 0 & 0 & -1 & 0 & b_3 \end{bmatrix}$$

to the Hermite normal form

$$\begin{bmatrix} 1 & 0 & 0 & -1 & 0 & b_3 \\ 0 & 1 & 0 & 1 & 0 & b_2 \\ 0 & 0 & 1 & 1 & 1 & b_1 - b_3 \end{bmatrix}.$$

This means the solution ξ is represented by

$$(b_3, b_2, b_1 - b_3, 0, 0) + x_4(1, -1, -1, 1, 0) + x_5(0, 0, -1, 0, 1).$$

$$(b_3, b_2, b_1 - b_3, 0, 0) = b_1(0, 0, 1, 0, 0) + b_2(0, 1, 0, 0, 0) + b_3(1, 0, -1, 0, 0)$$

is a particular solution and a convenient basis for a subspace W complementary to K is $\{(0, 0, 1, 0, 0), (0, 1, 0, 0, 0), (1, 0, -1, 0, 0)\}$. σ maps $b_1(0, 0, 1, 0, 0) + b_2(0, 1, 0, 0, 0) + b_3(1, 0, -1, 0, 0)$ onto (b_1, b_2, b_3). Hence, W is mapped isomorphically onto R^3.

This example also provides an opportunity to illustrate the working of the first homomorphism theorem. For any $(x_1, x_2, x_3, x_4, x_5) \in R^5$.

$$\begin{aligned}(x_1, x_2, x_3, x_4 + x_5) = &(x_1 + x_3 + x_5)(0, 0, 1, 0, 0) \\ &+ (x_2 + x_4)(0, 1, 0, 0, 0) \\ &+ (x_1 - x_4)(1, 0, -1, 0, 0) \\ &+ x_4(1, -1, -1, 1, 0) + x_5(0, 0, -1, 0, 1).\end{aligned}$$

Thus $(x_1, x_2, x_3, x_4, x_5)$ is mapped onto the coset $(x_1 + x_3 + x_5)(0, 0, 1, 0, 0) + (x_2 + x_4)(0, 1, 0, 0, 0) + (x_1 - x_4)(1, 0, -1, 0, 0) + K$ under the natural homomorphism onto R^5/K. This coset is then mapped isomorphically onto $(x_1 + x_3 + x_5, x_2 + x_4, x_1 - x_y) \in R^3$. However, it is somewhat contrived to

work out an example of this type. The main importance of the first homomorphism theorem is theoretical and not computational.

*11 | Hom(U, V)

Let U and V be vector spaces over F. We have already observed in Section 1
that the set of all linear transformations of U into V can be made into a vector
space over F by defining addition and scalar multiplication appropriately.
In this section we will explore some of the elementary consequences of this
observation. We shall call this vector space Hom(U, V), "The space of all
homomorphisms of U into V."

Theorem 11.1. *If* dim $U = n$ *and* dim $V = m$, *then* dim Hom(U, V) $= mn$.

PROOF.   Let $\{\alpha_1, \ldots, \alpha_n\}$ be a basis of U and let $\{\beta_1, \ldots, \beta_m\}$ be a basis of
V. Define the linear transformation of σ_{ij} by the rule

$$\sigma_{ij}(\alpha_k) = \delta_{jk}\beta_i$$

$$= \sum_{r=1}^{m} \delta_{ri}\delta_{jk}\beta_r \tag{11.1}$$

Thus σ_{ij} is represented by the matrix $[\delta_{ri}\delta_{jk}] = A_{ij}$. A_{ij} has a zero in every
position except for a 1 in row i column j.

The set $\{\sigma_{ij}\}$ is linearly independent. For if a linear relation existed among
the σ_{ij} it would be of the form

$$\sum_{i,j} a_{ij}\sigma_{ij} = 0.$$

This means $\sum_{i,j} a_{ij}\sigma_{ij}(\alpha_k) = 0$ for all α_k. But $\sum_{i,j} a_{ij}\sigma_{ij}(\alpha_k) = \sum_{i,j} a_{ij}\delta_{jk}\beta_i =$
$\sum_i a_{ik}\beta_i = 0$. Since $\{\beta_i\}$ is a lineary independent set, $a_{ik} = 0$ for $i = 1$,
$2, \ldots, m$. Since this is true for each k, all $a_{ij} = 0$ and $\{\sigma_{ij}\}$ is linearly
independent.

If $\sigma \in$ Hom(U, V) and $\sigma(\alpha_k) = \sum_{i=1}^{m} a_{ik}\beta_i$,
then

$$\sigma(\alpha_k) = \sum_{i=1}^{m} \left(\sum_{j=1}^{n} a_{ij}\delta_{jk} \right) \beta_i$$

$$= \sum_{i=1}^{m} \sum_{j=1}^{n} a_{ij}\sigma_{ij}(\alpha_k)$$

$$= \left(\sum_{i=1}^{m} \sum_{j=1}^{n} a_{ij}\sigma_{ij} \right)(\alpha_k).$$

Thus $\{\sigma_{ij}\}$ spans Hom(U, V), which is therefore of dimension mn. $\square$

If V_1 is a subspace of V, every linear transformation of U into V_1 also defines
a mapping of U into V. This mapping of U into V is a linear transformation of

U into V. Thus, with each element of $\text{Hom}(U, V_1)$ there is associated in a natural way an element of $\text{Hom}(U, V)$. We can identify $\text{Hom}(U, V_1)$ with a subset of $\text{Hom}(U, V)$. With this identification $\text{Hom}(U, V_1)$ is a subspace of $\text{Hom}(U, V)$.

Now let U_1 be a subspace of U. In this case we cannot consider $\text{Hom}(U_1, V)$ to be a subset of $\text{Hom}(U, V)$ since a linear transformation in $\text{Hom}(U_1, V)$ is not necessarily defined on all of U. But any linear transformation in $\text{Hom}(U, V)$ is certainly defined on U_1. If $\sigma \in \text{Hom}(U, V)$ we shall consider the mapping obtained by applying σ only to elements in U_1 to be a new function and denote it by $R(\sigma)$. $R(\sigma)$ is called the *restriction* of σ to U_1. We can consider $R(\sigma)$ to be an element of $\text{Hom}(U_1, V)$.

It may happen that different linear transformations defined on U produce the same restriction on U_1. We say that σ_1 and σ_2 are equivalent on U_1 if and only if $R(\sigma_1) = R(\sigma_2)$. It is clear that $R(\sigma + \tau) = R(\sigma) + R(\tau)$ and $R(a\sigma) = aR(\sigma)$ so that the mapping of $\text{Hom}(U, V)$ into $\text{Hom}(U_1, V)$ is linear. We call this mapping R, the *restriction mapping*.

The kernel of R is clearly the set of all linear transformations in $\text{Hom}(U, V)$ that vanish on U_1. Let us denote this kernel by U_1^*.

If $\underline{\sigma}$ is any linear transformation belonging to $\text{Hom}(U_1, V)$, it can be extended to a linear transformation belonging to $\text{Hom}(U, V)$ in many ways. If $\{\alpha_1, \ldots, \alpha_n\}$ is a basis of U such that $\{\alpha_1, \ldots, \alpha_r\}$ is a basis of U_1, then let $\sigma(\alpha_j) = \underline{\sigma}(\alpha_j)$ for $j = 1, \ldots, r$, and let $\sigma(\alpha_j)$ be defined arbitrarily for $j = r + 1, \ldots, n$. Since $\underline{\sigma}$ is then the restriction of σ, we see that R is an epimorphism of $\text{Hom}(U, \overline{V})$ onto $\text{Hom}(U_1, V)$. Since $\text{Hom}(U, V)$ is of dimension mn and $\text{Hom}(U_1, V)$ is of dimension mr, U_1^* is of dimension $m(n - r)$.

Theorem 11.2. $\text{Hom}(U_1, V)$ *is canonically isomorphic to* $\text{Hom}(U, V)/U_1^*$. $\square$

Note: It helps the intuitive understanding of this theorem to examine the method by which we obtained an extension of $\underline{\sigma}$ on U_1, to σ on U. U_1^* is the set of all extensions of σ when $\underline{\sigma}$ is the zero mapping, and one can see directly that the dimension of U_1^* is $(n - r)m$.

chapter **III** | # Determinants, eigenvalues, and similarity transforma-tions

This chapter is devoted to the study of matrices representing linear transformations of a vector space into itself. We have seen that if A represents a linear transformation σ of V into itself with respect to a basis A, and P is the matrix of transition from A to a new basis A', then $P^{-1}AP = A'$ is the matrix representing σ with respect to A'. In this case A and A' are said to be similar and the mapping of A onto $A' = P^{-1}AP$ is called a similarity transformation (on the set of matrices, not on V).

Given σ, we seek a basis for which the matrix representing σ is particularly simple. In practice σ is given only implicitly by giving a matrix A representing σ. The problem, then, is to determine the matrix of transition P so that $P^{-1}AP$ has the desired form. The matrix representing σ has its simplest form whenever σ maps each basis vector onto a multiple of itself; that is, whenever for each basis vector α there exists a scalar λ such that $\sigma(\alpha) = \lambda\alpha$. It is not always possible to find such a basis, but there are some rather general conditions under which it is possible. These conditions include most cases of interest in the applications of this theory to physical problems.

The problem of finding non-zero α such that $\sigma(\alpha) = \lambda\alpha$ is equivalent to the problem of finding non-zero vectors in the kernel of $\sigma - \lambda$. This is a linear problem and we have given practical methods for solving it. But there is no non-zero solution to this problem unless $\sigma - \lambda$ is singular. Thus we are faced with the problem of finding those λ for which $\sigma - \lambda$ is singular. The values of λ for which $\sigma - \lambda$ is singular are called the eigenvalues of σ, and the non-zero vectors α for which $\sigma(\alpha) = \lambda\alpha$ are called eigenvectors of σ.

We introduce some topics from the theory of determinants solely for the purpose of finding the eigenvalues of a linear transformation. Were it not for this use of determinants we would not discuss them in this book. Thus, the treatment given them here is very brief.

85

Whenever a basis of eigenvectors exists, the use of determinants will provide a method for finding the eigenvalues and, knowing the eigenvalues, use of the Hermite normal form will enable us to find the eigenvectors. This method is convenient only for vector spaces of relatively small dimension. For numerical work with large matrices other methods are required.

The chapter closes with a discussion of what can be done if a basis of eigenvectors does not exist.

1 | Permutations

To define determinants and handle them we have to know something about permutations. Accordingly, we introduce permutations in a form most suitable for our purposes and develop their elementary properties.

A *permutation* π of a set S is a one-to-one mapping of S onto itself. We are dealing with permutations of finite sets and we take S to be the set of the first n integers; $S = \{1, 2, \ldots, n\}$. Let $\pi(i)$ denote the element which π associates with i. Whenever we wish to specify a particular permutation we describe it by writing the elements of S in two rows; the first row containing the elements of S in any order and the second row containing the element $\pi(i)$ directly below the element i in the first row. Thus for $S = \{1, 2, 3, 4\}$, the permutation π for which $\pi(1) = 2$, $\pi(2) = 4$, $\pi(3) = 3$, and $\pi(4) = 1$, can conveniently be described by the notations

$$\pi = \begin{pmatrix} 1 & 2 & 3 & 4 \\ 2 & 4 & 3 & 1 \end{pmatrix} \quad \text{or} \quad \begin{pmatrix} 2 & 4 & 1 & 3 \\ 4 & 1 & 2 & 3 \end{pmatrix} \quad \text{or} \quad \begin{pmatrix} 4 & 1 & 3 & 2 \\ 1 & 2 & 3 & 4 \end{pmatrix}.$$

Two permutations acting on the same set of elements can be combined as functions. Thus, if π and σ are two permutations, $\sigma\pi$ will denote that permutation mapping i onto $\sigma[\pi(i)]$; $(\sigma\pi)(i) = \sigma[\pi(i)]$. As an example, let π denote the permutation described above and let

$$\sigma = \begin{pmatrix} 1 & 2 & 3 & 4 \\ 1 & 3 & 4 & 2 \end{pmatrix}.$$

Then

$$\sigma\pi = \begin{pmatrix} 1 & 2 & 3 & 4 \\ 3 & 2 & 4 & 1 \end{pmatrix}.$$

Notice particularly that $\sigma\pi \neq \pi\sigma$.

If π and σ are two given permutations, there is a unique permutation ρ such that $\rho\pi = \sigma$. Since ρ must satisfy the condition that $\rho[\pi(i)] = \sigma(i)$, ρ can be described in our notation by writing the elements $\pi(i)$ in the first row and the elements $\sigma(i)$ in the second row. For the π and σ described above,

$$\rho = \begin{pmatrix} 2 & 4 & 3 & 1 \\ 1 & 3 & 4 & 2 \end{pmatrix}.$$

The permutation that leaves all elements of S fixed is called the *identity permutation* and will be denoted by ϵ. For a given π the unique permutation π^{-1} such that $\pi^{-1}\pi = \epsilon$ is called the *inverse of* π.

If for a pair of elements $i < j$ in S we have $\pi(i) > \pi(j)$, we say that π performs an *inversion*. Let $k(\pi)$ denote the total number of inversions performed by π; we then say that π contains $k(\pi)$ inversions. For the permutation π described above, $k(\pi) = 4$. The number of inversions in π^{-1} is equal to the number of inversions in π.

For a permutation π, let sgn π denote the number $(-1)^{k(\pi)}$. "*Sgn*" is an abbreviation for "signum" and we use the term "sgn π" to mean "the sign of π." If sgn $\pi = 1$, we say that π is *even*; if sgn $\pi = -1$, we say that π is *odd*.

Theorem 1.1. Sgn $\sigma\pi = $ sgn $\sigma \cdot$ sgn π.

PROOF. σ can be represented in the form

$$\sigma = \begin{pmatrix} \cdots & \pi(i) & \cdots & \pi(j) & \cdots \\ \cdots & \sigma\pi(i) & \cdots & \sigma\pi(j) & \cdots \end{pmatrix}$$

because every element of S appears in the top row. Thus, in counting the inversions in σ it is sufficient to compare $\pi(i)$ and $\pi(j)$ with $\sigma\pi(i)$ and $\sigma\pi(j)$. For a given $i < j$ there are four possibilities:
1. $i < j$; $\pi(i) < \pi(j)$; $\sigma\pi(i) < \sigma\pi(j)$: no inversions.
2. $i < j$; $\pi(i) < \pi(j)$; $\sigma\pi(i) > \sigma\pi(j)$: one inversion in σ, one in $\sigma\pi$.
3. $i < j$; $\pi(i) > \pi(j)$; $\sigma\pi(i) > \sigma\pi(j)$: one inversion in π, one in $\sigma\pi$.
4. $i < j$; $\pi(i) > \pi(j)$; $\sigma\pi(i) < \sigma\pi(j)$: one inversion in π, one in σ, and none in $\sigma\pi$.

Examination of the above table shows that $k(\sigma\pi)$ differs from $k(\sigma) + k(\pi)$ by an even number. Thus sgn $\sigma\pi = $ sgn $\sigma \cdot$ sgn π. $\square$

Theorem 1.2. *If a permutation π leaves an element of S fixed, the inversions involving that element need not be considered in determining whether π is even or odd.*

PROOF. Suppose $\pi(j) = j$. There are $j - 1$ elements of S less than j and $n - j$ elements of S larger than j. For $i < j$ an inversion occurs if and only if $\pi(i) > \pi(j) = j$. Let k be the number of elements i in S preceding j for which $\pi(i) > j$. Then there must also be exactly k elements i of S following j for which $\pi(i) < j$. It follows that there are $2k$ inversions involving j. Since their number is even they may be ignored in determining sgn π. $\square$

Theorem 1.3. *A permutation which interchanges exactly two elements of S and leaves all other elements of S fixed is an odd permutation.*

PROOF. Let π be a permutation which interchanges the elements i and j and leaves all other elements of S fixed. According to Theorem 1.2, in determining sgn π we can ignore the inversions involving all elements of S other than i and j. There is just one inversion left to consider and sgn $\pi = -1$. $\square$

Among other things, this shows that there is at least one odd permutation. In addition, there is at least one even permutation. From this it is but a step to show that the number of odd permutations is equal to the number of even permutations.

Let σ be a fixed odd permutation. If π is an even permutation, $\sigma\pi$ is odd. Furthermore, σ^{-1} is also odd so that to each odd permutation τ there corresponds an even permutation $\sigma^{-1}\tau$. Since $\sigma^{-1}(\sigma\pi) = \pi$, the mapping of the set of even permutations into the set of odd permutations defined by $\pi \to \sigma\pi$ is one-to-one and onto. Thus the number of odd permutations is equal to the number of even permutations.

EXERCISES

1. Show that there are $n!$ permutations of n objects.

2. There are six permutations of three objects. Determine which of them are even and which are odd.

3. There are 24 permutations of four objects. By use of Theorem 1.2 and Exercise 2 we can determine the parity (evenness or oddness) of 15 of these permutations without counting inversions. Determine the parity of these 15 permutations by this method and the parity of the remaining nine by any other method.

4. The nine permutations of four objects that leave no object fixed can be divided into two types of permutations, those that interchange two pairs of objects and those that permut the four objects in some cyclic order. There are three permutations of the first type and six of the second. Find them. Knowing the parity of the 15 permutations that leave at least one object fixed, as in Exercise 3, and that exactly half of the 24 permutations must be even, determine the parity of these nine.

5. By counting the inversions determine the parity of

$$\pi = \begin{pmatrix} 1 & 2 & 3 & 4 & 5 \\ 2 & 4 & 5 & 1 & 3 \end{pmatrix}.$$

Notice that π permutes the objects in $\{1, 2, 4\}$ among themselves and the objects in $\{3, 5\}$ among themselves. Determine the parity of π on each of these subsets separately and deduce the parity of π on all of S.

2 | Determinants

Let $A = [a_{ij}]$ be a square $n \times n$ matrix. We wish to associate with this matrix a scalar that will in some sense measure the "size" of A and tell us whether or not A is non-singular.

Definition. The *determinant* of the matrix $A = [a_{ij}]$ is defined to be the scalar $\det A = |a_{ij}|$ computed according to the rule

$$\det A = |a_{ij}| = \sum_{\pi} (\text{sgn } \pi) a_{1\pi(1)} a_{2\pi(2)} \cdots a_{n\pi(n)}, \qquad (2.1)$$

where the sum is taken over all permutations of the elements of $S = \{1, \ldots, n\}$. Each term of the sum is a product of n elements, each taken from a different row of A and from a different column of A, and sgn π. The number n is called the *order* of the determinant.

As a direct application of this definition we see that

$$\begin{vmatrix} a_{11} & a_{12} \\ a_{21} & a_{22} \end{vmatrix} = a_{11}a_{22} - a_{12}a_{21}. \qquad (2.2)$$

$$\begin{vmatrix} a_{11} & a_{12} & a_{13} \\ a_{21} & a_{22} & a_{23} \\ a_{31} & a_{32} & a_{33} \end{vmatrix} = \begin{aligned} & a_{11}a_{22}a_{33} + a_{12}a_{23}a_{31} + a_{13}a_{21}a_{32} - a_{12}a_{21}a_{33} \\ & \quad - a_{13}a_{22}a_{31} - a_{11}a_{23}a_{32}. \end{aligned} \qquad (2.3)$$

In general, a determinant of order n will be the sum of $n!$ products. As n increases, the amount of computation increases astronomically. Thus it is very desirable to develop more efficient ways of handling determinants.

Theorem 2.1. $\det A^T = \det A$.

PROOF. In the expansion of $\det A$ each term is of the form

$$(\text{sgn } \pi) a_{1\pi(1)} a_{2\pi(2)} \cdots a_{n\pi(n)}.$$

The factors of this term are ordered so that the indices of the rows appear in the usual order and the column indices appear in a permuted order. In the expansion of $\det A^T$ the same factors will appear but they will be ordered according to the row indices of A^T, that is, according to the column indices of A. Thus this same product will appear in the form

$$(\text{sgn } \pi^{-1}) a_{\pi^{-1}(1),1} a_{\pi^{-1}(2),2} \cdots a_{\pi^{-1}(n),n}.$$

But since sgn $\pi^{-1} =$ sgn π, this term is identical to the one given above. Thus, in fact, all the terms in the expansion of $\det A^T$ are equal to corresponding terms in the expansion of $\det A$, and $\det A^T = \det A$. $\square$

A consequence of this discussion is that any property of determinants developed in terms of the rows (or columns) of A will also imply a corresponding property in terms of the columns (or rows) of A.

Theorem 2.2. *If A' is the matrix obtained from A by multiplying a row (or column) of A by a scalar c, then $\det A' = c \det A$.*

PROOF.   Each term of the expansion of $\det A$ contains just one element from each row of A. Thus multiplying a row of A by c introduces the factor c into each term of $\det A$. Thus $\det A' = c \det A$. □

Theorem 2.3.   *If A' is the matrix obtained from A by interchanging any two rows (or columns) of A, then $\det A' = -\det A$.*

PROOF.   Interchanging two rows of A has the effect of interchanging two row indices of the elements appearing in A. If σ is the permutation interchanging these two indices, this operation has the effect of replacing each permutation π by the permutation $\pi\sigma$. Since σ is an odd permutation, this has the effect of changing the sign of every term in the expansion of $\det A$. Therefore, $\det A' = -\det A$. □

Theorem 2.4.   *If A has two equal rows, $\det A = 0$.*

PROOF.   The matrix obtained from A by interchanging the two equal rows is identical to A, and yet, by Theorem 2.3, this operation must change the sign of the determinant. Since the only number equal to its negative is 0 $\det A = 0$. □

Note: There is a minor point to be made here. If $1 + 1 = 0$, the proof of this theorem is not valid, but the theorem is still true. To see this we return our attention to the definition of a determinant. Sgn $\pi = 1$ for both even and odd permutations. Then the terms in (2.1) can be grouped into pairs of equal terms. Since the sum of each pair is 0, the determinant is 0.

Theorem 2.5.   *If A' is the matrix obtained from A by adding a multiple of one row (or column) to another, then $\det A' = \det A$.*

PROOF.   Let A' be the matrix obtained from A by adding c times row k to row j. Then

$$\det A' = \sum_\pi (\text{sgn } \pi) a_{1\pi(1)} \cdots (a_{j\pi(j)} + c a_{k\pi(j)}) \cdots a_{k\pi(k)} \cdots a_{n\pi(n)}$$

$$= \sum_\pi (\text{sgn } \pi) a_{1\pi(1)} \cdots a_{j\pi(j)} \cdots a_{k\pi(k)} \cdots a_{n\pi(n)}$$

$$+ c \sum_\pi (\text{sgn } \pi) a_{1\pi(1)} \cdots a_{k\pi(j)} \cdots a_{k\pi(k)} \cdots a_{n\pi(n)}. \tag{2.4}$$

The second sum on the right side of this equation is, in effect, the determinant of a matrix in which rows j and k are equal. Thus it is zero. The first term is just the expansion of $\det A$. Therefore, $\det A' = \det A$. □

It is evident from the definition that, if I is the identity matrix, $\det I = 1$.

If E is an elementary matrix of type I, det $E = c$ where c is the scalar factor employed in the corresponding elementary operation. This follows from Theorem 2.2 applied to the identity matrix.

If E is an elementary matrix of type II, det $E = 1$. This follows from Theorem 2.5 applied to the identity matrix.

If E is an elementary matrix of type III, det $E = -1$. This follows from Theorem 2.3 applied to the identity matrix.

Theorem 2.6. *If E is an elementary matrix and A is any matrix, then* det $EA = $ det $E \cdot$ det $A = $ det AE.

PROOF. This is an immediate consequence of Theorems 2.2, 2.5, 2.3, and the values of the determinants of the corresponding elementary matrices. □

Theorem 2.7. det $A = 0$ *if and only if A is singular.*

PROOF. If A is non-singular, it is a product of elementary matrices (see Chapter II, Theorem 6.1). Repeated application of Theorem 2.6 shows that det A is equal to the product of the determinants of the corresponding elementary matrices, and hence is non-zero.

If A is singular, the rows are linearly dependent and one row is a linear combination of the others. By repeated application of elementary operations of type II we can obtain a matrix with a row of zeros. The determinant of this matrix is zero, and by Theorem 2.5 so also is det A. □

Theorem 2.8. *If A and B are any two matrices of order n, then* det $AB = $ det $A \cdot$ det $B = $ det BA.

PROOF. If A and B are non-singular, the theorem follows by repeated application of Theorem 2.6. If either matrix is singular, then AB and BA are also singular and all terms are zero. □

EXERCISES

1. If all elements of a matrix below the main diagonal are zero, the matrix is said to be in *superdiagonal* form; that is, $a_{ij} = 0$ for $i > j$. If $A = [a_{ij}]$ is in superdiagonal form, compute det A.

2. Theorem 2.6 provides an effective and convenient way to evaluate determinants. Verify the following sequence of steps.

$$\begin{vmatrix} 3 & 2 & 2 \\ 1 & 4 & 1 \\ -2 & -4 & -1 \end{vmatrix} = - \begin{vmatrix} 1 & 4 & 1 \\ 3 & 2 & 2 \\ -2 & -4 & -1 \end{vmatrix} = - \begin{vmatrix} 1 & 4 & 1 \\ 0 & -10 & -1 \\ 0 & 4 & 1 \end{vmatrix}$$

$$= - \begin{vmatrix} 1 & 4 & 1 \\ 0 & -2 & 1 \\ 0 & 4 & 1 \end{vmatrix} = - \begin{vmatrix} 1 & 4 & 1 \\ 0 & -2 & 1 \\ 0 & 0 & 3 \end{vmatrix}.$$

Now use the results of Exercise 1 to evaluate the last determinant.

3. Actually, to compute a determinant there is no need to obtain a superdiagonal form. And elementary column operations can be used as well as elementary row operations. Any sequence of steps that will result in a form with a large number of zero elements will be helpful. Verify the following sequence of steps.

$$\begin{vmatrix} 3 & 2 & 2 \\ 1 & 4 & 1 \\ -2 & -4 & -1 \end{vmatrix} = \begin{vmatrix} 3 & 2 & 2 \\ 1 & 4 & 1 \\ -1 & 0 & 0 \end{vmatrix} = \begin{vmatrix} 3 & 0 & 2 \\ 1 & 3 & 1 \\ -1 & 0 & 0 \end{vmatrix}.$$

This last determinant can be evaluated by direct use of the definition by computing just one product. Evaluate this determinant.

4. Evaluate the determinants:

$$(a) \begin{vmatrix} 1 & -2 & 2 \\ -1 & 3 & 1 \\ 2 & 5 & -1 \end{vmatrix} \qquad (b) \begin{vmatrix} 1 & 2 & 0 & 1 \\ 1 & 3 & 4 & 0 \\ 0 & 1 & 5 & 6 \\ 1 & 2 & 3 & 4 \end{vmatrix}$$

5. Consider the real plane R^2. We agree that the two points (a_1, a_2), (b_1, b_2) suffice to describe a quadrilateral with corners at $(0, 0)$, (a_1, a_2), (b_1, b_2), and $(a_1 + b_1, a_2 + b_2)$. (See Fig. 2.) Show that the area of this quadrilateral is

$$\begin{vmatrix} a_1 & a_2 \\ b_1 & b_2 \end{vmatrix}$$

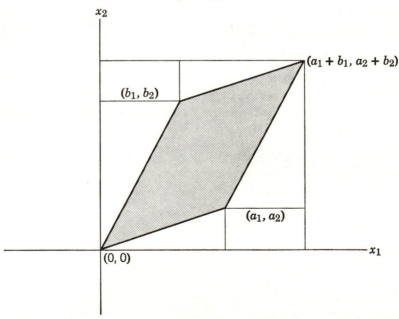

Fig. 2

Notice that the determinant can be positive or negative, and that it changes sign if the first and second rows are interchanged. To interpret the value of the determinant as an area, we must either use the absolute value of the determinant or give an interpretation to a negative area. We make the latter choice since to take the absolute value is to discard information. Referring to Fig. 2, we see that the direction of rotation from (a_1, a_2) to (b_1, b_2) across the enclosed area is the same as the direction of rotation from the positive x_1-axis to the positive x_2-axis. To interchange (a_1, a_2) and (b_1, b_2) would be to change the sense of rotation and the sign of the determinant. Thus the sign of the determinant determines an orientation of the quadrilateral on the coordinate system. Check the sign of the determinant for choices of (a_1, a_2) and (b_1, b_2) in various quadrants and various orientations.

6. (Continuation) Let E be an elementary transformation of R^2 onto itself. E maps the vertices of the given quadrilateral onto the vertices of another quadrilateral. Show that the area of the new quadrilateral is det E times the area of the old quadrilateral.

7. Let $x_1, \ldots, x_n$ be a set of indeterminates. The determinant

$$
V = \begin{vmatrix}
1 & x_1 & x_1^2 & \cdots & x_1^{n-1} \\
1 & x_2 & x_2^2 & \cdots & x_2^{n-1} \\
 & \cdot & \cdot & \cdots & \cdot \\
 & \cdot & \cdot & \cdots & \cdot \\
 & \cdot & \cdot & \cdots & \cdot \\
1 & x_n & x_n^2 & \cdots & x_n^{n-1}
\end{vmatrix}
$$

is called the Vandermonde determinant of order n.

(a) Show that V is a polynomial of degree $n - 1$ in each indeterminate separately and of degree $n(n - 1)/2$ in all the indeterminates together.

(b) Show that, for each $i < j$, V is divisible by $x_j - x_i$.

(c) Show that $\prod_{1 \le i < j \le n} (x_j - x_i)$ is a polynomial of degree $n - 1$ in each indeterminate separately, and of degree $n(n - 1)/2$ in all the indeterminates together.

(d) Show that $V = \prod_{1 \le i < j \le n} (x_j - x_i)$.

3 | Cofactors

For a given pair i, j, consider in the expansion for det A those terms which have a_{ij} as a factor. Det A is of the form det $A = a_{ij}A_{ij} +$ (terms which do not contain a_{ij} as a factor). The scalar A_{ij} is called the *cofactor* of a_{ij}.

In particular, we see that $A_{11} = \sum_\pi (\text{sgn } \pi) a_{2\pi(2)} \cdots a_{n\pi(n)}$ where this sum includes all permutations π that leave 1 fixed. Each such π defines a permutation π' on $S' = \{2, \ldots, n\}$ which coincides with π on S. Since no inversion of π involves the element 1, we see that sgn $\pi =$ sgn π'. Thus A_{ij} is a determinant, the determinant of the matrix obtained from A by crossing out the first row and the first column of A.

A similar procedure can be used to compute the cofactors A_{ij}. By a sequence of elementary row and column operations of type III we can obtain a matrix in which the element a_{ij} is moved into row 1, column 1. By applying the observation of the previous paragraph we see that the cofactor A_{ij} is essentially the determinant of the matrix obtained by crossing out the row and column containing the element a_{ij}. Furthermore, we can keep the other rows and columns in the same relative order if the sequence of operations we use interchanges only adjacent rows or columns. It takes $i - 1$ interchanges to move the element a_{ij} into the first row, and it takes $j - 1$ interchanges to move it into the first column. Thus A_{ij} is $(-1)^{i-1+j-1} = (-1)^{i+j}$ times the determinant of the matrix obtained by crossing out the ith row and the jth column of A.

Each term in the expansion of det A contains exactly one factor from each row and each column of A. Thus, for any given row of A each term of det A contains exactly one factor from that row. Hence, for any given i,

$$\det A = \sum_j a_{ij} A_{ij}. \tag{3.1}$$

Similarly, for any given column of A each term of det A contains exactly one factor from that column. Hence, for any given k,

$$\det A = \sum_j a_{jk} A_{jk}. \tag{3.2}$$

These expansions of a determinant according to the cofactors of a row or column reduce the problem of computing an nth order determinant to that of computing n determinants of order $n - 1$. We have already given explicit expansions for determinants of orders 2 and 3, and the technique of expansions according to cofactors enables us to compute determinants of higher orders. The labor of evaluating a determinant of even quite modest order is still quite formidable, however, and we make some suggestions as to how the work can be minimized.

First, observe that if any row or column has several zeros in it, expansion according to cofactors of that row or column will require the evaluation of only those cofactors corresponding to non-zero elements. It is clear that the presence of several zeros in any row or column would considerably reduce the labor. If we are not fortunate enough to find such a row or column, we can produce a row or column with a large number of zeros by applying some elementary operations of type II. For example, consider the determinant

$$\det A = \begin{vmatrix} 3 & 2 & -2 & 10 \\ 3 & 1 & 1 & 2 \\ -2 & 2 & 3 & 4 \\ 1 & 1 & 5 & 2 \end{vmatrix}.$$

If the numbers appearing in the array were unwieldy, there would be no choice but to wade in and make the best of it. The numbers in our example are all integers, and we will not introduce fractions if we take advantage of the 1's that appear in the array. By Theorem 2.5, a sequence of elementary operations of type II will not change the value of the determinant. Thus we can obtain

$$\det A = \begin{vmatrix} 0 & -1 & -17 & 4 \\ 0 & -2 & -14 & -4 \\ 0 & 4 & 13 & 8 \\ 1 & 1 & 5 & 2 \end{vmatrix} = - \begin{vmatrix} -1 & -17 & 4 \\ -2 & -14 & -4 \\ 4 & 13 & 8 \end{vmatrix}.$$

Now we face several options. We can expand the 3rd order determinants as it stands; we can try the same technique again; or we can try to remove a common factor from some row or column. We can remove the common factor -1 from the second row and the common factor 4 from the third column. Although 2 is factor of the second row, we cannot remove both a 2 from the second row and a 4 from the third column. Thus we can obtain

$$\det A = 4 \cdot \begin{vmatrix} -1 & -17 & 1 \\ 2 & 14 & 1 \\ 4 & 13 & 2 \end{vmatrix} = 4 \cdot \begin{vmatrix} -1 & -17 & 1 \\ 3 & 31 & 0 \\ 6 & 47 & 0 \end{vmatrix}$$

$$= 4 \cdot \begin{vmatrix} 3 & 31 \\ 6 & 47 \end{vmatrix} = -180.$$

If we multiply the elements in row i by the cofactors of the elements in row $k \neq i$, we get the same result as we would if the elements in row k were equal to the elements in row i. Hence,

$$\sum_j a_{ij}A_{kj} = 0 \qquad \text{for} \qquad i \neq k, \tag{3.3}$$

and

$$\sum_i a_{ij}A_{ik} = 0 \qquad \text{for} \qquad j \neq k. \tag{3.4}$$

The various relations we have developed between the elements of a matrix and their cofactors can be summarized in the form

$$\sum_j a_{ij}A_{kj} = \delta_{ik} \det A, \tag{3.5}$$

$$\sum_i a_{ij}A_{ik} = \delta_{jk} \det A. \tag{3.6}$$

If $A = [a_{ij}]$ is any square matrix and A_{ij} is the cofactor of a_{ij}, the matrix $[A_{ij}]^T = \text{adj } A$ is called the *adjunct* of A. What we here call the "adjunct" is traditionallly called the "adjoint." Unfortunately, the term "adjoint" is also used to denote a linear transformation that is not represented by the adjoint (or adjunct) matrix. A new term is badly needed. We shall have a use for the adjunct matrix only in this chapter. Thus, this unconventional terminology will cause only a minor inconvenience and help to avoid confusion.

Theorem 3.1. $A \cdot \operatorname{adj} A = (\operatorname{adj} A) \cdot A = (\det A) \cdot I.$

PROOF.

$$A \cdot \operatorname{adj} A = [a_{ij}] \cdot [A_{kl}]^T = \left[\sum_j a_{ij}A_{kj}\right] = (\det A) \cdot I. \qquad (3.7)$$

$$(\operatorname{adj} A) \cdot A = [A_{kl}]^T \cdot [a_{ij}] = \left[\sum_i A_{ik}a_{ij}\right] = (\det A) \cdot I. \;\square \qquad (3.8)$$

Theorem 3.1 provides us with an effective technique for computing the inverse of a non-singular matrix. However, it is effective only in the sense that the inverse can be computed by a prescribed sequence of steps. The number of steps is large for matrices of large order, and it is not sufficiently small for matrices of low order to make it a preferred technique. The method described in Section 6 of Chapter II is the best method that is developed in this text. In numerical analysis where matrices of large order are inverted, highly specialized methods are available. But a discussion of such methods is beyond the scope of this book.

A matrix A is non-singular if and only if $\det A \neq 0$, and in this case we can see from the theorem that

$$A^{-1} = \frac{1}{\det A} \operatorname{adj} A. \qquad (3.9)$$

This is illustrated in the following example.

$$A = \begin{bmatrix} 1 & 2 & 3 \\ 2 & 1 & 2 \\ -2 & 1 & -1 \end{bmatrix}, \quad \operatorname{adj} A = \begin{bmatrix} -3 & 5 & 1 \\ -2 & 5 & 4 \\ 4 & -5 & -3 \end{bmatrix},$$

$$A^{-1} = \tfrac{1}{5} \begin{bmatrix} -3 & 5 & 1 \\ -2 & 5 & 4 \\ 4 & -5 & -3 \end{bmatrix}.$$

The relations between the elements of a matrix and their cofactors lead to a method for solving a system of n simultaneous equations in n unknowns when the equations are independent. Suppose we are given the system of equations

$$\sum_{j=1}^{n} a_{ij}x_j = b_i, \qquad (i = 1, 2, \ldots, n). \tag{3.10}$$

The assumption that the equations are independent is expressed in the condition that $\det A \neq 0$, where $A = [a_{ij}]$. Let A_{ij} be the cofactor of a_{ij}. Then for a given k

$$\sum_{i=1}^{n} A_{ik}\left(\sum_{j=1}^{n} a_{ij}x_j\right) = \sum_{j=1}^{n}\left(\sum_{i=1}^{n} A_{ik}a_{ij}\right)x_j$$

$$= \sum_{j=1}^{n} \det A \, \delta_{kj}x_j$$

$$= \det A \, x_k = \sum_{i=1}^{n} A_{ik}b_i. \tag{3.11}$$

Since $\det A \neq 0$ we see that

$$x_k = \frac{\displaystyle\sum_{i=1}^{n} A_{ik}b_i}{\det A}. \tag{3.12}$$

The numerator can be interpreted as the cofactor expansion of the determinant of the matrix obtained by replacing the kth column of A by the column of the b_i. In this form the method is known as *Cramer's rule*.

Cramer's rule is convenient for systems of equations of low order, but it fails if the system of equations is dependent or the number of equations is different from the number of unknowns. Even in these cases Cramer's rule can be modified to provide solutions. However, the methods we have already developed are usually easier to apply, and the balance in their favor increases as the order of the system of equations goes up and the nullity increases.

EXERCISES

1. In the determinant

$$\begin{vmatrix} 2 & 7 & 5 & 8 \\ 7 & -1 & 2 & 5 \\ 1 & 0 & 4 & 2 \\ -3 & 6 & -1 & 2 \end{vmatrix},$$

find the cofactor of the "8"; find the cofactor of the "−3."

2. The expansion of a determinant in terms of a row or column, as in formulas (3.1) and (3.2), provides a convenient method for evaluating determinants. The

amount of work involved can be reduced if a row or column is chosen in which some of the elements are zeros. Expand the determinant

$$\begin{vmatrix} 1 & 3 & 4 & -1 \\ 2 & 2 & 0 & 1 \\ 0 & -1 & 1 & 3 \\ -3 & 0 & 1 & 2 \end{vmatrix}$$

in terms of the cofactors of the third row.

3. It is even more convenient to combine an expansion in terms of cofactors with the method of elementary row and column operations described in Section 2. Subtract appropriate multiples of column 2 from the other columns to obtain

$$\begin{vmatrix} 1 & 3 & 7 & 8 \\ 2 & 2 & 2 & 7 \\ 0 & -1 & 0 & 0 \\ -3 & 0 & 1 & 2 \end{vmatrix}$$

and expand this determinant in terms of cofactors of the third row.

4. Show that det (adj A) = (det A)$^{n-1}$.

5. Show that a matrix is non-singular if and only if its adj A is also non-singular.

6. Let $A = [a_{ij}]$ be an arbitrary $n \times n$ matrix and let adj A be the adjunct of A. If $X = (x_1, \ldots, x_n)$ and $Y = (y_1, \ldots, y_n)$ show that

$$Y^T(\text{adj } A)X = - \begin{vmatrix} a_{11} & \cdots & a_{1n} & x_1 \\ \cdot & \cdots & \cdot & \cdot \\ \cdot & \cdots & \cdot & \cdot \\ \cdot & \cdots & \cdot & \cdot \\ a_{n1} & \cdots & a_{nn} & x_n \\ y_1 & \cdots & y_n & 0 \end{vmatrix}$$

For notation see pages 42 and 55.

4 | The Hamilton-Cayley Theorem

Let $p(x) = a_m x^m + \cdots + a_0$ be a polynomial in an indeterminate x with scalar coefficients a_i. If A is an $n \times n$ matrix, by $p(A)$ we mean the matrix $a_m A^m + a_{m-1} A^{m-1} + \cdots + a_0 I$.   Notice particularly that the constant term a_0 must be replaced by $a_0 I$ so that each term of $p(A)$ will be a matrix. No particular problem is encountered with matric polynomials of this form since all powers of a single matrix commute with each other. Any polynomial identity will remain valid if the indeterminate is replaced by a matrix, provided any scalar terms are replaced by corresponding scalar multiples of the identity matrix.

We may also consider polynomials with matric coefficients. To make sense, all coefficients must be matrices of the same order. We consider only the possibility of substituting scalars for the indeterminate, and in all manipulations with such polynomials the matric coefficients commute with the powers of the indeterminate. Polynomials with matric coefficients can be added and multiplied in the usual way, but the order of the factors is important in multiplication since the coefficients may not commute. The algebra of polynomials of this type is not simple, but we need no more than the observation that two polynomials with matric coefficients are equal if and only if they have exactly the same coefficients.

We avoid discussing the complications that can occur for polynomials with matric coefficients in a matric variable.

Now we should like to consider matrices for which the elements are polynomials. If F is the field of scalars for the set of polynomials in the indeterminate x, let K be the set of all rational functions in x; that is, the set of all permissible quotients of polynomials in x. It is not difficult to show that K is a field. Thus a matrix with polynomial components is a special case of a matrix with elements in K.

From this point of view a polynomial with matric coefficients can be expressed as a single matrix with polynomial components. For example,

$$\begin{bmatrix} 1 & 0 \\ -1 & 2 \end{bmatrix} x^2 + \begin{bmatrix} 0 & 2 \\ -2 & 0 \end{bmatrix} x + \begin{bmatrix} 2 & -1 \\ 1 & 1 \end{bmatrix} = \begin{bmatrix} x^2 + 2 & 2x - 1 \\ -x^2 - 2x + 1 & 2x^2 + 1 \end{bmatrix}.$$

Conversely, a matrix in which the elements are polynomials in an indeterminate x can be expanded into a polynomial with matric coefficients. Since polynomials with matric coefficients and matrices with polynomial components can be converted into one another, we refer to both types of expressions as *polynomial matrices*.

Definition. If A is any square matrix, the polynomial matrix $A - xI = C$ is called the *characteristic matrix* of A.

C has the form

$$\begin{bmatrix} a_{11} - x & a_{12} & \cdots & a_{1n} \\ a_{21} & a_{22} - x & \cdots & a_{2n} \\ \cdot & \cdot & \cdots & \cdot \\ \cdot & \cdot & \cdots & \cdot \\ \cdot & \cdot & \cdots & \cdot \\ a_{n1} & a_{n2} & \cdots & a_{nn} - x \end{bmatrix} \tag{4.1}$$

The determinant of C is a polynomial det $C = f(x) = k_n x^n + k_{n-1} x^{n-1} + \cdots + k_0$ of degree n; it is called the *characteristic polynomial* of A. The equation $f(x) = 0$ is called the *characteristic equation* of A. First, we should observe that the coefficient of x^n in the characteristic polynomial is $(-1)^n$, the coefficient of x^{n-1} is $(-1)^{n-1} \sum_{i=1}^{n} a_{ii}$, and the constant term $k_0 = \det A$.

Theorem 4.1. (*Hamilton-Cayley theorem*). *If A is a square matrix and $f(x)$ is its characteristic polynomial, then $f(A) = 0$.*

PROOF.   Since C is of order n, adj C will contain polynomials in x of degree not higher than $n - 1$. Hence adj C can be expanded into a polynomial with matric coefficients of degree at most $n - 1$:

$$\text{adj } C = C_{n-1} x^{n-1} + C_{n-2} x^{n-2} + \cdots + C_1 x + C_0 \tag{4.2}$$

where each C_i is a matrix with scalar elements.

By Theorem 3.1 we have

$$\text{adj } C \cdot C = \det C \cdot I = f(x)I$$
$$= \text{adj } C \cdot (A - xI) = (\text{adj } C)A - (\text{adj } C)x. \tag{4.3}$$

Hence,

$$k_n I x^n + k_{n-1} I x^{n-1} + \cdots + k_1 I x + k_0 I$$
$$= -C_{n-1} x^n - C_{n-2} x^{n-1} - \cdots - C_0 x$$
$$+ C_{n-1} A x^{n-1} + \cdots + C_1 A x + C_0 A. \tag{4.4}$$

The expressions on the two sides of this equality are $n \times n$ polynomial matrices. Since two polynomial matrices are equal if and only if the corresponding coefficients are equal, (4.4) is equivalent to the following set of matric equations:

$$k_n I = -C_{n-1}$$
$$k_{n-1} I = -C_{n-2} + C_{n-1} A$$
$$\cdot$$
$$\cdot \tag{4.5}$$
$$\cdot$$
$$k_1 I = -C_0 + C_1 A$$
$$k_0 I = \qquad C_0 A.$$

Multiply each of these equations by A^n, A^{n-1}, ..., A, I from the right, respectively, and add them. The terms on the right side will cancel out leaving the zero matrix. The terms on the left add up to

$$k_n A^n + k_{n-1} A^{n-1} + \cdots + k_1 A + k_0 I = f(A) = 0. \; \square \tag{4.6}$$

The equation $m(x) = 0$ of lowest degree which A satisfies is called the *minimum equation* (or *minimal equation*) for A; $m(x)$ is called the *minimum polynomial* for A. Since A satisfies its characteristic equation the degree of $m(x)$ is not more than n. Since a linear transformation and any matrix representing it satisfy the same relations, similar matrices satisfy the same set of polynomial equations. In particular, similar matrices have the same minimum polynomials.

Theorem 4.2. *If $g(x)$ is any polynomial with coefficients in F such that $g(A) = 0$, then $g(x)$ is divisible by the minimum polynomial for A. The minimum polynomial is unique except for a possible non-zero scalar factor.*

PROOF.   Upon dividing $g(x)$ by $m(x)$ we can write $g(x)$ in the form

$$g(x) = m(x) \cdot q(x) + r(x), \tag{4.7}$$

where $q(x)$ is the quotient polynomial and $r(x)$ is the remainder, which is either identically zero or is a polynomial of degree less than the degree of $m(x)$. If $g(x)$ is a polynomial such that $g(A) = 0$, then

$$g(A) = 0 = m(A) \cdot q(A) + r(A) = r(A). \tag{4.8}$$

This would contradict the selection of $m(x)$ as the minimum polynomial for A unless the remainder $r(x)$ is identically zero. Since two polynomials of the same lowest degree must divide each other, they must differ by a scalar factor. □

As we have pointed out, the elements of adj C are polynomials of degree at most $n - 1$. Let $g(x)$ be the greatest common divisor of the elements of adj C. Since adj $C \cdot C = f(x)I$, $g(x)$ divides $f(x)$.

Theorem 4.3. $h(x) = \dfrac{f(x)}{g(x)}$ *is the minimum polynomial for A.*

PROOF.   Let adj $C = g(x)B$ where the elements of B have no non-scalar common factor. Since adj $C \cdot C = f(x)I$ we have $h(x) \cdot g(x)I = g(x)BC$. Since $g(x) \neq 0$ this yields

$$BC = h(x)I. \tag{4.9}$$

Using B in place of adj C we can repeat the argument used in the proof of the Hamilton-Cayley theorem to deduce that $h(A) = 0$. Thus $h(x)$ is divisible by $m(x)$.

On the other hand, consider the polynomial $m(x) - m(y)$. Since it is a sum of terms of the form $c_i(x^i - y^i)$, each of which is divisible by $y - x$, $m(x) - m(y)$ is divisible by $y - x$:

$$m(x) - m(y) = (y - x) \cdot k(x, y). \tag{4.10}$$

Replacing x by xI and y by A we have

$$m(xI) - m(A) = m(x)I = (A - xI) \cdot k(xI, A) = C \cdot k(xI, A). \tag{4.11}$$

Multiplying by adj C we have

$$m(x) \text{ adj } C = (\text{adj } C)C \cdot k(xI, A) = f(x) \cdot k(xI, A). \tag{4.12}$$

Hence,

$$m(x) \cdot g(x)B = h(x) \cdot g(x) \cdot k(xI, A), \tag{4.13}$$

or

$$m(x)B = h(x) \cdot k(xI, A). \tag{4.14}$$

Since $h(x)$ divides every element of $m(x)B$ and the elements of B have no non-scalar common factor, $h(x)$ divides $m(x)$. Thus, $h(x)$ and $m(x)$ differ at most by a scalar factor. $\square$

Theorem 4.4.   *Each irreducible factor of the characteristic polynomial $f(x)$ of A is also an irreducible factor of the minimum polynomial $m(x)$.*
    PROOF.   As we have seen in the proof of the previous theorem

$$m(x)I = C \cdot k(xI, A).$$

Thus

$$\det m(x)I = [m(x)]^n = \det C \cdot \det k(xI, A)$$
$$= f(x) \cdot \det k(xI, A). \tag{4.15}$$

We see then that every irreducible factor of $f(x)$ divides $[m(x)]^n$, and therefore $m(x)$ itself. $\square$

Theorem 4.4 shows that a characteristic polynomial without repeated factors is also the minimum polynomial. As we shall see, it is the case in which the characteristic polynomial has repeated factors that generally causes trouble.

We now ask the converse question. Given the polynomial $f(x) = (-1)^n x^n + k_{n-1} x^{n-1} + \cdots + k_0$, does there exist an $n \times n$ matrix A for which $f(x)$ is the minimum polynomial?

Let $A = \{\alpha_1, \ldots, \alpha_n\}$ be any basis and define the linear transformation σ by the rules

$$\sigma(\alpha_i) = \alpha_{i+1} \qquad \text{for} \qquad i < n, \tag{4.16}$$

and

$$(-1)^n \sigma(\alpha_n) = -k_0 \alpha_1 - k_1 \alpha_2 - \cdots - k_{n-1} \alpha_n.$$

It follows directly from the definition of σ that

$$f(\sigma)(\alpha_1) = (-1)^n \sigma(\alpha_n) + k_{n-1} \alpha_n + \cdots + k_1 \alpha_2 + k_0 \alpha_1 = 0. \tag{4.17}$$

For any other basis element we have

$$f(\sigma)(\alpha_j) = f(\sigma)[\sigma^{j-1}(\alpha_1)] = \sigma^{j-1}[f(\sigma)(\alpha_1)] = 0. \tag{4.18}$$

Since $f(\sigma)$ vanishes on the basis elements $f(\sigma) = 0$ and any matrix representing σ satisfies the equation $f(x) = 0$.

On the other hand, σ cannot satisfy an equation of lower degree because the corresponding polynomial in σ applied to α_1 could be interpreted as a relation among the basis elements. Thus, $f(x)$ is a minimum polynomial for σ and for any matrix representing σ. Since $f(x)$ is of degree n, it must also be the characteristic polynomial of any matrix representing σ.

With respect to the basis A the matrix representing σ is

$$A = \begin{bmatrix} 0 & 0 & \cdots & 0 & -(-1)^n k_0 \\ 1 & 0 & \cdots & 0 & -(-1)^n k_1 \\ 0 & 1 & \cdots & 0 & -(-1)^n k_2 \\ \cdot & \cdot & \cdots & \cdot & \cdot \\ \cdot & \cdot & \cdots & \cdot & \cdot \\ \cdot & \cdot & \cdots & \cdot & \cdot \\ 0 & 0 & & 1 & -(-1)^n k_{n-1} \end{bmatrix} \tag{4.19}$$

A is called the *companion matrix* of $f(x)$.

Theorem 4.5.   *$f(x)$ is a minimum polynomial for its companion matrix.* ☐

EXERCISES

1. Show that $-x^3 + 39x - 90$ is the characteristic polynomial for the matrix

$$\begin{bmatrix} 0 & 0 & -90 \\ 1 & 0 & 39 \\ 0 & 1 & 0 \end{bmatrix}.$$

2. Find the characteristic polynomial for the matrix

$$\begin{bmatrix} 2 & -2 & 3 \\ 1 & 1 & 1 \\ 1 & 3 & -1 \end{bmatrix}$$

and show by direct substitution that this matrix satisfies its characteristic equation.

3. Find the minimum polynomial for the matrix

$$\begin{bmatrix} 3 & 2 & 2 \\ 1 & 4 & 1 \\ -2 & -4 & -1 \end{bmatrix}.$$

4. Write down a matrix which has $x^4 + 3x^3 + 2x^2 - x + 6 = 0$ as its minimum equation.

5. Show that if the matrix A satisfies the equation $x^2 + x + 1 = 0$, then A is non-singular and the inverse A^{-1} is expressible as a linear combination of A and I.

6. Show that no real 3×3 matrix satisfies $x^2 + 1 = 0$. Show that there are complex 3×3 matrices which do. Show that there are real 2×2 matrices that satisfy the equation.

7. Find a 2×2 matrix with integral elements satisfying the equation $x^3 - 1 = 0$, but not satisfying the equation $x - 1 = 0$.

8. Show that the characteristic polynomial of

$$\begin{bmatrix} 7 & 4 & -4 \\ 4 & -8 & -1 \\ -4 & -1 & -8 \end{bmatrix}$$

is not its minimum polynomial. What is the minimum polynomial?

5 | Eigenvalues and Eigenvectors

Let σ be a linear transformation of V into itself. It is often useful to find subspaces of V in which σ also acts as a linear transformation. If W is such a subspace, this means that $\sigma(W) \subset W$. A subspace with this property is called an *invariant subspace* of V under σ. Generally, the problem of determining the properties of σ on V can be reduced to the problem of determining the properties of σ on the invariant subspaces.

The simplest and most restricted case occurs when an invariant subspace W is of dimension 1. In that case, let $\{\alpha_1\}$ be a basis for W. Then, since $\sigma(\alpha_1) \in W$, there is a scalar λ_1 such that $\sigma(\alpha_1) = \lambda_1 \alpha_1$. Also for any $\alpha \in W$, $\alpha = a_1 \alpha_1$ and hence $\sigma(\alpha) = a_1 \sigma(\alpha_1) = a_1 \lambda_1 \alpha_1 = \lambda_1 \alpha$. In some sense the scalar λ_1 is characteristic of the invariant subspace W; σ stretches every vector in W by the factor λ_1.

In general, a problem of finding those scalars λ and associated vectors ξ for which $\sigma(\xi) = \lambda \xi$ is called an *eigenvalue problem*. A non-zero vector ξ is called an *eigenvector* of σ if there exists a scalar λ such that $\sigma(\xi) = \lambda \xi$. A scalar λ is called an *eigenvalue* of σ if there exists a non-zero vector ξ such that $\sigma(\xi) = \lambda \xi$. Notice that the equation $\sigma(\xi) = \lambda \xi$ is an equation in two variables, one of which is a vector and the other a scalar. The solution $\xi = 0$ and λ any scalar is a solution we choose to ignore since it will not lead to an invariant subspace of positive dimension. Without further conditions we have no assurance that the eigenvalue problem has any other solutions.

A typical and very important eigenvalue problem occurs in the solution of partial differential equations of the form

$$\frac{\partial^2 u}{\partial x^2} + \frac{\partial^2 u}{\partial y^2} = 0,$$

subject to the boundary conditions that $u(0, y) = u(\pi, y) = 0$,

$$\lim_{y \to \infty} u(x, y) = 0, \quad \text{and} \quad u(x, 0) = f(x)$$

where $f(x)$ is a given function. The standard technique of separation of variables leads us to try to construct a solution which is a sum of functions of the form XY where X is a function of x alone and Y is a function of y alone. For this type of function, the partial differential equation becomes

$$\frac{d^2X}{dx^2} \cdot Y + \frac{d^2Y}{dy^2} \cdot X = 0.$$

Since

$$\frac{1}{Y} \cdot \frac{d^2Y}{dy^2} = -\frac{1}{X} \cdot \frac{d^2X}{dx^2}$$

is a function of x alone and also a function of y alone, it must be a constant (scalar) which we shall call k^2. Thus we are trying to solve the equations

$$\frac{d^2X}{dx^2} = -k^2X, \quad \frac{d^2Y}{dy^2} = k^2Y.$$

These are eigenvalue problems as we have defined the term. The vector space is the space of infinitely differentiable functions over the real numbers and the linear transformation is the differential operator d^2/dx^2.

For a given value of $k^2(k > 0)$ the solutions would be

$$X = a_1 \cos kx + a_2 \sin kx,$$
$$Y = a_3 e^{-ky} + a_4 e^{ky}.$$

The boundary conditions $u(0, y) = 0$ and $\lim_{y \to \infty} u(x, y) = 0$ imply that $a_1 = a_4 = 0$. The most interesting condition for the purpose of this example is that the boundary condition $u(\pi, y) = 0$ implies that k is an integer. Thus, the eigenvalues of this eigenvalue problem are the integers, and the corresponding eigenfunctions (eigenvectors) are of the form $a_k e^{-ky} \sin kx$. The fourth boundary condition leads to a problem in Fourier series; the problem of determining the a_k so that the series

$$\sum_{n=1}^{\infty} a_k \sin kx$$

represents the given function $f(x)$ for $0 \leq x \leq \pi$.

Although the vector space in this example is of infinite dimension, we restrict our attention to the eigenvalue problem in finite dimensional vector spaces. In a finite dimensional vector space there exists a simple necessary and sufficient condition which determines the eigenvalues of an eigenvalue problem.

The eigenvalue equation can be written in the form $(\sigma - \lambda)(\xi) = 0$. We know that there exists a non-zero vector ξ satisfying this condition if and only if $\sigma - \lambda$ is singular. Let $A = \{\alpha_1, \ldots, \alpha_n\}$ be any basis of V and let $A = [a_{ij}]$ be the matrix representing σ with respect to this basis. Then $A - \lambda I = C(\lambda)$ is the matrix representing $\sigma - \lambda$. Since $A - \lambda I$ is singular if and only if det $(A - \lambda I) = f(\lambda) = 0$, we see that we have proved

Theorem 5.1.   *A scalar λ is an eigenvalue of σ if and only if it is a solution of the characteristic equation of a matrix representing σ.* □

Notice that Theorem 5.1 applies only to scalars. In particular a solution of the characteristic equation which is not a scalar is not an eigenvalue. For example, if the field of scalars is the field of real numbers, then non-real complex solutions of the characteristic equation are not eigenvalues. In the published literature on matrices the terms "proper values" and "characteristic values" are also used to denote what we have called eigenvalues. But, unfortunately, the same terms are often also applied to the solutions of the characteristic equation. We call the solutions of the characteristic equation *characteristic values*. Thus, a characteristic value is an eigenvalue if and only if it is also in the given field of scalars. This distinction between eigenvalues and characteristic values is not standard in the literature on matrices, but we hope this or some other means of distinguishing between these concepts will become conventional.

In abstract algebra a field is said to be *algebraically closed* if every polynomial with coefficients in the field factors into linear factors in the field. The field of complex numbers is algebraically closed. Though many proofs of this assertion are known, none is elementary. It is easy to show that algebraically closed fields exist, but it is not easy to show that a specific field is algebraically closed.

Since for most applications of concepts using eigenvalues or characteristic values the underlying field is either rational, real or complex, we content ourselves with the observation that the concepts, eigenvalue and characteristic, value, coincide if the underlying field is complex, and do not coincide if the underlying field is rational or real.

The procedure for finding the eigenvalues and eigenvectors of σ is fairly direct. For some basis $A = \{\alpha_1, \ldots, \alpha_n\}$, let A be the matrix representing σ. Determine the characteristic matrix $C(x) = A - xI$ and the characteristic equation det $(A - \lambda I) = f(x) = 0$. Solve the characteristic equation. (It is this step that presents the difficulties. The characteristic equation may have no solution in F. In that event the eigenvalue problem has no solution. Even in those cases where solutions exists, finding them can present practical difficulties.) For each solution λ of $f(x) = 0$, solve the system of homogeneous equations

$$(A - \lambda I)X = C(\lambda) \cdot X = 0. \tag{5.1}$$

Since this system of equations has positive nullity, non-zero solutions exist and we should use the Hermite normal form to find them. All solutions are the representations of eigenvectors corresponding to σ.

Generally, we are given the matrix A rather than σ itself, and in this case we regard the problem as solved when the eigenvalues and the representations of the eigenvectors are obtained. We refer to the eigenvalues and eigenvectors of σ as eigenvalues and eigenvectors, respectively, of A.

Theorem 5.2. *Similar matrices have the same eigenvalues and eigenvectors.*
PROOF. This follows directly from the definitions since the eigenvalues and eigenvectors are associated with the underlying linear transformation. □

Theorem 5.3. *Similar matrices have the same characteristic polynomial.*
PROOF. Let A and $A' = P^{-1}AP$ be similar. Then

$$\det (A' - xI) = \det (P^{-1}AP - xI) = \det \{P^{-1}(A - xI)P\} = \det P^{-1}$$
$$\det (A - xI) \det P = \det (A - xI) = f(x). □$$

We call the characteristic polynomial of any matrix representing σ the *characteristic polynomial* of σ. Theorem 5.3 shows that the characteristic polynomial of a linear transformation is uniquely defined.

Let $S(\lambda)$ be the set of all eigenvectors of σ corresponding to λ, together with 0.

Theorem 5.4. $S(\lambda)$ *is a subspace of* V.
PROOF. If α and $\beta \in S(\lambda)$, then

$$\sigma(a\alpha + b\beta) = a\sigma(\alpha) + b\sigma(\beta)$$
$$= a\lambda\alpha + b\lambda\beta$$
$$= \lambda(a\alpha + b\beta). \tag{5.2}$$

Hence, $a\alpha + b\beta \in S(\lambda)$ and $S(\lambda)$ is a subspace. □

We call $S(\lambda)$ *the eigenspace* of σ corresponding to λ, and any subspace of $S(\lambda)$ is called *an eigenspace* of σ.

The dimension of $S(\lambda)$ is equal to the nullity of $C(\lambda)$, the characteristic matrix of A with λ substituted for the indeterminate x. The dimension of $S(\lambda)$ is called the *geometric multiplicity* of λ. We have shown that λ is also a solution of the characteristic equation $f(x) = 0$. Hence, $(x - \lambda)$ is a factor of $f(x)$. If $(x - \lambda)^k$ is a factor of $f(x)$ while $(x - \lambda)^{k+1}$ is not, λ is a root of $f(x) = 0$ of multiplicity k. We refer to this multiplicity as the *algebraic multiplicity* of λ.

Theorem 5.5. *The geometric multiplicity of λ does not exceed the algebraic multiplicity of λ.*
PROOF. Since the geometric multiplicity of λ is defined independently of any matrix representing σ and the characteristic equation is the same for all

matrices representing σ it will be sufficient to prove the theorem for any particular matrix representing σ. We shall choose the matrix representing σ so that the assertion of the theorem is evident. Let r be the dimension of $S(\lambda)$ and let $\{\xi_1, \ldots, \xi_r\}$ be a basis of $S(\lambda)$. This linearly independent set can be extended to a basis $\{\xi_1, \ldots, \xi_n\}$ of V. Since $\sigma(\xi_i) = \lambda \xi_i$ for $i \leq r$, the matrix A representing σ with respect to this basis has the form

$$A = \begin{bmatrix} \lambda & 0 & \cdots & 0 & a_{1,r+1} & \cdots & a_{1n} \\ 0 & \lambda & \cdots & 0 & a_{2,r+1} & \cdots & a_{2n} \\ \cdot & \cdot & & \cdot & \cdot & & \cdot \\ \cdot & \cdot & & \cdot & \cdot & & \cdot \\ \cdot & \cdot & & \cdot & \cdot & & \cdot \\ 0 & 0 & \cdots & \lambda & a_{r,r+1} & \cdots & \cdot \\ 0 & 0 & \cdots & 0 & a_{r+1,r+1} & \cdots & \cdot \\ \cdot & \cdot & & \cdot & \cdot & & \cdot \\ \cdot & \cdot & & \cdot & \cdot & & \cdot \\ \cdot & \cdot & & \cdot & \cdot & & \cdot \\ 0 & 0 & \cdots & 0 & a_{n,r+1} & \cdots & a_{nn} \end{bmatrix}. \tag{5.3}$$

From the form of A it is evident that $\det (A - xI) = f(x)$ is divisible by $(x - \lambda)^r$. Therefore, the algebraic multiplicity of λ is at least r, which is the geometric multiplicity. $\square$

Theorem 5.6.   *If the eigenvalues $\lambda_1, \ldots, \lambda_s$ are all different and $\{\xi_1, \ldots, \xi_s\}$ is a set of eigenvectors, ξ_i corresponding to λ_i, then the set $\{\xi_1, \ldots, \xi_s\}$ is linearly independent.*

PROOF.   Suppose the set is dependent and that we have reordered the eigenvectors so that the first k eigenvectors are linearly independent and the last $s - k$ are dependent on them. Then

$$\xi_s = \sum_{i=1}^{k} a_i \xi_i$$

where the representation is unique. Not all $a_i = 0$ since $\xi_s \neq 0$. Upon applying the linear transformation σ we have

$$\lambda_s \xi_s = \sum_{i=1}^{k} a_i \lambda_i \xi_i.$$

There are two possibilities to be considered. If $\lambda_s = 0$, then none of the λ_i for $i \leq k$ is zero since the eigenvalues are distinct. This would imply that $\{\xi_1, \ldots, \xi_k\}$ is linearly dependent, contrary to assumption. If $\lambda_s \neq 0$, then

$$\xi_s = \sum_{i=1}^{k} a_i \frac{\lambda_i}{\lambda_s} \xi_i.$$

Since not all $a_i = 0$ and $\lambda_i/\lambda_s \neq 1$, this would contradict the uniqueness of the representation of ξ_s. Since we get a contradiction in any event, the set $\{\xi_1, \ldots, \xi_s\}$ must be linearly independent. $\square$

EXERCISES

1. Show that $\lambda = 0$ is an eigenvalue of a matrix A if and only if A is singular.

2. Show that if ξ is an eigenvector of σ, then ξ is also an eigenvector of σ^n for each $n \geq 0$. If λ is the eigenvalue of σ corresponding to ξ, what is the eigenvalue of σ^n corresponding to ξ?

3. Show that if ξ is an eigenvector of both σ and τ, then ξ is also an eigenvector of $a\sigma$ (for $a \in F$) and $\sigma + \tau$. If λ_1 is the eigenvalue of σ and λ_2 is the eigenvalue of τ corresponding to ξ, what are the eigenvalues of $a\sigma$ and $\sigma + \tau$?

4. Show, by producing an example, that if λ_1 and λ_2 are eigenvalues of σ_1 and σ_2, respectively, it is not necessarily true that $\lambda_1 + \lambda_2$ is an eigenvalue of $\sigma_1 + \sigma_2$.

5. Show that if ξ is an eigenvector of σ, then it is also an eigenvector of $p(\sigma)$ where $p(x)$ is a polynomial with coefficients in F. If λ is an eigenvalue of σ corresponding to ξ, what is the eigenvalue of $p(\sigma)$ corresponding to ξ?

6. Show that if σ is non-singular and λ is an eigenvalue of σ, then λ^{-1} is an eigenvalue of σ^{-1}. What is the corresponding eigenvector?

7. Show that if every vector in V is an eigenvector of σ, then σ is a scalar transformation.

8. Let P_n be the vector space of polynomials of degree at most $n - 1$, and let D be the differentiation operator; that is $D(t^k) = kt^{k-1}$. Determine the characteristic polynomial for D. From your knowledge of the differentiation operator and net using Theorem 4.3, determine the minimum polynomial for D. What kind of differential equation would an eigenvector of D have to satisfy? What are the eigenvectors of D?

9. Let $A = [a_{ij}]$. Show that if $\sum_j a_{ij} = c$ independent of i, then $\xi = (1, 1, \ldots, 1)$ is an eigenvector. What is the corresponding eigenvalue?

10. Let W be an invariant subspace of V under σ, and let $A = \{\alpha_1, \ldots, \alpha_n\}$ be a basis of V such that $\{\alpha_1, \ldots, \alpha_k\}$ is a basis of W. Let $A = [a_{ij}]$ be the matrix representing σ with respect to the basis A. Show that all elements in the first k columns below the kth row are zeros.

11. Show that if λ_1 and $\lambda_2 \neq \lambda_1$ are eigenvalues of σ_1 and ξ_1 and ξ_2 are eigenvectors corresponding to λ_1 and λ_2, respectively, then $\xi_1 + \xi_2$ is not an eigenvector.

12. Assume that $\{\xi_1, \ldots, \xi_{1r}\}$ are eigenvectors with distinct eigenvalues. Show that $\sum_{i=1}^{r} a_i \xi_i$ is never an eigenvector unless precisely one coefficient is non-zero.

13. Let A be an $n \times n$ matrix with eigenvalues $\lambda_1, \lambda_2, \ldots, \lambda_n$. Show that if Λ is the diagonal matrix

$$
\Lambda = \begin{bmatrix}
\lambda_1 & 0 & 0 & \cdots & 0 \\
0 & \lambda_2 & 0 & \cdots & 0 \\
0 & 0 & \lambda_2 & \cdots & 0 \\
& \cdot & \cdot & \cdot & \\
& \cdot & \cdot & \cdot & \\
0 & 0 & 0 & & \lambda_n
\end{bmatrix},
$$

and $P = [p_{ij}]$ is the matrix in which column j is the n-tuple representing an eigenvector corresponding to λ_j, then $AP = P\Lambda$.

14. Use the notation of Exercise 13. Show that if A has n linearly independent eigenvalues, then eigenvectors can be chosen so that P is non-singular. In this case $P^{-1}AP = \Lambda$.

6 | Some Numerical Examples

Since we are interested here mainly in the numerical procedures, we start with the matrices representing the linear transformations and obtain their eigenvalues and the representations of the eigenvectors.

Example 1. Let

$$
A = \begin{bmatrix}
-1 & 2 & 2 \\
2 & 2 & 2 \\
-3 & -6 & -6
\end{bmatrix}.
$$

The first step is to obtain the characteristic matrix

$$
C(x) = \begin{bmatrix}
-1 - x & 2 & 2 \\
2 & 2 - x & 2 \\
-3 & -6 & -6 - x
\end{bmatrix}
$$

and then the characteristic polynomial

$$
\det C(x) = -(x + 2)(x + 3)x.
$$

Thus the eigenvalues of A are $\lambda_1 = -2$, $\lambda_2 = -3$, and $\lambda_3 = 0$. The next steps are to substitute, successively, the eigenvalues for x in the characteristic matrix. Thus we have

$$
C(-2) = \begin{bmatrix}
1 & 2 & 2 \\
2 & 4 & 2 \\
-3 & -6 & -4
\end{bmatrix}.
$$

The Hermite normal form obtained from $C(-2)$ is

$$\begin{bmatrix} 1 & 2 & 0 \\ 0 & 0 & 1 \\ 0 & 0 & 0 \end{bmatrix}.$$

The components of the eigenvector corresponding to $\lambda_1 = -2$ are found by solving the equations

$$x_1 + 2x_2 = 0$$
$$ x_3 = 0.$$

Thus $(2, -1, 0)$ is the representation of an eigenvector corresponding to λ_1; for simplicity we shall write $\xi_1 = (2, -1, 0)$, identifying the vector with its representation.

In a similar fashion we obtain

$$C(-3) = \begin{bmatrix} 2 & 2 & 2 \\ 2 & 5 & 2 \\ -3 & -6 & -3 \end{bmatrix}.$$

From $C(-3)$ we obtain the Hermite normal form

$$\begin{bmatrix} 1 & 0 & 1 \\ 0 & 1 & 0 \\ 0 & 0 & 0 \end{bmatrix}$$

and hence the eigenvector $\xi_2 = (1, 0, -1)$.

Similarly, from

$$C(0) = \begin{bmatrix} -1 & 2 & 2 \\ 2 & 2 & 2 \\ -3 & -6 & -6 \end{bmatrix}$$

we obtain the eigenvector $\xi_3 = (0, 1, -1)$.

By Theorem 5.6 the three eigenvectors obtained for the three different eigenvalues are linearly independent.

Example 2. Let

$$A = \begin{bmatrix} 1 & 1 & -1 \\ -1 & 3 & -1 \\ -1 & 2 & 0 \end{bmatrix}.$$

From the characteristic matrix

$$C(x) = \begin{bmatrix} 1-x & 1 & -1 \\ -1 & 3-x & -1 \\ -1 & 2 & -x \end{bmatrix}$$

we obtain the characteristic polynomial $\det C(x) = -(x-1)^2(x-2)$. Thus we have just two distinct eigenvalues; $\lambda_1 = \lambda_2 = 1$ with algebraic multiplicity two, and $\lambda_3 = 2$.

Substituting λ_1 for x in the characteristic matrix we obtain

$$C(1) = \begin{bmatrix} 0 & 1 & -1 \\ -1 & 2 & -1 \\ -1 & 2 & -1 \end{bmatrix}.$$

The corresponding Hermite normal form is

$$\begin{bmatrix} 1 & 0 & -1 \\ 0 & 1 & -1 \\ 0 & 0 & 0 \end{bmatrix}.$$

Thus it is seen that the nullity of $C(1)$ is 1. The eigenspace $S(1)$ is of dimension 1 and the geometric multiplicity of the eigenvalue 1 is 1. This shows that the geometric multiplicity can be lower than the algebraic multiplicity. We obtain $\xi_1 = (1, 1, 1)$.

The eigenvector corresponding to $\lambda_3 = 2$ is $\xi_3 = (0, 1, 1)$.

EXERCISES

For each of the following matrices find all the eigenvalues and as many linearly independent eigenvectors as possible.

1. $\begin{bmatrix} 2 & 4 \\ 5 & 3 \end{bmatrix}$

2. $\begin{bmatrix} 3 & 2 \\ -2 & 3 \end{bmatrix}$

3. $\begin{bmatrix} 1 & 2 \\ 2 & -2 \end{bmatrix}$

4. $\begin{bmatrix} 1 & -\sqrt{2} \\ \sqrt{2} & 4 \end{bmatrix}$

5. $\begin{bmatrix} 4 & 9 & 0 \\ 0 & -2 & 8 \\ 0 & 0 & 7 \end{bmatrix}$

6. $\begin{bmatrix} 3 & 2 & 2 \\ 1 & 4 & 1 \\ -2 & -4 & -1 \end{bmatrix}$

7. $\begin{bmatrix} 7 & 4 & -4 \\ 4 & -8 & -1 \\ -4 & -1 & -8 \end{bmatrix}$

8. $\begin{bmatrix} 2 & -i & 0 \\ i & 2 & 0 \\ 0 & 0 & 3 \end{bmatrix}$

7 | Similarity

Generally, for a given linear transformation σ we seek a basis for which the matrix representing σ has as simple a form as possible. The simplest form is that in which the elements not on the main diagonal are zero, a *diagonal* matrix. Not all linear transformations can be represented by diagonal matrices, but relatively large classes of transformations can be represented by diagonal matrices, and we seek conditions under which such a representation exists.

Theorem 7.1. *A linear transformation σ can be represented by a diagonal matrix if and only if there exists a basis consisting of eigenvectors of σ.*

PROOF. Suppose there is a linearly independent set $X = \{\xi_1, \ldots, \xi_n\}$ of eigenvectors and that $\{\lambda_1, \ldots, \lambda_n\}$ are the corresponding eigenvalues. Then $\sigma(\xi_i) = \lambda_i \xi_i$ so that the matrix representing σ with respect to the basis X has the form

$$\begin{bmatrix} \lambda_1 & 0 & \cdots & 0 \\ 0 & \lambda_2 & \cdots & 0 \\ \cdot & \cdot & \cdots & \cdot \\ \cdot & \cdot & \cdot & \cdot \\ \cdot & \cdot & \cdot & \cdot \\ 0 & 0 & \cdots & \lambda_n \end{bmatrix}, \tag{7.1}$$

that is, σ is represented by a diagonal matrix.

Conversely, if σ is represented by a diagonal matrix, the vectors in that basis are eigenvectors. $\square$

Usually, we are not given the linear transformation σ directly. We are given a matrix A representing σ with respect to an unspecified basis. In this case Theorem 7.1 is usually worded in the form: A matrix A is similar to a diagonal matrix if and only if there exist n linearly independent eigenvectors of A. In this form a computation is required. We must find the matrix P such that $P^{-1}AP$ is a diagonal matrix.

Let the matrix A be given; that is, A represents σ with respect to some basis $A = \{\alpha_1, \ldots, \alpha_n\}$. Let $\xi_j = \sum_{i=1}^{n} p_{ij}\alpha_j$ be the representations of the eigenvectors of A with respect to A. Then the matrix A' representing σ with respect to the basis $X = \{\xi_1, \ldots, \xi_n\}$ is $P^{-1}AP = A'$. By Theorem 7.1, A' is a diagonal matrix.

In Example 1 of Section 6, the matrix

$$A = \begin{bmatrix} -1 & 2 & 2 \\ 2 & 2 & 2 \\ -3 & -6 & -6 \end{bmatrix}$$

has three linearly independent eigenvectors, $\xi_1 = (2, -1, 0)$, $\xi_2 = (1, 0, -1)$, and $\xi_3 = (0, 1, -1)$. The matrix of transition P has the components of these vectors written in its columns:

$$P = \begin{bmatrix} 2 & 1 & 0 \\ -1 & 0 & 1 \\ 0 & -1 & -1 \end{bmatrix}, \qquad P^{-1} = \begin{bmatrix} 1 & 1 & 1 \\ -1 & -2 & -2 \\ 1 & 2 & 1 \end{bmatrix}.$$

The reader should check that $P^{-1}AP$ is a diagonal matrix with the eigenvalues appearing in the main diagonal.

In Example 2 of Section 6, the matrix

$$A = \begin{bmatrix} 1 & 1 & -1 \\ -1 & 3 & -1 \\ -1 & 2 & 0 \end{bmatrix}$$

has one linearly independent eigenvector corresponding to each of its two eigenvalues. As there are no other eigenvalues, there does not exist a set of three linearly independent eigenvectors. Thus, the linear transformation represented by this matrix cannot be represented by a diagonal matrix; A is not similar to a diagonal matrix.

Corollary 7.2. *If σ can be represented by a diagonal matrix D, the elements in the main diagonal of D are the eigenvalues of σ.* $\square$

Theorem 7.3. *If an $n \times n$ matrix has n distinct eigenvalues, then A is similar to a diagonal matrix.*

PROOF. By Theorem 5.6 the n eigenvectors corresponding to the n eigenvalues of A are linearly independent and form a basis. By Theorem 7.1 the matrix representing the underlying linear transformation with respect to this basis is a diagonal matrix. Hence, A is similar to a diagonal matrix. $\square$

Theorem 7.3 is quite practical because we expect the eigenvalues of a randomly given matrix to be distinct; however, there are circumstances under which the theorem does not apply. There may not be n distinct eigenvalues, either because some have algebraic multiplicity greater than one or because the characteristic equation does not have enough solutions in the field. The most general statement that can be made without applying more conditions to yield more results is

Theorem 7.4. *A necessary and sufficient condition that a matrix A be similar to a diagonal matrix is that its minimum polynomial factor into distinct linear factors with coefficients in F.*

PROOF.   Suppose first that the matrix A is similar to a diagonal matrix D.
By Theorem 5.3, A and D have the same characteristic polynomial. Since
D is a diagonal matrix the elements in the main diagonal are the solutions
of the chracteristic equation and the characteristic polynomial must factor
into linear factors. By Theorem 4.4 the minimum polynomial for A must
contain each of the linear factors of $f(x)$, although possibly with lower
multiplicity. It can be seen, however, either from Theorem 4.3 or by direct
substitution, that D satisfies an equation without repeated factors. Thus,
the minimum polynomial for A has distinct linear factors.

On the other hand, suppose that the minimum polynomial for A is
$m(x) = (x - \lambda_1)(x - \lambda_2) \cdots (x - \lambda_p)$ with distinct linear factors. Let
M_i be the kernel of $\sigma - \lambda_i$. The non-zero vectors in M_i are the eigenvectors
of σ corresponding to λ_i. It follows from Theorem 5.6 that a non-zero
vector in M_i cannot be expressed as a sum of vectors in $\sum_{j \neq i} M_j$. Hence,
the sum $M_1 + M_2 + \cdots + M_p$ is direct.

Let $v_i = \dim M_i$, that is, v_i is the nullity of $\sigma - \lambda_i$. Since $M_1 \oplus \cdots \oplus$
$M_p \subset V$ we have $v_1 + \cdots + v_p \leq n$. By Theorem 1.5 of Chapter II
$\dim (\sigma - \lambda_i)V = n - v_i = \rho_i$, By another application of the same theorem
we have $\dim (\sigma - \lambda_j)\{(\sigma - \lambda_i)V\} \geq \rho_i - v_j = n - (v_i + v_j)$.

Finally, by repeated application of the same ideas we obtain $0 = \dim m(\sigma)V \geq n - (v_1 + \cdots + v_p)$. Thus, $v_1 + \cdots + v_p = n$. This shows
that $M_1 \oplus \cdots \oplus M_p = V$. Since every vector in V is a linear combination of
eigenvectors, there exists a basis of eigenvectors. Thus, A is similar to a
diagonal matrix. $\square$

Theorem 7.4 is important in the theory of matrices, but it does not provide
the most effective means for deciding whether a particular matrix is similar
to diagonal matrix. If we can solve the characteristic equation, it is easier
to try to find the n linearly independent eigenvectors than it is to use Theorem
7.4 to ascertain that they do or do not exist. If we do use this theorem and
are able to conclude that a basis of eigenvectors does exist, the work done in
making this conclusion is of no help in the attempt to find the eigenvectors.
The straightforward attempt to find the eigenvectors is always conclusive
On the other hand, if it is not necessary to find the eigenvectors, Theorem 7.4
can help us make the necessary conclusion without solving the characteristic
equation.

For any square matrix $A = [a_{ij}]$, $\text{Tr}(A) = \sum_{i=1}^{n} a_{ii}$ is called the *trace* of A.
It is the sum of the elements in the diagonal of A. Since $\text{Tr}(AB) =$
$\sum_{i=1}^{n}(\sum_{j=1}^{n} a_{ij}b_{ji}) = \sum_{j=1}^{n}(\sum_{i=1}^{n} b_{ji}a_{ij}) = \text{Tr}(BA)$,

$$\text{Tr}(P^{-1}AP) = \text{Tr}(APP^{-1}) = \text{Tr}(A). \qquad (7.2)$$

This shows that the trace is invariant under similarity transformations;

that is, similar matrices have the same trace. For a given linear transformation σ of V into itself, all matrices representing σ have the same trace. Thus we can define $\text{Tr}(\sigma)$ to be the trace of any matrix representing σ.

Consider the coefficient of x^{n-1} in the expansion of the determinant of the characteristic matrix,

$$\begin{bmatrix} a_{11} - x & a_{12} & \cdots & a_{1n} \\ a_{21} & a_{22} - x & \cdots & a_{2n} \\ \cdot & \cdot & \cdots & \cdot \\ \cdot & \cdot & \cdots & \cdot \\ \cdot & \cdot & \cdots & \cdot \\ a_{n1} & a_{n2} & \cdots & a_{nn} - x \end{bmatrix}. \tag{7.3}$$

The only way an x^{n-1} can be obtained is from a product of $n-1$ of the diagonal elements, multiplied by the scalar from the remaining diagonal element. Thus, the coefficient of x^{n-1} is $(-1)^{n-1} \sum_{i=1}^{n} a_{ii}$, or $(-1)^{n-1}\text{Tr}(A)$.

If $f(x) = \det(A - xI)$ is the characteristic polynomial of A, then $\det A = f(0)$ is the constant term of $f(x)$. If $f(x)$ is factored into linear factors in the form

$$f(x) = (-1)^n (x - \lambda_1)^{r_1}(x - \lambda_2)^{r_2} \cdots (x - \lambda_p)^{r_p}, \tag{7.4}$$

the constant term is $\prod_{i=1}^{p} \lambda_i^{r_i}$. Thus $\det A$ is the product of the characteristic values (each counted with the multiplicity with which it is a factor of the characteristic polynomials). In a similar way it can be seen that $\text{Tr}(A)$ is the sum of the characteristic values (each counted with multiplicity).

We have now shown the existence of several objects associated with a matrix, or its underlying linear transformation, which are independent of the coordinate system. For example, the characteristic polynomial, the determinant, and the trace are independent of the coordinate system. Actually, this list is redundant since $\det A$ is the constant term of the characteristic polynomial, and $\text{Tr}(A)$ is $(-1)^{n-1}$ times the coefficient of x^{n-1} of the characteristic polynomial. Functions of this type are of interest because they contain information about the linear transformation, or the matrix, and they are sometimes rather easy to evaluate. But this raises a host of questions. What information do these invariants contain? Can we find a complete list of non-redundant invariants, in the sense that any other invariant can be computed from those in the list? While some partial answers to these questions will be given, a systematic discussion of these questions is beyond the scope of this book.

Theorem 7.5. *Let V be a vector space with a basis consisting of eigen-vectors of σ. If W is any subspace of V invariant under σ, then W also has a basis consisting of eigenvectors of σ.*

PROOF.   Let α be any vector in W. Since V has a basis of eigenvectors of σ, α can be expressed as a linear combination of eigenvectors of σ. By disregarding terms with zero coefficients, combining terms corresponding to the same eigenvalue, and renaming a term like $a_i\xi_i$, where ξ_i is an eigen-vector and $a_i \neq 0$, as an eigenvector with coefficient 1, we can represent α in the form

$$\alpha = \sum_{i=1}^{r} \xi_i,$$

where the ξ_i are eigenvectors of σ with distinct eigenvalues. Let λ_i be the eigenvalue corresponding to ξ_i. We will show that each $\xi_i \in W$.

$(\sigma - \lambda_2)(\alpha - \lambda_3) \cdots (\sigma - \lambda_r)(\alpha)$ is in W since W is invariant under σ, and hence invariant under $\sigma - \lambda$ for any scalar λ. But then $(\sigma - \lambda_2)(\sigma - \lambda_3)$ $\cdots (\sigma - \lambda_r)(\alpha) = (\lambda_1 - \lambda_2)(\lambda_1 - \lambda_3) \cdots (\lambda_1 - \lambda_r)\xi_1 \in W$, and $\xi_1 \in W$ since $(\lambda_1 - \lambda_2)(\lambda_1 - \lambda_3) \cdots (\lambda_1 - \lambda_r) \neq 0$. A similar argument shows that each $\xi_i \in W$.

Since this argument applies to any $\alpha \in W$, W is spanned by eigenvectors of σ. Thus, W has a basis of eigenvectors of σ. □

Theorem 7.6. *Let V be a vector space over C, the field of complex numbers. Let σ be a linear transformation of V into itself. V has a basis of eigenvectors for σ if and only if for every subspace S invariant under σ there is a subspace T invariant under σ such that $V = S \oplus T$.*

PROOF.   The theorem is obviously true if V is of dimension 1. Assume the assertions of the theorem are correct for spaces of dimension less than n, where n is the dimension of V.

Assume first that for every subspace S invariant under σ there is a com-plementary subspace T also invariant under σ. Since V is a vector space over the complex numbers σ has at least one eigenvalue λ_1. Let α_1 be an eigenvector corresponding to λ_1. The subspace $S_1 = \langle \alpha_1 \rangle$ is then invariant under σ. By assumption there is a subspace T_1 invariant under σ such that $V = S_1 \oplus T_1$.

Every subspace S_2 of T_1 invariant under $R\sigma$ is also invariant under σ. Thus there exists a subspace T_2 of V invariant under σ such that $V = S_2 \oplus T_2$. Now $S_2 \subset T_1$ and $T_1 = S_2 \oplus (T_2 \cap T_1)$. (See Exercise 15, Section 1-4.) Since $T_2 \cap T_1$ is invariant under σ, and therefore under $R\sigma$, the induction assumption holds for the subspace T_1. Thus, T_1 has a basis of eigenvectors, and by adjoining α_1 to this basis we obtain a basis of eigenvectors of V.

Now assume there is a basis of V consisting of eigenvectors of σ. By theorem 7.5 any invariant subspace S has a basis of eigenvectors. The method of proof of Theorem 2.7 of Chapter I (the Steinitz replacement theorem) will yield a basis of V consisting of eigenvectors of σ, and this basis will contain the basis of S consisting of eigenvectors. The eigenvectors adjoined will span a subspace T, and this subspace will be invariant under σ and complementary to S. $\square$

EXERCISES

1. For each matrix A given in the exercises of Section 6 find, when possible, a non-singular matrix P for which $P^{-1}AP$ is diagonal.

2. Show that the matrix $\begin{bmatrix} 1 & c \\ 0 & 1 \end{bmatrix}$ where $c \neq 0$ is not similar to a diagonal matrix

3. Show that any 2×2 matrix satisfying $x^2 + 1 = 0$ is similar to the matrix
$$\begin{bmatrix} 0 & -1 \\ 1 & 0 \end{bmatrix}.$$

4. Show that if A is non-singular, then AB is similar to BA.

5. Show that any two projections of the same rank are similar.

*8 | The Jordan Normal Form

A normal form that is obtainable in general when the field of scalars is the field of complex numbers is known as the *Jordan normal form*. An application of the Jordan normal form to power series of matrices and systems of linear differential equations is given in the chapter on applications. Except for these applications this section can be skipped without penalty.

We assume that the field of scalars is the field of complex numbers. Thus for any square matrix A the characteristic polynomial $f(x)$ factors into linear factors, $f(x) = (x - \lambda_1)^{r_1}(x - \lambda_2)^{r_2} \cdots (x - \lambda_p)^{r_p}$ where $\lambda_i \neq \lambda_j$ for $i \neq j$ and r_i is the algebraic multiplicity of the eigenvalue λ_i. The minimum polynomial $m(x)$ for A is of the form $m(x) = (x - \lambda_1)^{s_1}(x - \lambda_2)^{s_2} \cdots (x - \lambda_p)^{s_p}$ where $1 \leq s_i \leq r_i$.

In the theorems about the diagonalization of matrices we sought bases made up of eigenvectors. Because we are faced with the possibility that such bases do not exist, we must seek proper generalizations of the eigenvectors. It is more fruitful to think of the eigenspaces rather than the eigenvectors themselves. An eigenvalue is a scalar λ for which the linear transformation $\sigma - \lambda$ is singular. An eigenspace is the kernel (of positive dimension) of the linear transformation $\sigma - \lambda$. The proper generalization of eigenspaces turns out to be the kernels of higher powers of $\sigma - \lambda$. For a given eigenvalue λ, let M^k be the kernel of $(\sigma - \lambda)^k$. Thus, $M^0 = \{0\}$ and M^1

is the eigenspace of λ. For $\alpha \in M^k$, $(\sigma - \lambda)^{k+1}(\alpha) = (\sigma - \lambda)(\sigma - \lambda)^k(\alpha) = (\sigma - \lambda)(0) = 0$. Hence, $M^k \subset M^{k+1}$. Also, for $\alpha \in M^{k+1}$, $(\sigma - \lambda)^k(\sigma - \lambda)(\alpha) = (\sigma - \lambda)^{k+1}(\alpha) = 0$ so that $(\sigma - \lambda)(\alpha) \in M^k$. Hence, $(\sigma - \lambda)M^{k+1} \subset M^k$.

Since all $M^k \subset V$ and V is finite dimensional, the sequence of subspaces $M^0 \subset M^1 \subset M^2 \subset \cdots$ must eventually stop increasing. Let t be the smallest index such that $M^k = M^t$ for all $k \geq t$, and denote M^t by $M_{(\lambda)}$. Let m_k be the dimension of M^k and m_t the dimension of $M_{(\lambda)}$.

Let $(\sigma - \lambda)^k V = W^k$. Then $W^{k+1} = (\sigma - \lambda)^{k+1}V = (\sigma - \lambda)^k\{(\sigma - \lambda)V\} \subset (\sigma - \lambda)^k V = W^k$. Thus, the subspaces W^k form a decreasing sequence $W^0 \supset W^1 \supset W^2 \supset \cdots$. Since the dimension of W^k is $n - m_k$, we see that $W^k = W^t$ for all $k \geq t$. Denote W^t by $W_{(\lambda)}$.

Theorem 8.1. *V is the direct sum of $M_{(\lambda)}$ and $W_{(\lambda)}$.*

PROOF. Since $(\sigma - \lambda)W^t = (\sigma - \lambda)^{t+1}V = W^{t+1} = W^t$ we see that $\sigma - \lambda$ is non-singular on $W^t = W_{(\lambda)}$. Now let α be any vector in V. Then $(\sigma - \lambda)^t(\alpha) = \beta$ is an element of $W_{(\lambda)}$. Because $(\sigma - \lambda)^t$ is non-singular on $W_{(\lambda)}$ there is a unique vector $\gamma \in W_{(\lambda)}$ such that $(\sigma - \lambda)^t(\gamma) = \beta$. Let $\alpha - \gamma$ be denoted by δ. It is easily seen that $\delta \in M_{(\lambda)}$. Hence $V = M_{(\lambda)} + W_{(\lambda)}$. Finally, since dim $M_{(\lambda)} = m_t$ and dim $W_{(\lambda)} = n - m_t$, the sum is direct. $\square$

In the course of defining M^k and W^k we have shown that

(1) $(\sigma - \lambda)M^{k+1} \subset M^k \subset M^{k+1}$,
(2) $(\sigma - \lambda)W^k = W^{k+1} \subset W^k$.

This shows that each M^k and W^k is invariant under $\sigma - \lambda$. It follows immediately that each is invariant under any polynomial in $\sigma - \lambda$, and hence also under any polynomial in σ. The use we wish to make of this observation is that if μ is any other eigenvalue, then $\sigma - \mu$ also maps $M_{(\lambda)}$ and $W_{(\lambda)}$ into themselves.

Let $\lambda_1, \ldots, \lambda_p$ be the distinct eigenvalues of σ. Let M_i be a simpler notation for the subspace $M_{(\lambda_i)}$ defined as above, and let W_i be a simpler notation for $W_{(\lambda_i)}$.

Theorem 8.2. *For $\lambda_i \neq \lambda_j$, $M_i \subset W_j$.*

PROOF. Suppose $\alpha \in M_i$ is in the kernel of $\sigma - \lambda_j$. Then

$$(\lambda_j - \lambda_i)^{t_i}\alpha = \{(\sigma - \lambda_i) - (\sigma - \lambda_j)\}^{t_i}(\alpha)$$

$$= (\sigma - \lambda_i)^{t_i}(\alpha) + \sum_{k=1}^{t_i}(-1)^k\binom{t_i}{k}(\sigma - \lambda_i)^{t_i-k}(\sigma - \lambda_j)^k(\alpha).$$

The first term is zero because $\alpha \in M_i$, and the others are zero because α is in the kernel of $\sigma - \lambda_j$. Since $\lambda_j - \lambda_i \neq 0$, it follows that $\alpha = 0$. This means that $\sigma - \lambda_j$ is non-singular on M_i; that is, $\sigma - \lambda_j$ maps M_i onto itself. Thus M_i is contained in the set of images under $(\sigma - \lambda_j)^{t_j}$, and hence $M_i \subset W_j$. □

Theorem 8.3. $V = M_1 \oplus M_2 \oplus \cdots \oplus M_p$.

PROOF. Since $V = M_2 \oplus W_2$ and $M_2 \subset W_1$, we have $V = M_1 \oplus W_1 = M_1 \oplus \{M_2 \oplus (W_1 \cap W_2)\}$. Continuing in the same fashion, we get $V = M_1 \oplus \cdots \oplus M_p \oplus \{W_1 \cap \cdots \cap W_p\}$. Thus the theorem will follow if we can show that $W = W_1 \cap \cdots \cap W_p = \{0\}$. By an extension of remarks already made $(\sigma - \lambda_1) \cdots (\sigma - \lambda_p) = q(\sigma)$ is non-singular on W; that is, $q(\sigma)$ maps W onto itself. For arbitrarily large k, $[q(\sigma)]^k$ also maps W onto itself. But $q(x)$ contains each factor of the characteristic polynomial $f(x)$ so that for large enough k, $[q(x)]^k$ is divisible by $f(x)$. This implies that $W = \{0\}$. □

Corollary 8.4. $t_i = s_i$ for $i = 1, \ldots, p$.

PROOF. Since $V = M_1 \oplus \cdots \oplus M_p$ and $(\sigma - \lambda_i)^{t_i}$ vanishes on M_i, it follows that $(\sigma - \lambda_1)^{t_1} \cdots (\sigma - \lambda_p)^{t_p}$ vanishes on all of V. Thus $(x - \lambda_1)^{t_1} \cdots (x - \lambda_p)^{t_p}$ is divisible by the minimum polynomial and $s_i \leq t_i$.

On the other hand, if for a single i we have $s_i < t_i$, there is an $\alpha \in M_i$ such that $(\sigma - \lambda_i)^{s_i}(\alpha) \neq 0$. For all $\lambda_j \neq \lambda_i$, $\sigma - \lambda_j$ is non-singular on M_i. Hence $m(\sigma) \neq 0$. This is a contradiction so that $t_i = s_i$. □

Let us return to the situation where, for the single eigenvalue λ, M^k is the kernel of $(\sigma - \lambda)^k$ and $W^k = (\sigma - \lambda)^k V$. In view of Corollary 8.4 we let s be the smallest index such that $M^k = M^s$ for all $k \geq s$. By induction we can construct a basis $\{\alpha_1, \ldots, \alpha_{m_s}\}$ of $M_{(\lambda)}$ such that $\{\alpha_1, \ldots, \alpha_{m_k}\}$ is a basis of M^k.

We now proceed step by step to modify this basis. The set $\{\alpha_{m_{s-1}+1}, \ldots, \alpha_{m_s}\}$ consists of those basis elements in M^s which are not in M^{s-1}. These elements do not have to be replaced, but for consistency of notation we change their names; let $\alpha_{m_{s-1}+v} = \beta_{m_{s-1}+v}$. Now set $(\sigma - \lambda)(\beta_{m_{s-1}+v}) = \beta_{m_{s-2}+v}$ and consider the set $\{\alpha_1, \ldots, \alpha_{m_{s-2}}\} \cup \{\beta_{m_{s-2}+1}, \ldots, \beta_{m_{s-2}+m_s-m_{s-1}}\}$. We wish to show that this set is linearly independent.

If this set were linearly dependent, a non-trivial relation would exist and it would have to involve at least one of the β_i with a non-zero coefficient since the set $\{\alpha_1, \ldots, \alpha_{m_{s-2}}\}$ is linearly independent. But then a non-trivial linear combination of the β_i would be an element of M^{s-2}, and $(\sigma - \lambda)^{s-2}$ would map this linear combination onto 0. This would mean that $(\sigma - \lambda)^{s-1}$ would map a non-trivial linear combination of $\{\alpha_{m_{s-1}+1}, \ldots, \alpha_{m_s}\}$ onto 0. Then this non-trivial linear combination would be in M^{s-1}, which would contradict the linear independence of $\{\alpha_1, \ldots, \alpha_{m_s}\}$. Thus the set $\{\alpha_1, \ldots, \alpha_{m_{s-2}}\} \cup \{\beta_{m_{s-2}+1}, \ldots, \beta_{m_{s-2}+m_s-m_{s-1}}\}$ is linearly independent.

This linearly independent subset of M^{s-1} can be expanded to a basis of M^{s-1}. We use β's to denote these additional elements of this basis, if any

additional elements are required. Thus we have the new basis $\{\alpha_1, \ldots,$
$\alpha_{m_{s-2}}\} \cup \{\beta_{m_{s-2}+1}, \ldots, \beta_{m_{s-1}}\}$ of M^{s-1}.

We now set $(\sigma - \lambda)(\beta_{m_{s-2}+\nu}) = \beta_{m_{s-3}+\nu}$ and proceed as before to obtain
a new basis $\{\alpha_1, \ldots, \alpha_{m_{s-3}}\} \cup \{\beta_{m_{s-3}+1}, \ldots, \beta_{m_{s-2}}\}$ of M^{s-2}.

Proceeding in this manner we finally get a new basis $\{\beta_1, \ldots, \beta_{m_s}\}$ of
$M_{(\lambda)}$ such that $\{\beta_1, \ldots, \beta_{m_k}\}$ is a basis of M^k and $(\sigma - \lambda)(\beta_{m_k+\nu}) = \beta_{m_{k-1}+\nu}$
for $k \geq 1$. This relation can be rewritten in the form

$$
\begin{array}{lll}
\sigma(\beta_{m_k+\nu}) = \lambda\beta_{m_k+\nu} + \beta_{m_{k-1}+\nu} & \text{for} & k \geq 1, \qquad (8.1) \\
\sigma(\beta_\nu) = \lambda\beta_\nu & \text{for} & \nu \leq m_1.
\end{array}
$$

Thus we see that in a certain sense $\beta_{m_k+\nu}$ is "almost" an eigenvector.

This suggests reordering the basis vectors so that $\{\beta_1, \beta_{m_1+1}, \ldots, \beta_{m_{s-1}+1}\}$
are listed first. Next we should like to list the vectors $\{\beta_2, \beta_{m_1+2}, \ldots\}$, etc.
The general idea is to list each of the first elements from each section of the
β's, then each of the second elements from each section, and continue until
a new ordering of the basis is obtained.

With the basis of $M_{(\lambda)}$ listed in this order (and assuming for the moment
that that $M_{(\lambda)}$ is all of V) the matrix representing σ takes the form

Theorem 8.5. *Let A be a matrix with characteristic polynomial $f(x) = (x - \lambda_1)^{r_1} \cdots (x - \lambda_p)^{r_p}$ and minimum polynomial $m(x) = (x - \lambda_1)^{s_1} \cdots (x - \lambda_p)^{s_p}$. A is similar to a matrix J with submatrices of the form*

$$
B_i = \begin{bmatrix}
\lambda_i & 1 & 0 & \cdots & 0 & 0 \\
0 & \lambda_i & 1 & \cdots & 0 & 0 \\
0 & 0 & \lambda_i & \cdots & 0 & 0 \\
\cdot & \cdot & \cdot & \cdots & \cdot & \cdot \\
\cdot & \cdot & \cdot & \cdots & \cdot & \cdot \\
\cdot & \cdot & \cdot & \cdots & \cdot & \cdot \\
0 & 0 & 0 & \cdots & \lambda_i & 1 \\
0 & 0 & 0 & \cdots & 0 & \lambda_i
\end{bmatrix}
$$

along the main diagonal. All other elements of J are zero. For each λ_i there at least one B_i of order s_i. All other B_i corresponding to this λ_i are of order less than or equal to s_i. The number of B_i corresponding to this λ_i is equal to the geometric multiplicity of λ_i. The sum of the orders of all the B_i corresponding to λ_i is r_i. While the ordering of the B_i along the main diagonal of J is not unique, the number of B_i of each possible order is uniquely determined by A. J is called the Jordan normal form corresponding to A.

PROOF.   From Theorem 8.3 we have $V = M_1 \oplus \cdots \oplus M_p$. In the discussion preceding the statement of Theorem 8.5 we have shown that each M_i has a basis of a special type. Since V is the sum of the M_i, the union of these bases spans V. Since the sum is direct, the union of these bases is linearly independent and, hence, a basis for V. This shows that a matrix J of the type described in Theorem 8.5 does represent σ and is therefore similar to A.

The discussion preceding the statement of the theorem also shows that the dimensions $m_{i,k}$ of the kernels M_i^k of the various $(\sigma - \lambda_i)^k$ determine the orders of the B_i in J. Since A determines σ and σ determines the subspace M_i^k independently of the bases employed, the B_i are uniquely determined.

Since the λ_i appear along the main diagonal of J and all other non-zero elements of J are above the main diagonal, the number of times $x - \lambda_i$ appears as a factor of the characteristic polynomial of J is equal to the number of times λ_i appears in the main diagonal. Thus the sum of the orders of the B_i corresponding to λ_i is exactly r_i. This establishes all the statements of Theorem 8.5. □

Let us illustrate the workings of the theorems of this section with some examples. Unfortunately, it is a little difficult to construct an interesting example of low order. Hence, we give two examples, The first example illustrates the choice of basis as described for the space $M_{(\lambda)}$. The second example illustrates the situation described by Theorem 8.3.

Example 1. Let

$$
A = \begin{bmatrix}
1 & 0 & -1 & 1 & 0 \\
-4 & 1 & -3 & 2 & 1 \\
-2 & -1 & 0 & 1 & 1 \\
-3 & -1 & -3 & 4 & 1 \\
-8 & -2 & -7 & 5 & 4
\end{bmatrix}.
$$

The first step is to obtain the characteristic matrix

$$
C(x) = \begin{bmatrix}
1-x & 0 & -1 & 1 & 0 \\
-4 & 1-x & -3 & 2 & 1 \\
-2 & -1 & -x & 1 & 1 \\
-3 & -1 & -3 & 4-x & 1 \\
-8 & -2 & -7 & 5 & 4-x
\end{bmatrix}.
$$

Although it is tedious work we can obtain the characteristic polynomial $f(x) = (x - 2)^5$. We have one eigenvalue with algebraic multiplicity 5. What is the geometric multiplicity and what is the minimum equation for A? Although there is an effective method for determining the minimum equation, it is less work and less wasted effort to proceed directly with determining the eigenvectors. Thus, from

$$
C(2) = \begin{bmatrix}
-1 & 0 & -1 & 1 & 0 \\
-4 & -1 & -3 & 2 & 1 \\
-2 & -1 & -2 & 1 & 1 \\
-3 & -1 & -3 & 2 & 1 \\
-8 & -2 & -7 & 5 & 2
\end{bmatrix},
$$

we obtain by elementary row operations the Hermite normal form

$$
\begin{bmatrix}
1 & 0 & 0 & 0 & 0 \\
0 & 1 & 0 & 1 & -1 \\
0 & 0 & 1 & -1 & 0 \\
0 & 0 & 0 & 0 & 0 \\
0 & 0 & 0 & 0 & 0
\end{bmatrix}.
$$

From this we learn that there are two linearly independent eigenvectors corresponding to 2. The dimension of M^1 is 2. Without difficulty we find the eigenvectors

$$\alpha_1 = (0, -1, 1, 1, 0)$$
$$\alpha_2 = (0, 1, 0, 0, 1).$$

Now we must compute $(A - 2I)^2 = (C(2))^2$, and obtain

$$(A - 2I)^2 = \begin{bmatrix} 0 & 0 & 0 & 0 & 0 \\ 0 & 0 & 0 & 0 & 0 \\ -1 & 0 & -1 & 1 & 0 \\ -1 & 0 & -1 & 1 & 0 \\ -1 & 0 & -1 & 1 & 0 \end{bmatrix}.$$

The rank of $(A - 2I)^2$ is 1 and hence M^2 is of dimension 4. The α_1 and α_2 we already have are in M^2 and we must obtain two more vectors in M^2 which, together with α_1 and α_2, will form an independent set. There is quite a bit of freedom for choice and

$$\alpha_3 = (0, 1, 0, 0, 0)$$
$$\alpha_4 = (-1, 0, 1, 0, 0)$$

appear to be as good as any.

Now $(A - 2I)^3 = 0$, and we know that the minimum polynomial for A is $(x - 2)^3$. We have this knowledge and quite a bit more more for less work than would be required to find the minimum polynomial directly. We see, then, that $M^3 = V$ and we have to find another vector independent of α_1, α_2, α_3, and α_4. Again, there are many possible choices. Some choices will lead to a simpler matrix of transition than will others, and there seems to be no very good way to make the choice that will result in the simplest matrix of transition. Let us take

$$\alpha_5 = (0, 0, 0, 1, 0).$$

We now have the basis of $\{\alpha_1, \alpha_2, \alpha_3, \alpha_4, \alpha_5\}$ such that $\{\alpha_1, \alpha_2\}$ is a basis of M^1, $\{\alpha_1, \alpha_2, \alpha_3, \alpha_4\}$ is a basis of M^2, and $\{\alpha_1, \alpha_2, \alpha_3, \alpha_4, \alpha_5\}$ is a basis of M^3. Following our instructions, we set $\beta_5 = \alpha_5$. Then

$$(A - 2I)\begin{bmatrix} 0 \\ 0 \\ 0 \\ 1 \\ 0 \end{bmatrix} = \begin{bmatrix} 1 \\ 2 \\ 1 \\ 1 \\ 5 \end{bmatrix}.$$

Hence, we set $\beta_3 = (1, 2, 1, 2, 5)$. Now we must choose β_4 so that $\{\alpha_1,$ $\alpha_2, \beta_3, \beta_4\}$ is a basis for M^2. We can choose $\beta_4 = (-1, 0, 1, 0, 0)$. Then

$$(A - 2I)\begin{bmatrix} 1 \\ 2 \\ 1 \\ 2 \\ 5 \end{bmatrix} = \begin{bmatrix} 0 \\ 0 \\ 1 \\ 1 \\ 1 \end{bmatrix}, \qquad (A - 2I)\begin{bmatrix} -1 \\ 0 \\ 1 \\ 0 \\ 0 \end{bmatrix} = \begin{bmatrix} 0 \\ 1 \\ 0 \\ 0 \\ 1 \end{bmatrix}.$$

Hence, we choose $\beta_1 = (0, 0, 1, 1, 1)$ and $\beta_2 = (0, 1, 0, 0, 1)$. Thus,

$$P = \begin{bmatrix} 0 & 1 & 0 & 0 & -1 \\ 0 & 2 & 0 & 1 & 0 \\ 1 & 1 & 0 & 0 & 1 \\ 1 & 2 & 1 & 0 & 0 \\ 1 & 5 & 0 & 1 & 0 \end{bmatrix}$$

is the matrix of transition that will transform A to the Jordan normal form

$$J = \begin{bmatrix} 2 & 1 & 0 & 0 & 0 \\ 0 & 2 & 1 & 0 & 0 \\ 0 & 0 & 2 & 0 & 0 \\ 0 & 0 & 0 & 2 & 1 \\ 0 & 0 & 0 & 0 & 2 \end{bmatrix}.$$

Example 2. Let

$$A = \begin{bmatrix} 5 & -1 & -3 & 2 & -5 \\ 0 & 2 & 0 & 0 & 0 \\ 1 & 0 & 1 & 1 & -2 \\ 0 & -1 & 0 & 3 & 1 \\ 1 & -1 & -1 & 1 & 1 \end{bmatrix}.$$

The characteristic polynomial is $f(x) = -(x - 2)^3(x - 3)^2$. Again we have repeated eigenvalues, one of multiplicity 3 and one of multiplicity 2.

$$C(2) = \begin{bmatrix} 3 & -1 & -3 & 2 & -5 \\ 0 & 0 & 0 & 0 & 0 \\ 1 & 0 & -1 & 1 & -2 \\ 0 & -1 & 0 & 1 & 1 \\ 1 & -1 & -1 & 1 & -1 \end{bmatrix},$$

from which we obtain the Hermite normal form

$$\begin{bmatrix} 1 & 0 & -1 & 0 & -2 \\ 0 & 1 & 0 & 0 & -1 \\ 0 & 0 & 0 & 1 & 0 \\ 0 & 0 & 0 & 0 & 0 \\ 0 & 0 & 0 & 0 & 0 \end{bmatrix}.$$

Again, the geometric multiplicity is less than the algebraic multiplicity. We obtain the eigenvectors

$$\alpha_1 = (1, 0, 1, 0, 0)$$
$$\alpha_2 = (2, 1, 0, 0, 1).$$

Now we must compute $(A - 2I)^2$. We find

$$(A - 2I)^2 = \begin{bmatrix} 1 & 0 & -1 & 0 & -2 \\ 0 & 0 & 0 & 0 & 0 \\ 0 & 0 & 0 & 0 & 0 \\ 1 & -2 & -1 & 2 & 0 \\ 1 & -1 & -1 & 1 & -1 \end{bmatrix},$$

from which we obtain the Hermite normal form

$$\begin{bmatrix} 1 & 0 & -1 & 0 & -2 \\ 0 & 1 & 0 & -1 & -1 \\ 0 & 0 & 0 & 0 & 0 \\ 0 & 0 & 0 & 0 & 0 \\ 0 & 0 & 0 & 0 & 0 \end{bmatrix}.$$

For the third basis vector we can choose

$$\alpha_3 = (0, 1, 0, 1, 0).$$

Then

$$(A - 2I)\begin{bmatrix} 0 \\ 1 \\ 0 \\ 1 \\ 0 \end{bmatrix} = \begin{bmatrix} 1 \\ 0 \\ 1 \\ 0 \\ 0 \end{bmatrix};$$

hence, we have $\beta_3 = \alpha_3$, $\beta_1 = \alpha_1$, and we can choose $\beta_2 = \alpha_2$.

In a similar fashion we find $\beta_4 = (-1, 0, 0, 1, 0)$ and $\beta_5 = (2, 0, 0, 0, 1)$ corresponding to the eigenvalue 3. β_4 is an eigenvector and $(A - 3I)\beta_5 = \beta_4$.

IV | Linear functionals, bilinear forms, quadratic forms

In this chapter we study scalar-valued functions of vectors. Linear functionals are linear transformations of a vector space into a vector space of dimension 1. As such they are not new to us. But because they are very important, they have been the subject of much investigation and a great deal of special terminology has accumulated for them.

For the first time we make use of the fact that the set of linear transformations can profitably be considered to be a vector space. For finite dimensional vector spaces the set of linear functionals forms a vector space of the same dimension, the dual space. We are concerned with the relations between the structure of a vector space and its dual space, and between the representations of the various objects in these spaces.

In Chapter V we carry the vector point of view of linear functionals one step further by mapping them into the original vector space. There is a certain aesthetic appeal in imposing two separate structures on a single vector space, and there is value in doing it because it motivates our concentration on the aspects of these two structures that either look alike or are symmetric. For clarity in this chapter, however, we keep these two structures separate in two different vector spaces.

Bilinear forms are functions of two vector variables which are linear in each variable separately. A quadratic form is a function of a single vector variable which is obtained by identifying the two variables in a bilinear form. Bilinear forms and quadratic forms are intimately tied together, and this is the principal reason for our treating bilinear forms in detail. In Chapter VI we give some applications of quadratic forms to physical problems.

If the field of scalars is the field of complex numbers, then the applications we wish to make of bilinear forms and quadratic forms leads us to modify the definition slightly. In this way we are led to study Hermitian forms. Aside from their definition they present little additional difficulty.

1 | Linear Functionals

Definition. Let V be a vector space over a field of constants F. A linear transformation ϕ of V into F is called a *linear form* or *linear functional* on V.

Any field can be considered to be a 1-dimensional vector space over itself (see Exercise 10, Section I-1). It is possible, for example, to imagine two copies of F, one of which we label U. We retain the operation of addition in U, but drop the operation of multiplication. We then define scalar multiplication in the obvious way: the product is computed as if both the scalar and the vector were in the same copy of F and the product taken to be an element of U. Thus the concept of a linear functional is not really something new. It is our familiar linear transformation restricted to a special case. Linear functionals are so useful, however, that they deserve a special name and particular study. Linear concepts appear throughout mathematics particularly in applied mathematics, and in all cases linear functionals play an important part. It is usually the case, however, that special terminology is used which tends to obscure the widespread occurrence of this concept.

The term "linear form" would be more consistent with other usage throughout this book and the history of the theory of matrices. But the term "linear functional" has come to be almost universally adopted.

Theorem 1.1. *If V is a vector space of dimension n over F, the set of all linear functionals on V is a vector space of dimension n.*

PROOF. If ϕ and ψ are linear functionals on V, by $\phi + \psi$ we mean the mapping defined by $(\phi + \psi)(\alpha) = \phi(\alpha) + \psi(\alpha)$ for all $\alpha \in V$. For any $a \in F$, by $a\phi$ we mean the mapping defined by $(a\phi)(\alpha) = a[\phi(\alpha)]$ for all $\alpha \in V$. We must then show that with these laws for vector addition and scalar multiplication of linear functionals the axioms of a vector space are satisfied.

These demonstrations are not difficult and they are left to the reader. (Remember that proving axioms $A1$ and $B1$ are satisfied really requires showing that $\phi + \psi$ and $a\phi$, as defined, are linear functionals.)

We call the vector space of all linear functionals on V the *dual or conjugate space of V* and denote it by $\hat{V}$ (pronounced "vee hat" or "vee caret"). We have yet to show that $\hat{V}$ is of dimension n. Let $A = \{\alpha_1, \alpha_2, \ldots, \alpha_n\}$ be a basis of V. Define ϕ_i by the rule that for any $\alpha = \sum_{j=1}^{n} a_j \alpha_j$, $\phi_i(\alpha) = a_i \in F$. We shall call ϕ_i the *ith coordinate function*.

For any $\beta = \sum_{i=1}^{n} b_i \alpha_i$ we have $\phi_i(\beta) = b_i$ and $\phi_i(\alpha + \beta) = \phi_i\{\sum_{j=1}^{n} a_j\alpha_j + \sum_{j=1}^{n} b_j\alpha_j\} = \phi_i\{\sum_{j=1}^{n} (a_j + b_j)\alpha_j\} = a_i + b_i = \phi_i(\alpha) + \phi_i(\beta)$. Also $\phi_i(a\alpha) = \phi_i\{a\sum_{j=1}^{n} a_j\alpha_j\} = \phi_i\{\sum_{j=1}^{n} aa_j\alpha_j\} = aa_i = a\phi_i(\alpha)$. Thus ϕ_i is a linear functional.

Suppose that $\sum_{j=1}^{n} b_j\phi_j = 0$. Then $(\sum_{j=1}^{n} b_j\phi_j)(\alpha) = 0$ for all $\alpha \in V$. In particular for α_i we have $(\sum_{j=1}^{n} b_j\phi_j)(\alpha_i) = \sum_{j=1}^{n} b_j\phi_j(\alpha_i) = b_i = 0$. Hence, all $b_i = 0$ and the set $\{\phi_1, \phi_2, \ldots, \phi_n\}$ must be linearly independent. On the other hand, for any $\phi \in \hat{V}$ and any $\alpha = \sum_{i=1}^{n} a_i\alpha_i \in V$, we have

$$\phi(\alpha) = \phi\left\{\sum_{i=1}^{n} a_i\alpha_i\right\} = \sum_{i=1}^{n} a_i\phi(\alpha_i). \tag{1.1}$$

If we let $\phi(\alpha_i) = b_i$, then for $\sum_{j=1}^{n} b_j\phi_j$ we have

$$\left(\sum_{j=1}^{n} b_j\phi_j\right)(\alpha) = \sum_{j=1}^{n} b_j\phi_j(\alpha) = \sum_{j=1}^{n} b_j a_j = \phi(\alpha). \tag{1.2}$$

Thus the set $\{\phi_1, \ldots, \phi_n\} = \hat{A}$ spans $\hat{V}$ and forms a basis of $\hat{V}$. This shows that $\hat{V}$ is of dimension n. $\square$

The basis $\hat{A}$ of $\hat{V}$ that we have constructed in the proof of Theorem 1.1 has a very special relation to the basis A. This relation is characterized by the equations

$$\phi_i(\alpha_j) = \delta_{ij}, \tag{1.3}$$

for all i, j. In the proof of Theorem 1.1 we have shown that a basis satisfying these conditions exists. For each i, the conditions in Equation (1.3) specify the values of ϕ_i on all the vectors in the basis A. Thus ϕ_i is uniquely determined as a linear functional. And thus $\hat{A}$ is uniquely determined by A and the conditions (1.3). We call $\hat{A}$ the basis *dual* to the basis A.

If $\phi = \sum_{i=1}^{n} b_i\phi_i$,

$$\phi(\alpha_j) = \sum_{i=1}^{n} b_i\phi_i(\alpha_j) = b_j$$

so that, as a linear transformation, ϕ is represented by the $1 \times n$ matrix $[b_1 \cdots b_n]$. For this reason we choose to represent the linear functionals in $\hat{V}$ by one-row matrices. With respect to the basis $\hat{A}$ in $\hat{V}$, $\phi = \sum_{i=1}^{n} b_i\phi_i$ will be represented by the row $[b_1 \cdots b_n] = B$. It might be argued that, since $\hat{V}$ is a vector space, the elements of $\hat{V}$ should be represented by columns. But the set of all linear transformations of one vector space into another also forms a vector space, and we can as justifiably choose to emphasize the aspect of $\hat{V}$ as a set of linear transformations. At most, the choice of a representing notation is a matter of taste and convenience. The choice we have made means that some adjustments will have to be made when using the matrix of transition to change the coordinates of a linear functional when the basis is changed. But no choice of representing notation seems to avoid all such difficulties and the choice we have made seems to offer the most advantages.

If the vector $\xi \in V$ is represented by the n-tuple $(x_1, \ldots, x_n) = X$, then we can compute $\phi(\xi)$ directly in terms of the representations.

$$\phi(\xi) = \left(\sum_{j=1}^{n} b_j \phi_j \right) \left(\sum_{i=1}^{n} x_i \alpha_i \right)$$

$$= \sum_{j=i}^{n} \sum_{i=1}^{n} b_j x_i \phi_j(\alpha_i)$$

$$= \sum_{j=1}^{n} b_j x_j$$

$$= [b_i \cdots b_n] \begin{bmatrix} x_1 \\ \cdot \\ \cdot \\ \cdot \\ x_n \end{bmatrix}$$

$$= BX. \tag{1.4}$$

EXERCISES

1. Let $A = \{\alpha_1, \alpha_2, \alpha_3\}$ be a basis in a 3-dimensional vector space V over R. Let $\hat{A} = \{\phi_1, \phi_2, \phi_3\}$ be the basis in $\hat{V}$ dual to A. Any vector $\xi \in V$ can be written in the form $\xi = x_1\alpha_1 + x_2\alpha_2 + x_3\alpha_3$. Determine which of the following functions on V are linear functionals. Determine the coordinates of those that are linear functionals in terms of the basis $\hat{A}$.

   (a) $\phi(\xi) = x_1 + x_2 + x_3$.
   (b) $\phi(\xi) = (x_1 + x_2)^2$.
   (c) $\phi(\xi) = \sqrt{2}x_1$.
   (d) $\phi(\xi) = x_2 - \frac{1}{2}x_1$.
   (e) $\phi(\xi) = x_2 - \frac{1}{2}$.

2. For each of the following bases of R^3 determine the dual basis in $\hat{R}^3$.

   (a) $\{(1, 0, 0), (0, 1, 0), (0, 0, 1)\}$.
   (b) $\{(1, 0, 0), (1, 1, 0), (1, 1, 1)\}$.
   (c) $\{(1, 0, -1), (-1, 1, 0), (0, 1, 1)\}$.

3. Let $V = P_n$, the space of polynomials of degree less than n over R. For a fixed $a \in R$, let $\phi(p) = p^{(k)}(a)$, where $p^{(k)}(x)$ is the kth derivative of $p(x) \in P_n$. Show that ϕ is a linear functional.

4. Let V be the space of real functions continuous on the interval $[0, 1]$, and let g be a fixed function in V. For each $f \in V$ define

$$L_g(f) = \int_0^1 f(t)g(t)\, dt.$$

Show that L_g is a linear functional on V. Show that if $L_g(f) = 0$ for every $g \in V$, then $f = 0$.

5. Let $A = \{\alpha_1, \ldots, \alpha_n\}$ be a basis of V and let $\hat{A} = \{\phi_1, \ldots, \phi_n\}$ be the basis of $\hat{V}$ dual to the basis A. Show that an arbitrary $\alpha \in V$ can be represented in the form

$$\alpha = \sum_{i=1}^{n} \phi_i(\alpha)\alpha_i.$$

6. Let V be a vector space of finite dimension $n \geq 2$ over F. Let α and β be two vectors in V such that $\{\alpha, \beta\}$ is linearly independent. Show that there exists a linear functional ϕ such that $\phi(\alpha) = 1$ and $\phi(\beta) = 0$.

7. Let $V = P_n$, the space of polynomials over F of degree less than $n(n > 1)$. Let $a \in F$ be any scalar. For each $p(x) \in P_n$, $p(a)$ is a scalar. Show that the mapping of $p(x)$ onto $p(a)$ is a linear functional on P_n (which we denote by σ_a). Show that if $a \neq b$ then $\sigma_a \neq \sigma_b$.

8. (Continuation) In Exercise 7 we showed that for each $a \in F$ there is defined a linear functional $\sigma_a \in \hat{P}_n$. Show that if $n > 1$, then not every linear functional in $\hat{P}_n$ can be obtained in this way.

9. (Continuation) Let $\{a_1, \ldots, a_n\}$ be a set of n distinct scalars. Let $f(x) = (x - a_1)(x - a_2) \cdots (x - a_n)$ and $h_k(x) = f(x) = f(x)/(x - a_k)$. Show that $h_k(a_j) = \delta_{ik}f'(a_j)$, where $f'(x)$ is the derivative of $f(x)$.

10. (Continuation) For the a_k given in Exercise 9, let

$$\sigma_j = \frac{1}{f'(a_j)}\, \sigma_{a_j}.$$

Show that $\{\sigma_1, \ldots, \sigma_n\}$ is linearly independent and a basis of $\hat{P}_n$. Show that $\{h_1(x), \ldots, h_n(x)\}$ is linearly independent and, hence, a basis of P_n. (*Hint:* Apply σ_j to $\sum_{k=1}^{n} b_k h_k(x)$.) Show that $\{\sigma_1, \ldots, \sigma_n\}$ is the basis dual to $\{h_1(x), \ldots, h_n(x)\}$

11. (Continuation) Let $p(x)$ be any polynomial in P_n. Show that $p(x)$ can be represented in the form

$$p(x) = \sum_{k=1}^{n} \frac{p(a_k)}{f'(a_k)}\, h_k(x).$$

(*Hint:* Use Exercise 5.) This formula is known as the *Lagrange interpolation formula*. It yields the polynomial of least degree taking on the n specified values $\{p(a_1), \ldots, p(a_n)\}$ at the points $\{a_1, \ldots, a_n\}$.

12. Let W be a proper subspace of the n-dimensional vector space V. Let α_0 be a vector in V but not in W. Show that there is a linear functional $\phi \in \hat{V}$ such that $\phi(\alpha_0) = 1$ and $\phi(\alpha) = 0$ for all $\alpha \in W$.

13. Let W be a proper subspace of the n-dimensional vector space V. Let ψ be a linear functional on W. It must be emphasized that ψ is an element of $\hat{W}$

and not an element of $\hat{V}$. Show that there is at least one element $\phi \in \hat{V}$ such that ϕ coincides with ψ on **W**.

14. Show that if $\alpha \neq 0$, there is a linear functional ϕ such that $\phi(\alpha) \neq 0$.

15. Let α and β be vectors such that $\phi(\beta) = 0$ implies $\phi(\alpha) = 0$. Show that α is a multiple of β.

2 | Duality

Until now, we have encouraged an unsymmetric point of view with respect to **V** and $\hat{V}$. Indeed, it is natural to consider $\phi(\alpha)$ for a chosen ϕ and a range of choices for α. However, there is no reason why we should not choose a fixed α and consider the expression $\phi(\alpha)$ for a range of choices for ϕ. Since $(b_1\phi_1 + b_2\phi_2)(\alpha) = (b_1\phi_1)(\alpha) + (b_2\phi_2)(\alpha)$, we see that α behaves like a linear functional on $\hat{V}$.

This leads us to consider the space $\hat{\hat{V}}$ of all linear functionals on $\hat{V}$. Corresponding to any $\alpha \in V$ we can define a linear functional $\tilde{\alpha}$ in $\hat{\hat{V}}$ by the rule $\tilde{\alpha}(\phi) = \phi(\alpha)$ for all $\phi \in \hat{V}$. Let the mapping defined by this rule be denoted by J, that is, $J(\alpha) = \tilde{\alpha}$. Since $J(a\alpha + b\beta)(\phi) = \phi(a\alpha + b\beta) = a\phi(\alpha) + b\phi(\beta) = aJ(\alpha)(\phi) + bJ(\beta)(\phi) = [aJ(\alpha) + bJ(\beta)](\phi)_0$ we see that J is a linear transformation mapping **V** into $\hat{\hat{V}}$.

Theorem 2.1. *If* **V** *is finite dimensional, the mapping* J *of* **V** *into* $\hat{\hat{V}}$ *is a one-to-one linear transformation of* **V** *onto* $\hat{\hat{V}}$.

PROOF. Let **V** be of dimension n. We have already shown that J is linear and into. If $J(\alpha) = 0$ then $J(\alpha)(\phi) = 0$ for all $\phi \in \hat{V}$. In particular, $J(\alpha)(\phi_i) = 0$ for the basis of coordinate functions. Thus if $\alpha = \sum_{i=1}^{n} a_i\alpha_i$ we see that

$$J(\alpha)(\phi_i) = \phi_i(\alpha) = \sum_{j=1}^{n} a_j\phi_i(\alpha_j) = a_i = 0$$

for each $i = 1, \ldots, n$. Thus $\alpha = 0$ and the kernel of J is $\{0\}$, that is, $J(V)$ is of dimension n. On the other hand, if **V** is of dimension n, then $\hat{V}$ and $\hat{\hat{V}}$ are also of dimension n. Hence $J(V) = \hat{\hat{V}}$ and the mapping is onto. $\square$

If the mapping J of **V** into $\hat{\hat{V}}$ is actually onto **V** we say that **V** is *reflexive*. Thus Theorem 2.1 says that a finite dimensional vector space is reflexive. Infinite dimensional vector spaces are not reflexive, but a proof of this assertion is beyond the scope of this book. Moreover, infinite dimensional vector spaces of interest have a topological structure in addition to the algebraic structure we are studying. This additional condition requires a more restricted definition of a linear functional. With this restriction the dual space is smaller than our definition permits. Under these condition it is again possible for the dual of the dual to be covered by the mapping J.

Since J is onto, we identify V and $J(V)$, and consider V as the space of linear functionals on $\hat{V}$. Thus V and $\hat{V}$ are considered in a symmetrical position and we speak of them as *dual spaces*. We also drop the parentheses from the notation, except when required for grouping, and write $\phi\alpha$ instead of $\phi(\alpha)$. The bases $\{\alpha_1, \ldots, \alpha_n\}$ and $\{\phi_1, \ldots, \phi_n\}$ are *dual bases* if and only if $\phi_i\alpha_j = \delta_{ij}$.

EXERCISES

1. Let $A = \{\alpha_1, \ldots, \alpha_n\}$ be a basis of V, and let $\hat{A} = \{\phi_1, \ldots, \phi_n\}$ be the basis of $\hat{V}$ dual to the basis A. Show that an arbitrary $\phi \in \hat{V}$ can be represented in the form

$$\phi = \sum_{i=1}^{n} \phi(\alpha_i)\phi_i.$$

2. Let V be a vector space of finite dimension $n \geq 2$ over F. Let ϕ and ψ be two linear functionals in $\hat{V}$ such that $\{\phi, \psi\}$ is linearly independent. Show that there exists a vector α such that $\phi(\alpha) = 1$ and $\psi(\alpha) = 0$.

3. Let ϕ_0 be a linear functional not in the subspace S of the space of linear functionals $\hat{V}$. Show that there exists a vector α such that $\phi_0(\alpha) = 1$ and $\phi(\alpha) = 0$ for all $\phi \in S$.

4. Show that if $\phi \neq 0$, there is a vector α such that $\phi(\alpha) \neq 0$.

5. Let ϕ and ψ be two linear functionals such that $\phi(\alpha) = 0$ implies $\psi(\alpha) = 0$. Show that ψ is a multiple of ϕ.

3 | Change of Basis

If the basis $A' = \{\alpha_1', \alpha_2', \ldots, \alpha_n'\}$ is used instead of the basis $A = \{\alpha_1, \alpha_2, \ldots, \alpha_n\}$, we ask how the dual basis $\hat{A}' = \{\phi_1', \ldots, \phi_n'\}$ is related to the dual basis $\hat{A} = \{\phi_1, \ldots, \phi_n\}$. Let $P = [p_{ij}]$ be the matrix of transition from the basis A to the basis A'. Thus $\alpha_j' = \sum_{i=1}^{n} p_{ij}\alpha_i$. Since $\phi_i(\alpha_j') = \sum_{k=1}^{n} p_{kj}\phi_i(\alpha_k) = p_{ij}$ we see that $\phi_i = \sum_{j=1}^{n} p_{ij}\phi_j'$. This means that P^T is the matrix of transition from the basis $\hat{A}'$ to the basis $\hat{A}$. Hence, $(P^T)^{-1} = (P^{-1})^T$ is the matrix of transition from $\hat{A}$ to $\hat{A}'$.

Since linear functionals are represented by row matrices instead of column matrices, the matrix of transition appears in the formulas for change of coordinates in a slightly different way. Let $B = [b_1 \cdots b_n]$ be the representation of a linear functional ϕ with respect to the basis $\hat{A}$ and $B' = [b_1' \cdots b_n']$

be its representation with respect to the basis $\hat{A}'$. Then

$$\phi = \sum_{i=1}^{n} b_i \phi_i$$

$$= \sum_{i=1}^{n} b_i \left(\sum_{j=1}^{n} p_{ij} \phi_j' \right)$$

$$= \sum_{j=1}^{n} \left(\sum_{i=1}^{n} b_i p_{ij} \right) \phi_j'. \tag{3.1}$$

Thus,

$$B' = BP. \tag{3.2}$$

We are looking at linear functionals from two different points of view. Considered as a linear transformation, the effect of a change of coordinates is given by formula (4.5) of Chapter II, which is identical with (3.2) above. Considered as a vector, the effect of a change of coordinates is given by formula (4.3) of Chapter II. In this case we would represent ϕ by B^T, since vectors are represented by column matrices. Then, since $(P^{-1})^T$ is the matrix of transition, we would have

$$B^T = (P^{-1})^T B'^T = (B' P^{-1})^T,$$

or $\tag{3.3}$

$$B = B' P^{-1},$$

which is equivalent to (3.2). Thus the end result is the same from either point of view. It is this two-sided aspect of linear functionals which has made them so important and their study so fruitful.

Example 1. In analytic geometry, a hyperplane passing through the origin is the set of all points with coordinates $(x_1, x_2, \ldots, x_n)$ satisfying an equation of the form $b_1 x_1 + b_2 x_2 + \cdots + b_n x_n = 0$. Thus the n-tuple $[b_1 b_2 \cdots b_n]$ can be considered as representing the hyperplane. Of course, a given hyperplane can be represented by a family of equations, so that there is not a one-to-one correspondence between the hyperplanes through the origin and the n-tuples. However, we can still profitably consider the space of hyperplanes as dual to the space of points.

Suppose the coordinate system is changed so that points now have the coordinates $(y_1, \ldots, y_n)$ where $x_i = \sum_{j=1}^{n} a_{ij} y_j$. Then the equation of the hyperplane becomes

$$\sum_{i=1}^{n} b_i x_i = \sum_{i=1}^{n} b_i \left[\sum_{j=1}^{n} a_{ij} y_j \right]$$

$$= \sum_{j=1}^{n} \left[\sum_{i=1}^{n} b_i a_{ij} \right] y_j$$

$$= \sum_{j=1}^{n} c_j y_j = 0. \tag{3.4}$$

Thus the equation of the hyperplane is transformed by the rule $c_j = \sum_{i=1}^{n} b_i a_{ij}$. Notice that while we have expressed the old coordinates in terms of the new coordinates we have expressed the new coefficients in terms of the old coefficients. This is typical of related transformations in dual spaces.

Example 2. A much more illuminating example occurs in the calculus of functions of several variables. Suppose that w is a function of the variables $x_1, x_2, \ldots, x_n$, $w = f(x_1, x_2, \ldots, x_n)$. Then it is customary to write down formulas of the following form:

$$dw = \frac{\partial w}{\partial x_1} dx_1 + \frac{\partial w}{\partial x_2} dx_2 + \cdots + \frac{\partial w}{\partial x_n} dx_n, \tag{3.5}$$

and

$$\nabla w = \left(\frac{\partial w}{\partial x_1}, \frac{\partial w}{\partial x_2}, \ldots, \frac{\partial w}{\partial x_n} \right). \tag{3.6}$$

dw is usually called the *differential* of w, and ∇w is usually called the *gradient* of w. It is also customary to call ∇w a vector and to regard dw as a scalar, approximately a small increment in the value of w.

The difficulty in regarding ∇w as a vector is that its coordinates do not follow the rules for a change of coordinates of a vector. For example, let us consider $(x_1, x_2, \ldots, x_n)$ as the coordinates of a vector in a linear vector space. This implies the existence of a basis $\{\alpha_1, \ldots, \alpha_n\}$ such that the linear combination

$$\xi = \sum_{i=1}^{n} x_i \alpha_i \tag{3.7}$$

is the vector with coordinates $(x_1, x_2, \ldots, x_n)$. Let $\{\beta_1, \ldots, \beta_n\}$ be a new basis with matrix of transition $P = [p_{ij}]$ where

$$\beta_j = \sum_{i=1}^{n} p_{ij} \alpha_i. \tag{3.8}$$

Then, if $\xi = \sum_{j=1}^{n} y_j \beta_j$ is the representation of ξ in the new coordinate system, we would have

$$x_i = \sum_{j=1}^{n} p_{ij} y_j, \tag{3.9}$$

or

$$x_i = \sum_{j=1}^{n} \frac{\partial x_i}{\partial y_j} y_j. \tag{3.10}$$

Let us contrast this with the formulas for changing the coordinates of ∇w. From the calculus of functions of several variables we know that

$$\frac{\partial w}{\partial y_j} = \sum_{i=1}^{n} \frac{\partial w}{\partial x_i} \cdot \frac{\partial x_i}{\partial y_j}. \tag{3.11}$$

This formula corresponds to (3.2). Thus ∇w changes coordinates as if it were in the dual space.

In vector analysis it is customary to call a vector whose coordinates change according to formula (3.10) a *contravariant vector*, and a vector whose coordinates change according to formula (3.11) a *covariant vector*. The reader should verify that if $P = \left[\dfrac{\partial x_i}{\partial y_j} = p_{ij}\right]$, then $\left[\dfrac{\partial y_j}{\partial x_i}\right] = (P^T)^{-1}$. Thus Thus (3.11) is equivalent to the formula

$$\frac{\partial w}{\partial x_i} = \sum_{j=1}^{n} \frac{\partial y_j}{\partial x_i} \cdot \frac{\partial w}{\partial y_j}. \tag{3.12}$$

From the point of view of linear vector spaces it is a mistake to regard both types of vectors as being in the same vector space. As a matter of fact, their sum is not defined. It is clearer and more fruitful to consider the covariant and contravariant vectors to be taken from a pair of dual spaces.

This point of view is now taken in modern treatments of advanced calculus and vector analysis. Further details in developing this point of view are given in Chapter VI, Section 4.

In traditional discussions of these topics, all quantities that are represented by n-tuples are called vectors.

In fact, the n-tuples themselves are called vectors. Also, it is customary to restrict the discussion to coordinate changes in which both covariant and contravariant vectors transform according to the same formulas. This amounts to having P, the matrix of transition, satisfy the condition $(P^{-1})^T = P$. While this does simplify the discussion it makes it almost impossible to understand the foundations of the subject.

Let $A = \{\alpha_1, \ldots, \alpha_n\}$ be a basis of V and let $\hat{A} = \{\phi_1, \ldots, \phi_n\}$ be the dual basis in $\hat{V}$. Let $B = \{\beta_1, \ldots, \beta_n\}$ be any new basis of V. We are asked to find the dual basis $\hat{\beta}$ in $\hat{V}$. This problem is ordinarily posed by giving the representation of the β_j with respect to the basis A and expecting the representations of the elements of the dual basis with respect of $\hat{A}$. Let the β_j be represented with respect to A in the form

$$\beta_j = \sum_{i=1}^{n} p_{ij}\alpha_i, \tag{3.13}$$

and let

$$\psi_i = \sum_{j=1}^{n} q_{ij}\phi_j \tag{3.14}$$

be the representations of the elements of the dual bases $\hat{B} = \{\psi_1, \ldots, \psi_n\}$.

Then

$$\delta_{kl} = \psi_k \beta_l = \left(\sum_{i=1}^{n} q_{ki}\phi_i \right)\left(\sum_{j=1}^{n} p_{jl}\alpha_j \right)$$

$$= \sum_{i=1}^{n} \sum_{j=1}^{n} q_{ki}p_{jl}\phi_i\alpha_j$$

$$= \sum_{i=1}^{n} q_{ki}p_{il}. \tag{3.15}$$

In matrix form, (3.15) is equivalent to

$$I = QP. \tag{3.16}$$

Q is the inverse of P. Because of (3.15), the ψ_i are represented by the rows of Q. Thus, to find the dual basis, we write the representation of the basis B in the columns of P, find the inverse matrix P^{-1}, and read out the representations of the basis $\hat{B}$ in the rows of P^{-1}.

EXERCISES

1. Let $A = \{(1, 0, \ldots, 0), (0, 1, \ldots, 0), \ldots, (0, 0, \ldots, 1)\}$ be a basis of R^n. The basis of $\widehat{R^n}$ dual to A has the same coordinates. It is of interest to see if there are other bases of R^n for which the dual basis has excatly the same coordinates. Let A' be another basis of R^n with matrix of transition P. What condition should P satisfy in order that the elements of the basis dual to A' have the same coordinates as the corresponding elements of the basis A'?

2. Let $A = \{\alpha_1, \alpha_2, \alpha_3\}$ be a basis of a 3-dimensional vector space V, and let $\hat{A} = \{\phi_1, \phi_2, \phi_3\}$ be the basis of $\hat{V}$ dual to A. Then let $A' = \{(1, 1, 1), (1, 0, 1), (0, 1, -1)\}$ be another basis of V (where the coordinates are given in terms of the basis A). Use the matrix of transition to find the basis $\hat{A}'$ dual to A'.

3. Use the matrix of transition to find the basis dual to $\{(1, 0, 0), (1, 1, 0), (1, 1, 1)\}$.

4. Use the matrix of transition to find the basis dual to $\{(1, 0, -1), (-1, 1, 0), (0, 1, 1)\}$.

5. Let B represent a linear functional ϕ, and X a vector ξ with respect to dual bases, so that BX is the value $\phi\xi$ of the linear functional. Let P be the matrix of transition to a new basis so that if X' is the new representation of ξ, then $X = PX'$. By substituting PX' for X in the expression for the value of $\phi\xi$ obtain another proof that BP is the representation of ϕ in the new dual coordinate system.

4 | Annihilators

Definition. Let V be an n-dimensional vector space and $\hat{V}$ its dual. If, for an $\alpha \in V$ and a $\phi \in \hat{V}$, we have $\phi\alpha = 0$, we say that ϕ and α are *orthogonal*.

Since ϕ and α are from different vector spaces, it should be clear that we do not intend to say that the ϕ and α are at "right angles."

Definition. Let W be a subset (not necessarily a subspace) of V. The set of all linear functionals ϕ such that $\phi\alpha = 0$ for all $\alpha \in W$ is called *the annihilator* of W, and we denote it by $W^{\perp}$. Any $\phi \in W^{\perp}$ is called *an annihilator of W*.

Theorem 4.1. *The annihilator $W^{\perp}$ of W is a subspace of $\hat{V}$. If W is a subspace of dimension ρ, then $W^{\perp}$ is of dimension $n - \rho$.*

PROOF. If ϕ and ψ are in $W^{\perp}$, then $(a\phi + b\psi)\alpha = a\phi\alpha + b\psi\alpha = 0$ for all $\alpha \in W$. Hence, $W^{\perp}$ is a subspace of $\hat{V}$.

Suppose W is a subspace of V of dimension ρ, and let $A = \{\alpha_1, \ldots, \alpha_n\}$ be a basis of V such that $\{\alpha_1, \ldots, \alpha_\rho\}$ is a basis of W. Let $\hat{A} = \{\phi_1, \ldots, \phi_n\}$ be the dual basis of A. For $\{\phi_{\rho+1}, \ldots, \phi_n\}$ we see that $\phi_{\rho+k}\alpha_i = 0$ for all $i \leq \rho$. Hence, $\{\phi_{\rho+1}, \ldots, \phi_n\}$ is a subset of the annihilator of W. On the other hand, if $\phi = \sum_{j=1}^{n} b_j\phi_j$ is an annihilator of W, we have $\phi\alpha_i = 0$ for each $i \leq \rho$. But $\phi\alpha_i = \sum_{j=1}^{n} b_j\phi_j\alpha_i = b_i$. Hence, $b_i = 0$ for $i \leq \rho$ and the set $\{\phi_{\rho+1}, \ldots, \phi_n\}$ spans $W^{\perp}$. Thus $\{\phi_{\rho+1}, \ldots, \phi_n\}$ is a basis for $W^{\perp}$, and $W^{\perp}$ is of dimension $n - \rho$. The dimension of $W^{\perp}$ is called the *co-dimension* of W. $\square$

It should also be clear from this argument that W is exactly the set of all $\alpha \in V$ annihilated by all $\phi \in W^{\perp}$. Thus we have

Theorem 4.2. *If S is any subset of $\hat{V}$, the set of all $\alpha \in V$ annihilated by all $\phi \in S$ is a subspace of V, denoted by $S^{\perp}$. If S is a subspace of dimension r, then $S^{\perp}$ is a subspace of dimension $n - r$.* $\square$

Theorem 4.2 is really Theorem 1.16 of Chapter II in a different form. If a linear transformation of V into another vector space W is represented by a matrix A, then each row of A can be considered as representing a linear functional on V. The number r of linearly independent rows of A is the dimension of the subspace S of $\hat{V}$ spanned by these linear functionals. $S^{\perp}$ is the kernel of the linear transformation and its dimension is $n - r$.

The symmetry in this discussion should be apparent. If $\phi \in W^{\perp}$, then $\phi\alpha = 0$ for all $\alpha \in W$. On the other hand, for $\alpha \in W$, $\phi\alpha = 0$ for all $\phi \in W^{\perp}$.

Theorem 4.3. *If W is a subspace, $(W^{\perp})^{\perp} = W$.*

PROOF. By definition, $(W^{\perp})^{\perp} = W^{\perp\perp}$ is the set of $\alpha \in V$ such that $\phi\alpha = 0$ for all $\phi \in W^{\perp}$. Clearly, $W \subset W^{\perp\perp}$. Since dim $W^{\perp\perp} = n -$ dim $W^{\perp} = $ dim W, $W^{\perp\perp} = W$. $\square$

This also leads to a reinterpretation of the discussion in Section II-8.

A subspace W of V of dimension ρ can be characterized by giving its annihilator $W^{\perp} \subset \hat{V}$ of dimension $r = n - \rho$.

Theorem 4.4. *If W_1 and W_2 are two subspaces of V, and $W_1^\perp$ and $W_2^\perp$ are their respective annihilators in $\hat{V}$, the annihilator of $W_1 + W_2$ is $W_1^\perp \cap W_2^\perp$ and the annihilator of $W_1 \cap W_2$ is $W_1^\perp + W_2^\perp$.*

PROOF. If ϕ is an annihilator of $W_1 + W_1$, then ϕ annihilates all $\alpha \in W_1$ and all $\beta \in W_2$ so that $\phi \in W_1^\perp \cap W_2^\perp$. If $\phi \in W_1^\perp \cap W_2^\perp$, then for all $\alpha \in W_1$ and $\beta \in W_2$ we have $\phi\alpha = 0$ and $\phi\beta = 0$. Hence, $\phi(a\alpha + b\beta) = a\phi\alpha + b\phi\beta = 0$ so that ϕ annihilates $W_1 + W_2$. This shows that $(W_1 + W_2)^\perp = W_1^\perp \cap W_2^\perp$.

The symmetry between the annihilator and the annihilated means that the second part of the theorem follows immediately from the first. Namely, since $(W_1 + W_2)^\perp = W_1^\perp \cap W_2^\perp$, we have by substituting $W_1^\perp$ and $W_2^\perp$ for W_1 and W_2, $(W_1^\perp + W_2^\perp)^\perp = (W_1^\perp)^\perp \cap (W_2^\perp)^\perp = W_1 \cap W_2$. Hence, $(W_1 \cap W_2)^\perp = W_1^\perp + W_2^\perp$. □

Now the mechanics for finding the sum of two subspaces is somewhat simpler than that for finding the intersection. To find the sum we merely combine the two bases for the two subspaces and then discard dependent vectors until an independent spanning set for the sum remains. It happens that to find the intersection $W_1 \cap W_2$ it is easier to find $W_1^\perp$ and $W_2^\perp$ and then $W_1^\perp + W_2^\perp$ and obtain $W_1 \cap W_2$ as $(W_1^\perp + W_2^\perp)^\perp$, than it is to find the intersection directly.

The example in Chapter II-8, page 71, is exactly this process carried out in detail. In the notation of this discussion $E_1 = W_1^\perp$ and $E_2 = W_2^\perp$.

Let V be a vector space, $\hat{V}$ the corresponding dual vector space, and let W be a subspace of V. Since $W \subset V$, is there any simple relation between $\hat{W}$ and $\hat{V}$? There is a relation but it is fairly sophisticated. Any function defined on all of V is certainly defined on any subset. A linear functional $\phi \in \hat{V}$, therefore, defines a function on W, which we have called the restriction of ϕ to W. This does not mean that $\hat{V} \subset \hat{W}$; it means that the restriction defines a mapping of $\hat{V}$ into $\hat{W}$.

Let us denote the restriction of ϕ to W by $\bar{\phi}$, and denote the mapping of ϕ onto $\bar{\phi}$ by R. We call R the restriction mapping. It is easily seen that R is linear. The kernel of R is the set of all $\phi \in \hat{V}$ such that $\phi(\alpha) = 0$ for all $\alpha \in W$. Thus $K(R) = W^\perp$. Since $\dim \hat{W} = \dim W = n - \dim W^\perp = n - \dim K(R)$, the restriction map is an epimorphism. Every linear functional on W is the restriction of a linear functional on V.

Since $K(R) = W^\perp$, we have also shown that $\hat{W}$ and $\hat{V}/W^\perp$ are isomorphic. But two vector spaces of the same dimension are isomorphic in many ways. We have done more than show that $\hat{W}$ and $\hat{V}/W^\perp$ are isomorphic. We have shown that there is a canonical isomorphism that can be specified in a natural way independent of any coordinate system. If $\bar{\phi}$ is a residue class in $\hat{V}/W^\perp$, and ϕ is any element of this residue class, then $\bar{\phi}$ and $R(\phi)$ correspond under this natural isomorphism. If η denotes the natural homomorphism of $\hat{V}$ onto $\hat{V}/W^{\perp}$, and τ denotes the mapping of $\bar{\phi}$ onto $R(\phi)$ defined above, then $R = \tau\eta$, and τ is uniquely determined by R and η and this relation.

Theorem 4.5. *Let W be a subspace of V and let $W^{\perp}$ be the annihilator of W in $\hat{V}$. Then $\hat{W}$ is isomorphic to $\hat{V}/W^{\perp}$. Furthermore, if R is the restriction map of $\hat{V}$ onto $\hat{W}$, if η is the natural homomorphism of $\hat{V}$ onto $\hat{V}/W^{\perp}$, and τ is the unique isomorphism of $\hat{V}/W^{\perp}$ onto $\hat{W}$ characterized by the condition $R = \tau\eta$, then $\tau(\bar{\phi}) = R(\phi)$ where ϕ is any linear functional in the residue class $\bar{\phi} \in \hat{V}/W^{\perp}$.* $\square$

EXERCISES

1. (*a*) Find a basis for the annihilator of $W = \langle(1, 0, -1), (1, -1, 0), (0, 1, -1)\rangle$. (*b*) Find a basis for the annihilator of $W = \langle(1, 1, 1, 1, 1), (1, 0, 1, 0, 1), (0, 1, 1, 1, 0), (2, 0, 0, 1, 1), (2, 1, 1, 2, 1), (1, -1, -1, -2, 2), (1, 2, 3, 4, -1)\rangle$. What are the dimensions of W and $W^{\perp}$?

2. Find a non-zero linear functional which takes on the same non-zero value for $\xi_1 = (1, 2, 3)$, $\xi_2 = (2, 1, 1)$, and $\xi_3 = (1, 0, 1)$.

3. Use an argument based on the dimension of the annihilator to show that if $\alpha \neq 0$, there is a $\phi \in \hat{V}$ such that $\phi\alpha \neq 0$.

4. Show that if $S \subset T$, then $S^{\perp} \supset T^{\perp}$.

5. Show that $\langle S \rangle = S^{\perp\perp}$.

6. Show that if S and T are subsets of V each containing 0, then

$$(S + T)^{\perp} \subset S^{\perp} \cap T^{\perp},$$

and

$$S^{\perp} + T^{\perp} \subset (S \cap T)^{\perp}.$$

7. Show that if S and T are subspaces of V, then

$$(S + T)^{\perp} = S^{\perp} \cap T^{\perp},$$

and

$$S^{\perp} + T^{\perp} = (S \cap T)^{\perp}.$$

8. Show that if S and T are subspaces of V such that the sum $S + T$ is direct, then $S^{\perp} + T^{\perp} = \hat{V}$.

9. Show that if S and T are subspaces of V such that $S + T = V$, then $S^{\perp} \cap T^{\perp} = \{0\}$.

10. Show that if S and T are subspaces of V such that $S \oplus T = V$, then $\hat{V} = S^{\perp} \oplus T^{\perp}$. Show that $S^{\perp}$ is isomorphic to $\hat{T}$ and that $T^{\perp}$ is isomorphic to $\hat{S}$.

11. Let V be vector space over the real numbers, and let ϕ be a non-zero linear functional on V. We refer to the subspace S of V annihilated by ϕ as a *hyperplane* of V. Let $S^{+} = \{\alpha \mid \phi(\alpha) > 0\}$, and $S^{-} = \{\alpha \mid \phi(\alpha) < 0\}$. We call S^{+} and S^{-} the two

sides of the hyperplane S. If α and β are two vectors, the line segment joining α and β is defined to be the set $\{t\alpha + (1 - t)\beta \mid 0 \le t \le 1\}$, which we denote by $\overline{\alpha\beta}$. Show that if α and β are both in the same side of S, then each vector in $\overline{\alpha\beta}$ is also in the same side. And show that if α and β are in opposite sides of S, then $\overline{\alpha\beta}$ contains a vector in S.

5 | The Dual of a Linear Transformation

Let U and V be vector spaces and let σ be a linear transformation mapping U into V. Let $\hat{V}$ be the dual space of V and let ϕ be a linear functional on V. For each $\alpha \in U$, $\sigma(\alpha) \in V$ so that ϕ can be applied to $\sigma(\alpha)$. Thus $\phi[\sigma(\alpha)] \in F$ and $\phi\sigma$ can be considered to be a mapping which maps U into F. For $\alpha, \beta \in U$ and $a, b \in F$ we have $\phi[\sigma(a\alpha + b\beta)] = \phi[a\sigma(\alpha) + b\sigma(\beta)] = a\phi\sigma(\alpha) + b\phi\sigma(\beta)$ so that we have shown

Theorem 5.1. *For σ a linear transformation of U into V, and $\phi \in \hat{V}$, the mapping $\phi\sigma$ defined by $\phi[\sigma(\alpha)] = \phi\sigma(\alpha)$ is a linear functional on U; that is, $\phi\sigma \in \hat{U}$.* $\square$

Theorem 5.2. *For a given linear transformation σ mapping U into V, the mapping of $\hat{V}$ into $\hat{U}$ defined by making $\phi \in \hat{V}$ correspond to $\phi\sigma \in \hat{U}$ is a linear transformation of $\hat{V}$ into $\hat{U}$.*

PROOF. For $\phi_1, \phi_2 \in \hat{V}$ and $a, b \in F$, $(a\phi_1 + b\phi_2)\sigma(\alpha) = a\phi_1\sigma(\alpha) + b\phi_2\sigma(\alpha)$ for all $\alpha \in U$ so that $a\phi_1 + b\phi_2$ in $\hat{V}$ is mapped onto $a\phi_1\sigma + b\phi_2\sigma \in \hat{U}$ and the mapping defined is linear. $\square$

Definition. The mapping of $\phi \in \hat{V}$ onto $\phi\sigma \in \hat{U}$ is called the *dual* of σ and is denoted by $\hat{\sigma}$. Thus $\hat{\sigma}(\phi) = \phi\sigma$.

Let A be the matrix representing σ with respect to the bases A in U and B in V. Let $\hat{A}$ and $\hat{B}$ be the dual bases in $\hat{U}$ and $\hat{V}$, respectively. The question now arises: "How is the matrix representing $\hat{\sigma}$ with respect to the bases $\hat{B}$ and $\hat{A}$ related to the matrix representing σ with respect to the bases A and B?"

For $A = \{\alpha_1, \ldots, \alpha_m\}$ and $B = \{\beta_1, \ldots, \beta_n\}$ we have $\sigma(\alpha_j) = \sum_{i=1}^{n} a_{ij}\beta_i$. Let $\{\phi_1, \ldots, \phi_m\}$ be the basis of $\hat{U}$ dual to A and let $\{\psi_1, \ldots, \psi_n\}$ be the basis of $\hat{V}$ dual to B. Then for $\psi_i \in \hat{V}$ we have

$$[\hat{\sigma}(\psi_i)](\alpha_j) = (\psi_i\sigma)(\alpha_j) = \psi_i\sigma(\alpha_j)$$

$$= \psi_i\left(\sum_{k=1}^{n} a_{kj}\beta_k\right)$$

$$= \sum_{k=1}^{n} a_{kj}\psi_i\beta_k$$

$$= a_{ij}. \tag{5.1}$$

The linear functional on U which has the effect $[\hat{\sigma}(\psi_i)](\alpha_j) = a_{ij}$ is $\hat{\sigma}(\psi_i) = \sum_{k=1}^{m} a_{ik}\phi_k$. If $\psi = \sum_{i=1}^{n} b_i\psi_i$, then $\hat{\sigma}(\psi) = \sum_{i=1}^{n} b_i(\sum_{k=1}^{m} a_{ik}\phi_k) = \sum_{k=1}^{m} (\sum_{i=1}^{n} b_i a_{ik})\phi_k$. Thus the representation of $\hat{\sigma}(\psi)$ is BA. To follow absolutely the notational conventions for representing a liner transformation as given in Chapter II, (2.2), $\hat{\sigma}$ should be represented by A^T. However, because we have chosen to represent ψ by the row matrix B, and because $\hat{\sigma}(\psi)$ is represented by BA, we also use A to represent $\hat{\sigma}$. We say that A represents $\hat{\sigma}$ with respect to $\hat{B}$ in $\hat{V}$ and $\hat{A}$ in $\hat{U}$.

In most texts the convention to represent $\hat{\sigma}$ by A^T is chosen. The reason we have chosen to represent $\hat{\sigma}$ by A in this: in Chapter V we define a closely related linear transformation σ^*, the adjoint of σ. The adjoint is not represented by A^T; it is represented by $A^* = \bar{A}^T$, the conjugate complex of the transpose. If we chose to represent $\hat{\sigma}$ by A^T, we would have σ represented by A, $\hat{\sigma}$ by A^T in both the real and complex case, and σ^* represented by A^T in the real case and $\bar{A}^T$ in the complex case. Thus, the fact that the adjoint is represented by A^T in the real case does not, in itself, provide a compelling reason for representing the dual by A^T. There seems to be less confusion if both σ and $\hat{\sigma}$ are represented by A, and σ^* is represented by A^* (which reduces to A^T in the real case). In a number of other respects our choice results in simplified notation.

If $\xi \in U$, then $\psi(\sigma(\xi)) = \hat{\sigma}(\psi)(\xi)$, by definition of $\hat{\sigma}(\psi)$. If ξ is represented by X, then $\psi(\sigma(\xi)) = B(AX) = (BA)X = \hat{\sigma}(\psi)(\xi)$. Thus the representation convention we are using allows us to interpret taking the dual of a linear transformation as equivalent to the associative law. The interpretation could be made to look better if we considered σ as a left operator on U and a right operator on V. In other words, write $\sigma(\xi)$ as $\sigma\xi$ and $\hat{\sigma}(\psi)$ as $\psi\sigma$. Then $\psi(\sigma\xi) = (\psi\sigma)\xi$ would correspond to passing to the dual.

Theorem 5.3.　$K(\hat{\sigma})^\perp = \text{Im}(\sigma)$.

PROOF.　If $\psi \in K(\hat{\sigma}) \subset \hat{V}$, then for all $\alpha \in U$, $\psi(\sigma)\alpha)) = \hat{\sigma}(\psi)(\alpha) = 0$. Thus $\psi \in \text{Im}(\sigma)^\perp$. If $\psi \in \text{Im}(\sigma)^\perp$, then for all $\alpha \in U$, $\hat{\sigma}(\psi)(\alpha) = \psi(\sigma(\alpha)) = 0$. Thus $\psi \in K(\hat{\sigma})$ and $K(\hat{\sigma}) = \text{Im}(\sigma)^\perp$. $\square$

Corollary 5.4.　*A necessary and sufficient condition for the solvability of the linear problem $\sigma(\xi) = \beta$ is that $\beta \in K(\hat{\sigma})^\perp$.* $\square$

The ideas of this section provide a simple way of proving a very useful theorem concerning the solvability of systems of linear equations. The theorem we prove, worded in terms of linear functionals and duals, may not at first appear to have much to do with with linear equations. But, when worded in terms of matrices, it is identical to Theorem 7.2 of Chapter II.

Theorem 5.5.   *Let σ be a linear transformation of U into V and let β be any vector in V.   Either there is a $\xi \in U$ such that*

(1) $\sigma(\xi) = \beta$,

or there is a $\phi \in \hat{V}$ such that

(2) $\hat{\sigma}(\phi) = 0$ *and* $\phi\beta = 1$.

PROOF.   Condition (1) means that $\beta \in \text{Im}(r)$ and condition (2) means that $\beta \notin K(\hat{\sigma})^{\perp}$.   Thus the assertion of the Theorem follows directly from Theorem 5.3. $\square$

Theorem 5.5 is also equivalent to Theorem 7.2 of Chapter 2.

In matrix notation Theorem 5.5 reads: Let A be an $m \times n$ matrix and B an $m \times 1$ matrix.   Either there is an $n \times 1$ matrix X such that

(1) $AX = B$,

or there is a $1 \times m$ matrix C such that

(2) $CA = 0$ and $CB = 1$.

Theorem 5.6.   *σ and $\hat{\sigma}$ have the same rank.*

PROOF.   By Theorems 5.3 and 4.1, $v(\hat{\sigma}) = n - \rho(\sigma) = v(\sigma)$. $\square$

Theorem 5.7.   *Let W be a subspace of V invariant under σ.   Then $W^{\perp}$ is a subspace of $\hat{V}$ invariant under $\hat{\sigma}$.*

PROOF.   Let $\phi \in W^{\perp}$.   For any $\alpha \in W$ we have $\hat{\sigma}\phi(\alpha) = \phi\sigma(\alpha) = 0$, since $\sigma(\alpha) \in W$.   Thus $\hat{\sigma}\phi \in W^{\perp}$. $\square$

Theorem 5.8.   *The dual of a scalar transformation is also a scalar transformation generated by the same scalar.*

PROOF.   If $\sigma(\alpha) = a\alpha$ for all $\alpha \in V$, then for each $\phi \in \hat{V}$, $(\hat{\sigma}\phi)(\alpha) = \phi\sigma(\alpha) = \phi a\alpha = a\phi\alpha$. $\square$

Theorem 5.9.   *If λ is an eigenvalue for σ, then λ is also an eigenvalue for $\hat{\sigma}$.*

PROOF.   If λ is an eigenvalue for σ, then $\sigma - \lambda$ is singular.   The dual of $\sigma - \lambda$ is $\hat{\sigma} - \lambda$ and it must also be singular by Theorem 5.6.   Thus λ is an eigenvalue of $\hat{\sigma}$. $\square$

Theorem 5.10.   *Let V have a basis consisting of eigenvectors of σ.   Then $\hat{V}$ has a basis consisting of eigenvectors of $\hat{\sigma}$.*

PROOF.   Let $\{\alpha_1, \alpha_2, \ldots, \alpha_n\}$ be a basis of V, and assume that α_i is an eigenvector of σ with eigenvalue λ_i.   Let $\{\phi_1, \phi_2, \ldots, \phi_n\}$ be the corresponding dual basis.   For all α_j, $\hat{\sigma}\phi_i(\alpha_j) = \phi_i\sigma(\alpha_j) = \phi_i\lambda_j\alpha_j = \lambda_j\phi_i\alpha_j = \lambda_j\delta_{ij} = \lambda_i\delta_{ij}$. Thus $\hat{\sigma}\phi_i = \lambda_i\phi_i$ and ϕ_i is an eigenvector of $\hat{\sigma}$ with eigenvalue λ_i. $\square$

EXERCISES

1. Show that $\widehat{\sigma\tau} = \hat{\tau}\hat{\sigma}$.

2. Let σ be a linear transformation of R^2 into R^3 represented by

$$A = \begin{bmatrix} 1 & -1 \\ 2 & -4 \\ 2 & 2 \end{bmatrix}.$$

Find a basis for $(\sigma(R^2))^\perp$. Find a linear functional that does not annihilate $(1, 2, 1)$. Show that $(1, 2, 1) \notin \sigma(R^2)$.

3. The following system of linear equations has no solution. Find the linear functional whose existence is asserted in Theorem 5.5.

$$3x_1 + x_2 = 2$$
$$x_1 + 2x_2 = 1$$
$$-x_1 + 3x_2 = 1.$$

*6 | Duality of Linear Transformations

In Section 5 we have defined the dual of a linear transformation. What is the dual of the dual? In considering this question we restrict our attention to finite dimensional vector spaces. In this case, the mapping J of V into $\hat{\hat{V}}$, defined in Section 2, is an isomorphism. Since $\hat{\hat{\sigma}}$, the dual of $\hat{\sigma}$, is a mapping of $\hat{\hat{V}}$ into itself, the isomorphism J allows us to define a corresponding linear transformation on V. For convenience, we also denote this linear transformation by $\hat{\hat{\sigma}}$. Thus,

$$\hat{\hat{\sigma}}(\alpha) = J^{-1}[\hat{\hat{\sigma}}(J(\alpha))]. \tag{6.1}$$

where the $\hat{\hat{\sigma}}$ on the left is the mapping of V into itself defined by the expression on the right.

Theorem 6.1. *The relation between σ and $\hat{\sigma}$ is symmetric; that is, σ is the dual of $\hat{\sigma}$.*

PROOF. By definition,

$$\hat{\hat{\sigma}}(J(\alpha))(\phi) = J(\alpha)\hat{\sigma}(\phi) = \hat{\sigma}(\phi)(\alpha) = \phi\sigma(\alpha) = J(\sigma(\alpha))(\phi).$$

Thus $\hat{\hat{\sigma}}(J(\alpha)) = J(\sigma(\alpha))$. By (6.1) this means $\hat{\hat{\sigma}}(\alpha) = J^{-1}[\hat{\hat{\sigma}}(J(\alpha))] = J^{-1}[J(\sigma(\alpha))] = \sigma(\alpha)$. Hence, σ is the dual of $\hat{\sigma}$. $\square$

The reciprocal nature of duality allows us to establish dual forms of theorems without a new proof. For example, the dual form of Theorem 5.3 asserts that $K(\sigma)^\perp = \text{Im}(\hat{\sigma})$. We exploit this principle systematically in this section.

Theorem 6.2. *The dual of a monomorphism is an epimorphism. The dual of an epimorphism is a monomorphism.*

PROOF. By Theorem 5.3, $\text{Im}(\sigma) = K(\hat{\sigma})^{\perp}$. If σ is an epimorphism, $\text{Im}(\sigma) = V$ so that $K(\hat{\sigma}) = V^{\perp} = \{0\}$. Dually, $\text{Im}(\hat{\sigma}) = K(\sigma)^{\perp}$. If σ is a monomorphism, $K(\sigma) = \{0\}$ and $\text{Im}(\hat{\sigma}) = U$. $\square$

ALTERNATE PROOF. By Theorem 1.15 and 1.16 of Chapter II, σ is an epimorphism if and only if $\tau\sigma = 0$ implies $\tau = 0$. Thus $\hat{\sigma}\hat{\tau} = 0$ implies $\hat{\tau} = 0$ if and only if σ is an epimorphism. Thus σ is an epimorphism if and only if $\hat{\sigma}$ is a monomorphism. Dually, τ is a monomorphism if and only if $\hat{\tau}$ is an epimorphism. $\square$

Actually, a much more precise form of this theorem can be established. If W is a subspace of V, the mapping ι of W into V that maps $\alpha \in W$ onto $\alpha \in V$ is called the *injection* of W into V.

Theorem 6.3. *Let W be a subspace of V and let ι be the injection mapping of W into V. Let R be the restriction map of $\hat{V}$ onto $\hat{W}$. Then ι and R are dual mappings.*

PROOF. Let $\phi \in \hat{V}$. For any $\alpha \in W$, $R(\phi)(\alpha) = \phi\iota(\alpha) = \hat{\iota}(\phi)(\alpha)$. Thus $R(\phi) = \hat{\iota}(\phi)$ for each ϕ. Hence, $R = \hat{\iota}$. $\square$

Theorem 6.4. *If π is a projection of U onto S along T, the dual $\hat{\pi}$ is a projection of $\hat{U}$ onto $T^{\perp}$ along $S^{\perp}$.*

PROOF. A projection is characterized by the property $\pi^2 = \pi$. By Theorem 5.7, $\hat{\pi}^2 = \widehat{\pi^2} = \hat{\pi}$ so that $\hat{\pi}$ is also a projection. By Theorem 5.3, $K(\hat{\pi}) = \text{Im}(\pi)^{\perp} = S^{\perp}$ and $\text{Im}(\hat{\pi}) = K(\pi)^{\perp} = T^{\perp}$. $\square$

A careful comparison of Theorems 6.2 and 6.4 should reveal the perils of being careless about the domain and codomain of a linear transformation. A projection π of U onto the proper subspace S is not an epimorphism because the codomain of π is U, not S. Since $\hat{\pi}$ is a projection with the same rank as π, $\hat{\pi}$ cannot be a monomorphism, which it would be if π were an epimorphism.

Theorem 6.5. *Let σ be a linear transformation of U into V and let τ be a linear transformation of V into W. Let $\hat{\sigma}$ and $\hat{\tau}$ be the corresponding dual transformations. If $\text{Im}(\sigma) = K(\tau)$, then $\text{Im}(\hat{\tau}) = K(\hat{\sigma})$.*

PROOF. Since $\text{Im}(\sigma) \subset K(\tau)$, $\tau\sigma(\alpha) = 0$ for all $\alpha \in U$; that is, $\tau\sigma = 0$. Since $\hat{\sigma}\hat{\tau} = \widehat{\tau\sigma} = 0$, $\text{Im}(\hat{\tau}) \subset K(\hat{\sigma})$. Now dim $\text{Im}(\hat{\tau}) = $ dim $\text{Im}(\tau)$ since τ and $\hat{\tau}$ have the same rank. Thus dim $\text{Im}(\hat{\tau}) = $ dim $V - $ dim $K(\tau) = $ dim $V - $ dim $\text{Im}(\sigma) = $ dim $\hat{V} - $ dim $\text{Im}(\hat{\sigma}) = $ dim $K(\hat{\sigma})$. Thus $K(\hat{\sigma}) = \text{Im}(\hat{\tau})$. $\square$

Definition. Experience has shown that the condition $\text{Im}(\sigma) = K(\tau)$ is very useful because it is preserved under a variety of conditions, such as the taking of duals in Theorem 6.5. Accordingly, this property is given a special name. We say the sequence of mappings

$$U \xrightarrow{\sigma} V \xrightarrow{\tau} W \tag{6.1}$$

is *exact* at V if $\text{Im}(\sigma) = K(\tau)$. A sequence of mappings of any length is said to be exact if it is exact at every place where the above condition can apply. In these terms, Theorem 6.5 says that if the sequence (6.1) is exact at V, the sequence

$$\hat{U} \xleftarrow{\hat{\tau}} \hat{V} \xleftarrow{\hat{\sigma}} \hat{W} \tag{6.2}$$

is exact at $\hat{V}$. We say that (6.1) and (6.2) are dual sequences of mappings.

Consider the linear transformation σ of U into V. Associated with σ is the following sequence of mappings

$$0 \longrightarrow K(\sigma) \xrightarrow{\iota} U \xrightarrow{\sigma} V \xrightarrow{\eta} V/\text{Im}(\sigma) \longrightarrow 0, \tag{6.3}$$

where ι is the injection mapping of $K(\sigma)$ into U, and η is the natural homomorphism of V onto $V/\text{Im}(\sigma)$. The two mappings at the ends are the only ones they could be, zero mappings. It is easily seen that this sequence is exact.

Associated with $\hat{\sigma}$ is the exact sequence

$$0 \xleftarrow{} \hat{U}/\text{Im}(\hat{\sigma}) \xleftarrow{\eta} \hat{U} \xleftarrow{\hat{\sigma}} \hat{V} \xleftarrow{\iota} K(\hat{\sigma}) \xleftarrow{} 0. \tag{6.4}$$

By Theorem 6.3 the restriction map R is the dual of ι, and by Theorem 4.5 R an η differ by a natural isomorphism. With the understanding that $\hat{U}/\text{Im}(\hat{\sigma})$ is isomoprhic to $\widehat{K(\sigma)}$, and $V/\text{Im}(\sigma)$ is isomorphic to $\widehat{K(\hat{\sigma})}$, the sequences (6.3) and (6.4) are dual to each other.

*7 | Direct Sums

Definition. If A and B are any two sets, the set of pairs, (a, b), where $a \in A$ and $b \in B$, is called the *product set* of A and B, and is denoted by $A \times B$. If $\{A_i \mid i = 1, 2, \ldots, n\}$ is a finite indexed collection of sets, the product set of the $\{A_i\}$ is the set of all n-tuples, $(a_1, a_2, \ldots, a_n)$, where $a_i \in A_i$. This product set is denoted by $X_{i=1}^n A_i$. If the index set is not ordered, the description of the product set is a little more complicated. To see the appropriate generalization, notice that an n-tuple in $X_{i=1}^n A_i$, in effect, selects one element from each of the A_i. Generally, if $\{A_\mu \mid \mu \in M\}$ is an indexed collection of sets, an element of the product set $X_{\mu \in M} A_\mu$ selects for each index μ an element of A_μ. Thus, an element of $X_{\mu \in M} A_\mu$ is a function defined on M which associates with each $\mu \in M$ an element $a_\mu \in A_\mu$.

Let $\{V_i \mid i = 1, 2, \ldots, n\}$ be a collection of vector spaces, all defined over the same field of scalars F. With appropriate definitions of addition and scalar multiplication it is possible to make a vector space over F out of the product set $X_{i=1}^n V_i$. We define addition and scalar multiplication as follows:

$$(\alpha_1, \ldots, \alpha_n) + (\beta_1, \ldots, \beta_n) = (\alpha_1 + \beta_1, \ldots, \alpha_n + \beta_n) \qquad (7.1)$$

$$a(\alpha_1, \ldots, \alpha_n) = (a\alpha_1, \ldots, a\alpha_n). \qquad (7.2)$$

It is not difficult to show that the axioms of a vector space over F are satisfied, and we leave this to the reader.

Definition.   The vector space constructed from the product set $X_{i=1}^n V_i$ by the definitions given above is called the *external direct sum* of the V_i and is denoted by $V_1 \oplus V_2 \oplus \cdots \oplus V_n = \oplus_{i=1}^n V_i$.

If $D = \oplus_{i=1}^n V_i$ is the external direct sum of the V_i, the V_i are not subspaces of D (for $n > 1$). The elements of D are n-tuples of vectors while the elements of any V_i are vectors. For the direct sum defined in Chapter I, Section 4, the summand spaces were subspaces of the direct sum. If it is necessary to distinguish between these two direct sums, the direct sum defined in Chapter I will be called the *internal* direct sum.

Even though the V_i are not subspaces of D it is possible to map the V_i monomorphically into D in such a way that D is an internal direct sum of these images. Associate with $\alpha_k \in V_k$ the element $(0, \ldots, 0, \alpha_k, 0, \ldots, 0) \in D$, in which α_k appears in the kth position. Let us denote this mapping by ι_k. ι_k is a monomorphism of V_k into D, and it is called an *injection*. It provides an embedding of V_k in D. If $V_k' = \text{Im}(\iota_k)$ it is easily seen that D is an internal direct sum of the V_k'.

It should be emphasized that the embedding of V_k in D provided by the injection map ι_k is entirely arbitrary even though it looks quite natural. There are actually infinitely many ways to embed V_k in D. For example, let σ be any linear transformation of V_k into V_1 (we assume $k \neq 1$). Then define a new mapping ι_k' of V_k into D in which $\alpha_k \in V_k$ is mapped onto $(\sigma(\alpha_k), 0, \ldots, 0, \alpha_n, 0, \ldots, 0) \in D$. It is easily seen that ι_k' is also a monomorphism of V_k into D.

Theorem 7.1.   *If* $\dim U = m$ *and* $\dim V = n$, *then* $\dim U \oplus V = m + n$.
  PROOF.   Let $A = \{\alpha_1, \ldots, \alpha_m\}$ be a basis of U and let $B = \{\beta_1, \ldots, \beta_n\}$ be a basis of V. Then consider the set $\{(\alpha_1, 0), \ldots, (\alpha_m, 0), (0, \beta_1), \ldots, (0, \beta_n)\} = (A, B)$ in $U \oplus V$. If $\alpha = \sum_{i=1}^m a_i \alpha_i$ and $\beta = \sum_{j=1}^n b_j \beta_j$, then

$$(\alpha, \beta) = \sum_{i=1}^m a_i(\alpha_i, 0) + \sum_{j=1}^n b_j(0, \beta_j)$$

and hence (A, B) spans $U \oplus V$. If we have a relation of the form

$$\sum_{i=1}^m a_i(\alpha_i, 0) + \sum_{j=1}^n b_j(0, \beta_j) = 0,$$

then

$$\left(\sum_{i=1}^{m} a_i\alpha_i, \sum_{j=1}^{n} b_j\beta_j \right) = 0,$$

and hence $\sum_{i=1}^{m} a_i\alpha_i = 0$ and $\sum_{j=1}^{n} b_j\rho_j = 0$. Since A and B are linearly independent, all $a_i = 0$ and all $b_i = 0$. Thus (A, B) is a basis of $U \oplus V$ and $U \oplus V$ is of dimension $m + n$. $\square$

It is easily seen that the external direct sum $\oplus_{i=1}^{n} V_i$, where dim $V_i = m_i$, is of dimension $\sum_{i=1}^{n} m_i$.

We have already noted that we can consider the field F to be a 1-dimensional vector space over itself. With this starting point we can construct the external direct sum $F \oplus F$, which is easily seen to be equivalent to the 2-dimensional coordinate space F^2. Similarly, we can extend the external direct sum to include more summands, and consider F^n to be equivalent to $F \oplus \cdots \oplus F$, where this direct sum includes n summands.

We can define a mapping π_k of D onto V_k by the rule $\pi_k(\alpha_1, \ldots, \alpha_n) = \alpha_k$. π_k is called a *projection* of D onto the kth component. Actually, π_k is not a projection in the sense of the definition given in Section II-1, because here the domain and codomain of π_k are different and π_k^2 is not defined. However, $(\iota_k \pi_k)^2 = \iota_k \pi_k \iota_k \pi_k = \iota_k 1 \pi_k = \iota_k \pi_k$ so that $\iota_k \pi_k$ is a projection. Let W_k denote kernel of π_k. It is easily seen that

$$W_k = V_1 \oplus \cdots \oplus V_{k-1} \oplus \{0\} \oplus V_{k+1} \oplus \cdots \oplus V_n. \tag{7.3}$$

The injections and projections defined are related in simple but important ways. It is readily established that

$$\pi_k \iota_k = 1_{V_k}, \tag{7.4}$$

$$\pi_i \iota_k = 0 \quad \text{for} \quad i \neq k, \tag{7.5}$$

$$\iota_1 \pi_1 + \cdots + \iota_n \pi_n = 1_D. \tag{7.6}$$

The mappings $\iota_k \pi_i$ for $i \neq k$ are not defined since the domain of ι_k does not include the codomain of π_i.

Conversely, the relation (7.4), (7.5), and (7.6), are sufficient to define the direct sum. Starting with the V_k, the monomorphisms ι_k embed the V_k in D. Let $V'_k = \text{Im}(\iota_k)$. Let $D' = V'_1 + \cdots + V'_n$. Conditions (7.4) and (7.5) imply that D' is a direct sum of the V'_k. For if $0 = \alpha'_1 + \cdots + \alpha'_n$, with $\alpha'_k \in V'_k$, there exist $\alpha_k \in V_k$ such that $\iota_k(\alpha_k) = \alpha'_k$. Then $\pi_k(0) = \pi_k(\alpha'_1) + \cdots + \pi_k(\alpha'_n) = \pi_k \iota_1(\alpha_1) + \cdots + \pi_k \iota_n(\alpha_n) = \alpha_k = 0$. Thus $\alpha'_k = 0$ and the sum is direct. Condition (7.6) implies that $D' = D$.

Theorem 7.2. *The dual space of $U \oplus V$ is naturally isomorphic to $\hat{U} \oplus \hat{V}$.*

PROOF. First of all, if dim $U = m$ and dim $V = n$, then dim $\widehat{U \oplus V} = m + n$ and dim $\hat{U} \oplus \hat{V} = m + n$. Since $\widehat{U \oplus V}$ and $\hat{U} \oplus \hat{V}$ have the same dimension, there exists an isomorphism between them. The real content of this theorem, however, is that this isomorphism can be specified in a natural way independent of any coordinate system.

For $(\phi, \psi) \in \hat{U} \oplus \hat{V}$ and $(\alpha, \beta) \in U \oplus V$, define

$$(\phi, \psi)(\alpha, \beta) = \phi\alpha + \psi\beta. \tag{7.7}$$

It is easy to verify that this mapping of $(\alpha, \beta) \in U \oplus V$ onto $\phi\alpha + \psi\beta \in F$ is linear and, therefore, corresponds to a linear functional, an element of $\widehat{U \oplus V}$. It is also easy to verify that the mapping of $\hat{U} \oplus \hat{V}$ into $\widehat{U \oplus V}$ that this defines is a linear mapping. Finally, if (ϕ, ψ) corresponds to the zero linear functional, then $(\phi, \psi)(\alpha, 0) = \phi\alpha = 0$ for all $\alpha \in U$. This implies that $\phi = 0$. In a similar way we can conclude that $\psi = 0$. This shows that the mapping of $\hat{U} \oplus \hat{V}$ into $\widehat{U \oplus V}$ has kernel $\{(0, 0)\}$. Thus the mapping is an isomorphism. $\square$

Corollary 7.3. *The dual space to $V_1 \oplus \cdots \oplus V_n$ is naturally isomorphic to $\hat{V}_1 \oplus \cdots \oplus \hat{V}_n$.* $\square$

The direct sum of an infinite number of spaces is somewhat more complicated. In this case an element of the product set $P = X_{\mu \in M} V_\mu$ is a function on the index set M. For $\alpha \in X_{\mu \in M} V_\mu$, let $\alpha_\mu = \alpha(\mu)$ denote the value of this function in V_μ. Then we can define $\alpha + \beta$ and $a\alpha$ (for $a \in F$) by the rules

$$(\alpha + \beta)(\mu) = \alpha_\mu + \beta_\mu, \tag{7.8}$$

$$(a\alpha)(\mu) = a\alpha_\mu. \tag{7.9}$$

It is easily seen that these definitions convert the product set into a vector space. As before, we can define injective mappings ι_μ of V_μ into P. However, P is not the direct sum of these image spaces because, in algebra, we permit sums of only finitely many summands.

Let D be the subset of P consisting of those functions that vanish on all but a finite number of elements of M. With the operations of vector addition and scalar multiplication defined in P, D is a subspace. Both D and P are useful concepts. To distinguish them we call D the *external direct sum* and P the *direct product*. These terms are not universal and the reader of any mathematical literature should be careful about the intended meaning of these or related terms. To indicate the summands in P and D, we will denote P by $X_{\mu \in M} V_\mu$ and D by $\bigoplus_{\mu \in M} V_\mu$.

In a certain sense, the external direct sum and the direct product are dual concepts. Let ι_μ denote the injection of V_μ into P and let π_μ denote the projection of P onto V_μ. It is easily seen that we have

$$\pi_\mu \iota_\mu = 1_{V_\mu},$$

and

$$\pi_\nu \iota_\mu = 0 \quad \text{for} \quad \nu \neq \mu.$$

These mappings also have meaning in reference to D. Though we use the same notation, π_μ requires a restriction of the domain and ι_μ requires a restriction of the codomain. For D the analog of (7.6) is correct,

$$\sum_{\mu \in M} \iota_\mu \pi_\mu = 1_D. \tag{7.6}'$$

Even though the left side of (7.6)' involves an infinite number of terms, when applied to an element $\alpha \in D$,

$$\left(\sum_{\mu \in M} \iota_\mu \pi_\mu \right)(\alpha) = \sum_{\mu \in M} (\iota_\mu \pi_\mu)(\alpha) = \sum_{\mu \in M} \iota_\mu \alpha_\mu = \alpha \tag{7.10}$$

involves only a finite number of terms. An analog of (7.6) for the direct product is not available.

Consider the diagram of mappings

$$V_\mu \xrightarrow{\iota_\mu} D \xrightarrow{\pi_\nu} V_\nu, \tag{7.11}$$

and consider the dual diagram

$$\hat{V}_\mu \xleftarrow{\hat{\iota}_\mu} \hat{D} \xleftarrow{\hat{\pi}_\nu} \hat{V}_\nu. \tag{7.12}$$

For $\nu \neq \mu$, $\pi_\nu \iota_\mu = 0$. Thus $\hat{\iota}_\mu \hat{\pi}_\nu = \widehat{\pi_\nu \iota_\mu} = 0$. For $\nu = \mu$, $\hat{\iota}_\mu \hat{\pi}_\mu = \widehat{\pi_\mu \iota_\mu} = \hat{1} = 1$. By Theorem 6.2, $\hat{\iota}_\mu$ is an epimorphism and $\hat{\pi}_\mu$ is a monomorphism. Thus $\hat{\pi}_\mu$ is an injection of $\hat{V}_\mu$ into $\hat{D}$, and $\hat{\iota}_\mu$ is a projection of $\hat{D}$ onto $\hat{V}_\mu$.

Theorem 7.4. *If D is the external direct sum of the indexed collection $\{V_\mu \mid \mu \in M\}$, $\hat{D}$ is isomorphic to the direct product of the indexed collection $\{\hat{V}_\mu \mid \mu \in M\}$.*

PROOF. Let $\phi \in \hat{D}$. For each $\mu \in M$, $\phi \iota_\mu$ is a linear functional defined on V_μ; that is, $\phi \iota_\mu$ corresponds to an element in $\hat{V}_\mu$. In this way we define a function on M which has at $\mu \in M$ the value $\phi \iota_\mu \in \hat{V}_\mu$. By definition, this is an element in $X_{\mu \in M} \hat{V}_\mu$. It is easy to check that this mapping of $\hat{D}$ into the direct product $X_{\mu \in M} \hat{V}_\mu$ is linear.

If $\phi \neq 0$, there is an $\alpha \in D$ such that $\phi\alpha \neq 0$. Since $\phi\alpha = \phi[(\sum_{\mu \in M} \iota_\mu \pi_\mu)(\alpha)] = \sum_{\mu \in M} \phi \iota_\mu \pi_\mu(\alpha) \neq 0$, there is a $\mu \in M$ such that $\phi \iota_\mu \pi_\mu(\alpha) \neq 0$. Since $\pi_\mu(\alpha) \in V_\mu$, $\phi \iota_\mu \neq 0$. Thus, the kernel of the mapping of $\hat{D}$ into $X_{\mu \in M} \hat{V}_\mu$ is zero.

Finally, we show that this mapping is an epimorphism. Let $\psi \in X_{\mu \in M} \hat{V}_\mu$. Let $\psi_\mu = \psi(\mu) \in \hat{V}_\mu$ be the value of ψ at μ. For $\alpha \in D$, define $\phi\alpha = \sum_{\mu \in M} \psi_\mu(\pi_\mu \alpha)$. This sum is defined since $\pi_\mu \alpha = 0$ for all but finitely many μ.

For $\alpha_\nu \in V_\nu$,

$$\phi\iota_\nu(\alpha_\nu) = \phi(\iota_\nu\alpha_\nu)$$

$$= \sum_{\mu \in M} \psi_\mu(\pi_\mu\iota_\nu\alpha_\nu)$$

$$= \psi_\nu(\alpha_\nu). \tag{7.13}$$

This shows that ψ is the image of ϕ. Hence, $\hat{D}$ and $X_{\mu \in M} \hat{V}_\mu$ are ismorphic. □

While Theorem 7.4 shows that the direct product $\hat{D}$ is the dual of the external direct sum D, the external direct sum is generally not the dual of the direct product. This conclusion follows from a fact (not proven in this book) that infinite dimensional vector spaces are not reflexive. However, there is more symmetry in this relationship than this negative assertion seems to indicate. This is brought out in the next two theorems.

Theorem 7.5. *Let $\{V_\mu \mid \mu \in M\}$ be an indexed collection of vector spaces over F and let $\{\sigma_\mu \mid \mu \in M\}$ be an indexed collection of linear transformations, where σ_μ has domain V_μ and codomain U for all μ. Then there is a unique linear transformation σ of $\oplus_{\mu \in M} V_\mu$ into U such that $\sigma_\mu = \sigma\iota_\mu$ for each μ.*

PROOF.   Define

$$\sigma = \sum_{\mu \in M} \sigma_\mu\pi_\mu. \tag{7.14}$$

For each $\alpha \in \oplus_{\mu \in M} V_\mu$, $\sigma(\alpha) = \sum_{\mu \in M} \sigma_\mu\pi_\mu(\alpha)$ is well defined since only a finite number of terms on the right are non-zero. Then, for $\alpha_\nu \in V_\nu$,

$$\sigma\iota_\nu(\sigma_\nu) = \sum_{\mu \in M} \sigma_\mu\pi_\mu(\iota_\nu\alpha_\nu)$$

$$= \sum_{\mu \in M} \sigma_\mu(\pi_\mu\iota_\nu)(\alpha_\nu)$$

$$= \sigma_\nu\alpha_\nu. \tag{7.15}$$

Thus $\sigma\iota_\nu = \sigma_\nu$.

If σ' is another linear transformation of $\oplus_{\mu \in M} V_\mu$ into U such that $\sigma_\mu = \sigma'\iota_\mu$, then

$$\sigma' = \sigma'1_D$$

$$= \sigma' \sum_{\mu \in M} \iota_\mu\pi_\mu$$

$$= \sum_{\mu \in M} \sigma'\iota_\mu\pi_\mu$$

$$= \sum_{\mu \in M} \sigma_\mu\pi_\mu$$

$$= \sigma.$$

Thus, the σ with the desired property is unique. □

Theorem 7.6. *Let* $\{V_\mu \mid \mu \in M\}$ *be an indexed collection of vector spaces over F and let* $\{\tau_\mu \mid \mu \in M\}$ *be an indexed collection of linear transformations where* τ_μ *has domain W and codomain* V_μ *for all* μ. *Then there is a linear transformation* τ *of W into* $X_{\mu \in M} V_\mu$ *such that* $\tau_\mu = \pi_\mu \tau$ *for each* μ.

PROOF. Let $\alpha \in W$ be given. Since $\tau(\alpha)$ is supposed to be in $X_{\mu \in M} V_\mu$, $\tau(\alpha)$ is a function on M which for $\mu \in M$ has a value in V_μ. Define

$$\tau(\alpha)(\mu) = \tau_\mu(\alpha). \qquad (7.16)$$

Then

$$\pi_\mu \tau(\alpha) = \tau(\alpha)(\mu) = \tau_\mu(\alpha), \qquad (7.17)$$

so that $\pi_\mu \tau = \tau_\mu$. $\square$

The distinction between the external direct sum and the direct product is that the external direct sum is too small to replace the direct product in Theorem 7.6. This replacement could be done only if the indexed collection of linear transformations were restricted so that for each $\alpha \in W$ only finitely many mappings have non-zero values $\tau_\mu(\alpha)$.

The properties of the external direct sum and the direct product established in Theorems 7.5 and 7.6 are known as "universal factoring" properties. In Theorem 7.5 we have shown that any collection of mappings of V_μ into a space U can be factored through D. In Theorem 7.6 we have shown that any collection of mappings of W into the V_μ can be factored through P. Theorems 7.7 and 7.8 show that D and P are the smallest spaces with these properties.

Theorem 7.7. *Let W be a vector space over F with an indexed collection of linear transformations* $\{\lambda_\mu \mid \mu \in M\}$ *where each* λ_μ *has domain* V_μ *and co-domain W. Suppose that, for any indexed collection of linear transformations* $\{\sigma_\mu \mid \mu \in M\}$ *with domain* V_μ *and codomain U, there exists a linear transformation* λ *of W into U such that* $\sigma_\mu = \lambda \lambda_\mu$. *Then there exists a monomorphism of D into W.*

PROOF. By assumption, there exists a linear transformation λ of W into D such that $\iota_\mu = \lambda \lambda_\mu$. By Theorem 7.5 there is a unique linear transformation σ of D into W such that $\lambda_\mu = \sigma \iota_\mu$. Then

$$1 = \sum_{\mu \in M} \iota_\mu \pi_\mu$$

$$= \sum_{\mu \in M} \lambda \lambda_\mu \pi_\mu$$

$$= \sum_{\mu \in M} \lambda \sigma \iota_\mu \pi_\mu$$

$$= \lambda \sigma \sum_{\mu \in M} \iota_\mu \pi_\mu$$

$$= \lambda \sigma. \qquad (7.18)$$

This means that σ is a monomorphism and λ is an epimorphism. $\square$

Theorem 7.8. *Let Y be a vector space over F with an indexed collection of linear transformations $\{\theta_\mu \mid \mu \in M\}$ where each θ_μ has domain Y and codomain V_μ. Suppose that, for any indexed collection of linear transformations $\{\tau_\mu \mid \mu \in M\}$ with domain W and codomain V_μ, there exists a linear transformation θ of W into Y such that $\tau_\mu = \theta_\mu \theta$. Then P is isomorphic to a subspace of Y.*

PROOF. With P in place of W and π_μ in place of τ_μ, the assumptions of the theorem say there is a linear transformation θ of P into Y such that $\pi_\mu = \theta_\mu \theta$ for each μ. By Theorem 7.6 there is a linear transformation τ of Y into P such that $\theta_\mu = \pi_\mu \tau$ for each μ. Then

$$\pi_\mu = \theta_\mu \theta = \pi_\mu \tau \theta.$$

Recall that $\alpha \in P$ is a function defined on M that has at $\mu \in M$ a value α_μ in V_μ. Thus α is uniquely defined by its values. For $\mu \in M$

$$\pi_\mu(\tau\theta(\alpha)) = \pi_\mu(\alpha) = \alpha_\mu.$$

Thus $\tau\theta(\alpha) = \alpha$ and $\tau\theta = 1_P$. This means that θ is a monomorphism and τ is an epimorphism and P is isomorphic to $\text{Im}(\theta)$. □

Theorem 7.9. *Suppose a space D' is given with an indexed collection of monomorphisms $\{\iota'_\mu \mid \mu \in M\}$ of V_μ into D' and an indexed collection of epimorphisms $\{\pi'_\mu \mid \mu \in M\}$ of D' onto V_μ such that*

$$\pi'_\mu \iota'_\mu = 1_{V_\mu}$$

$$\pi'_\nu \iota'_\mu = 0 \qquad \text{for} \qquad \nu \neq \mu,$$

$$\sum_{\mu \in M} \iota'_\mu \pi'_\mu = 1_{D'}.$$

Then D and D' are isomorphic.

This theorem says, in effect, that conditions (7.4), (7.5), and (7.6)′ characterize the external direct sum.

PROOF. For $\alpha \in D'$ let $\alpha_\mu = \pi'_\mu(\alpha)$. We wish to show first that for a given $\alpha \in D'$ only finitely many α_μ are non-zero. By (7.6)′ $\alpha = 1_{D'}(\alpha) = \sum_{\mu \in M} \iota'_\mu \pi'_\mu(\alpha) = \sum_{\mu \in M} \iota'_\mu \alpha_\mu$. Thus, only finitely many of the $\iota'_\mu \alpha_\mu$ are non-zero. Since ι'_μ is a monomorphism, only finitely many of the α_μ are non-zero.

Now suppose that $\{\sigma_\mu \mid \mu \in M\}$ is an indexed collection of linear transformations with domain V_μ and codomain U. Define $\lambda = \sum_{\mu \in M} \sigma_\mu \pi'_\mu$. For $\alpha \in D'$, $\lambda(\alpha) = \sum_{\mu \in M} \sigma_\mu \pi'_\mu(\alpha) = \sum_{\mu \in M} \sigma_\mu \alpha_\mu$ is defined in U since only finitely many α_μ are non-zero. Also, $\lambda \iota'_\mu = (\sum_{\nu \in M} \sigma_\nu \pi'_\nu)\iota'_\mu = \sigma_\mu$. Thus D' satisfies the conditions of W in Theorem 7.7.

Repeating the steps of the proof of Theorem 7.7, we have a monomorphism σ of D into D' and an epimorphism λ of D' onto D such that $1_D = \lambda\sigma$. But we also have

$$1_{D'} = \sum_{\mu \in M} \iota'_\mu \pi'_\mu$$

$$= \sum_{\mu \in M} \sigma \iota_\mu \pi'_\mu$$

$$= \sum_{\mu \in M} \sigma \lambda \iota'_\mu \pi'_\mu$$

$$= \sigma \lambda \sum \iota'_\mu \pi'_\mu$$

$$= \sigma \lambda.$$

Since σ is both a monomorphism and an epimorphism, D and D' are isomorphic. $\square$

The direct product cannot be characterized quite so neatly. Although the direct product has a collection of mappings satisfying (7.4) and (7.5), (7.6)′ is not satisfied for this collection if M is an infinite set. The universal factoring property established for direct products in Theorem 7.6 is independent of (7.4) and (7.5), since direct sums satisfy (7.4) and (7.5) but not the universal factoring property of Theorem 7.6. We can combine these three conditions and state the following theorem.

Theorem 7.10. *Let P′ be a vector space over F with an indexed collection of monomorphisms $\{\iota'_\mu \mid \mu \in M\}$ of V_μ into P′ and an indexed collection of epimorphisms $\{\pi'_\mu \mid \mu \in M\}$ of P′ onto V_μ such that*

$$\pi'_\mu \iota'_\mu = 1_{V_\mu}$$

$$\pi'_\nu \iota'_\mu = 0 \qquad \text{for} \qquad \nu \neq \mu$$

and such that if $\{\rho_\mu \mid \mu \in M\}$ is any indexed collection of linear transformations with domain W and codomain V_μ, there is a linear transformation ρ of W into P′ such that $\rho_\mu = \pi'_\mu \rho$ for each μ. If P′ is minimal with respect to these three properties, then P and P′ are isomorphic.

When we say that P′ is minimal with respect to these three properties we mean: Let P″ be a subspace of P′ and let π''_μ be the restriction of π'_μ to P″. If there exists an indexed collection of monomorphisms $\{\iota''_\mu \mid \mu \in M\}$ with domain V_μ and codomain P″ such that (7.4), (7.5) and the universal factoring properties are satisfied with ι''_μ in place of ι'_μ and π''_μ in place of π'_μ, then P″ = P′.

PROOF.   By Theorem 7.8, P is isomorphic to a subspace of P′. Let θ be the isomorphism and let $P'' = \text{Im}(\theta)$. With appropriate changes in notation (P′ in place of Y and π'_μ in place of θ_μ), the proof of Theorem 7.8 yields the relations

$$\pi_\mu = \pi'_\mu \theta,$$

$$\pi'_\mu = \pi_\mu \tau,$$

where τ is an epimorphism of P' onto P. Thus, if π''_μ is the restriction of π'_μ to P'', we have

$$\pi_\mu = \pi''_\mu \theta.$$

This shows that π''_μ is an epimorphism.

Now let $\iota''_\mu = \theta \iota_\mu$.

$$\pi''_\mu \iota''_\mu = \pi''_\mu \theta \iota_\mu = \pi_\mu \iota_\mu = 1_{V\mu},$$

and

$$\pi''_\nu \iota''_\mu = \pi''_\mu \theta \iota_\mu = \pi_\nu \iota_\mu = 0 \qquad \text{for} \qquad \nu \neq \mu.$$

Since P has the universal factoring property, let τ be a linear transformation of W into P such that $\rho_\mu = \pi_\mu \tau$ for each μ. Then

$$\rho_\mu = \pi_\mu \tau = \pi''_\mu \theta \tau = \pi''_\mu \tau''$$

for each μ, where $\tau'' = \theta \tau$. This shows that P'' has universal factoring property of Theorem 7.6. Since we have assumed P' is minimal, we have $P'' = P'$ so that P and P' are isomorphic. $\square$

8 | Bilinear Forms

Definition. Let U and V be two vector spaces with the same field of scalars F. Let f be a mapping of pairs of vectors, one from U and one from V, into the field of scalars such that $f(\alpha, \beta)$, where $\alpha \in U$ and $\beta \in V$, is a linear function of α and β separately. Thus,

$$\begin{aligned}
f(a_1\alpha_1 + a_2\alpha_2, b_1\beta_1 + b_2\beta_2) &= a_1 f(\alpha_1, b_1\beta_1 + b_2\beta_2) + a_2 f(\alpha_2, b_1\beta_1 + b_2\beta_2) \\
&= a_1 b_1 f(\alpha_1, \beta_1) + a_1 b_2 f(\alpha_1, \beta_2) \\
&\quad + a_2 b_1 f(\alpha_2, \beta_1) + a_2 b_2 f(\alpha_2, \beta_2).
\end{aligned} \qquad (8.1)$$

Such a mapping is called a *bilinear form*. In most cases we shall have $U = V$.

(1) Take $U = V = R_n$ and $F = R$. Let $A = \{\alpha_1, \ldots, \alpha_n\}$ be a basis in R_n. For $\xi = \sum_{i=1}^n x_i\alpha_i$ and $\eta = \sum_{i=1}^n y_i\alpha_i$ we may define $f(\xi, \eta) = \sum_{i=1}^n x_i y_i$. This is a bilinear form and it is known as the inner, or dot, product.

(2) We can take $F = R$ and $U = V =$ space of continuous real-valued functions on the interval $[0, 1]$. We may then define $f(\alpha, \beta) = \int_0^1 \alpha(x)\beta(x) \, dx$. This is an infinite dimensional form of an inner product. It is a bilinear form.

As usual, we proceed to define the matrices representing bilinear forms with respect to bases in U and V and to see how these matrices are transformed when the bases are changed.

Let $A = \{\alpha_1, \ldots, \alpha_m\}$ be a basis in U and let $B = \{\beta_1, \ldots, \beta_n\}$ be a basis in V. Then, for any $\alpha \in U$, $\beta \in V$, we have $\alpha = \sum_{i=1}^m x_i\alpha_i$ and $\beta = \sum_{j=1}^n y_j\beta_j$

where x_i, $y_j \in F$. Then

$$f(\alpha, \beta) = f\left(\sum_{i=1}^{m} x_i \alpha_i, \beta\right)$$

$$= \sum_{i=1}^{m} x_i f\left(\alpha_i, \sum_{j=1}^{n} y_j \beta_j\right)$$

$$= \sum_{i=1}^{m} x_i \left(\sum_{j=1}^{n} y_j f(\alpha_i, \beta_j)\right)$$

$$= \sum_{i=1}^{m} \sum_{j=1}^{n} x_i y_j f(\alpha_i, \beta_j). \tag{8.2}$$

Thus we see that the value of the bilinear form is known and determined for any $\alpha \in U$, $\beta \in V$, as soon as we specify the mn values $f(\alpha_i, \beta_j)$. Conversely, values can be assigned to $f(\alpha_i, \beta_j)$ in an arbitrary way and $f(\alpha, \beta)$ can be defined uniquely for all $\alpha \in U$, $\beta \in V$, because A and B are bases in U and V, respectively.

We denote $f(\alpha_i, \beta_j)$ by b_{ij} and define $B = [b_{ij}]$ to be the matrix representing the bilinear form with respect to the bases A and B. We can use the m-tuple $X = (x_1, \ldots, x_m)$ to represent α and the n-tuple $Y = (y_1, \ldots, y_n)$ to represent β. Then

$$f(\alpha, \beta) = \sum_{i=1}^{m} \sum_{j=1}^{n} x_i b_{ij} y_j$$

$$= [x_1 \cdots x_m] B \begin{bmatrix} y_1 \\ \cdot \\ \cdot \\ \cdot \\ y_n \end{bmatrix}$$

$$= X^T B Y. \tag{8.3}$$

(Remember, our convention is to use an m-tuple $X = (x_1, \ldots, x_m)$ to represent an $m \times 1$ matrix. Thus X and Y are one-column matrices.)

Suppose, now, that $A' = \{\alpha_1', \ldots, \alpha_m'\}$ is a new basis of U with matrix of transition P, and that $B' = \{\beta_1', \ldots, \beta_n'\}$ is a new basis of V with matrix of transition Q. The matrix $B' = [b_{ij}']$ representing f with respect to these new bases is determined as follows:

$$b_{ij}' = f(\alpha_i', \beta_j') = f\left(\sum_{r=1}^{m} p_{ri} \alpha_r, \sum_{s=1}^{n} q_{sj} \beta_s\right)$$

$$= \sum_{r=1}^{m} p_{ri} \left[\sum_{s=1}^{n} q_{sj} f(\alpha_r, \beta_s)\right]$$

$$= \sum_{r=1}^{m} \sum_{s=1}^{n} p_{ri} b_{rs} q_{sj}. \tag{8.4}$$

Thus,

$$B' = P^T B Q. \tag{8.5}$$

From now on we assume that $U = V$. Then when we change from one basis to another, there is but one matrix of transition and $P = Q$ in the discussion above. Hence a change of basis leads to a new representation of f in the form

$$B' = P^T B P. \tag{8.6}$$

Definition. The matrices B and $P^T B P$, where P is non-singular, are said to be *congruent*.

Congruence is another equivalence relation among matrices. Notice that the particular kind of equivalence relation that is appropriate and meaningful depends on the underlying concept which the matrices are used to represent. Still other equivalence relations appear later. This occurs, for example, when we place restrictions on the types of bases we allow.

Definition. If $f(\alpha, \beta) = f(\beta, \alpha)$ for all $\alpha, \beta \in V$, we say that the bilinear form f is *symmetric*. Notice that for this definition to have meaning it is necessary that the bilinear form be defined on pairs of vectors from the same vector space, not from different vector spaces. If $f(\alpha, \alpha) = 0$ for all $\alpha \in V$, we say that the bilinear form f is *skew-symmetric*.

Theorem 8.1. *A bilinear form f is symmetric if and only if any matrix B representing f has the property $B^T = B$.*

PROOF. The matrix $B = [b_{ij}]$ is determined by $f(\alpha_i, \alpha_j)$. But $b_{ji} = f(\alpha_j, \alpha_i) = f(\alpha_i, \alpha_j) = b_{ij}$ so that $B^T = B$.

If $B^T = B$, we say the matrix B is *symmetric*. We shall soon see that symmetric bilinear forms and symmetric matrices are particularly important.

If $B^T = B$, then $f(\alpha_i, \alpha_j) = b_{ij} = b_{ji} = f(\alpha_j, \alpha_i)$. Thus $f(\alpha, \beta) = f(\sum_{i=1}^n a_i \alpha_i, \sum_{j=1}^n b_j \alpha_j) = \sum_{i=1}^n \sum_{j=1}^n a_i b_j f(\alpha_i, \alpha_j) = \sum_{i=1}^n \sum_{j=1}^n b_j a_i f(\alpha_j, \alpha_i) = f(\beta, \alpha)$. It then follows that any other matrix representing f will be symmetric; that is, if B is symmetric, then $P^T B P$ is also symmetric. $\square$ $\{$ non-sing

Theorem 8.2. *If a bilinear form f is skew-symmetric, then any matrix B representing f has the property $B^T = -B$.*

PROOF. For any $\alpha, \beta \in V$, $0 = f(\alpha + \beta, \alpha + \beta) = f(\alpha, \alpha) + f(\alpha, \beta) + f(\beta, \alpha) + f(\beta, \beta) = f(\alpha, \beta) + f(\beta, \alpha)$. From this it follows that $f(\alpha, \beta) = -f(\beta, \alpha)$ and hence $B^T = -B$. $\square$

Theorem 8.3. *If $1 + 1 \neq 0$ and the matrix B representing f has the property $B^T = -B$, then f is skew-symmetric.*

PROOF. Suppose that $B^T = -B$, or $f(\alpha, \beta) = -f(\beta, \alpha)$ for all $\alpha, \beta \in V$. Then $f(\alpha, \alpha) = -f(\alpha, \alpha)$, from which we have $f(\alpha, \alpha) + f(\alpha, \alpha) = (1 + 1)f(\alpha, \alpha) = 0$. Thus, if $1 + 1 \neq 0$, we can conclude that $f(\alpha, \alpha) = 0$ so that f is skew-symmetric. $\square$

If $B^T = -B$, we say the matrix B is *skew-symmetric*. The importance of symmetric and skew-symmetric bilinear forms is implicit in

Theorem 8.4. *If* $1 + 1 \neq 0$, *every bilinear form can be represented uniquely as a sum of a symmetric bilinear form and a skew-symmetric bilinear form.*

PROOF. Let f be the given bilinear form. Define $f_s(\alpha, \beta) = \frac{1}{2}[f(\alpha, \beta) + f(\beta, \alpha)]$ and $f_{ss}(\alpha, \beta) = \frac{1}{2}[f(\alpha, \beta) - f(\beta, \alpha)]$. (The assumption that $1 + 1 \neq 0$ is required to assure that the coefficient "$\frac{1}{2}$" has meaning.) It is clear that $f_s(\alpha, \beta) = f_s(\beta, \alpha)$ and $f_{ss}(\alpha, \alpha) = 0$ so that f_s is symmetric and f_{ss} is skew-symmetric.

We must yet show that this representation is unique. Thus, suppose that $f(\alpha, \beta) = f_1(\alpha, \beta) + f_2(\alpha, \beta)$ where f_1 is symmetric and f_2 is skew-symmetric. Then $f(\alpha, \beta) + f(\beta, \alpha) = f_1(\alpha, \beta) + f_2(\alpha, \beta) + f_1(\beta, \alpha) + f_2(\beta, \alpha) = 2f_1(\alpha, \beta)$. Hence $f_1(\alpha, \beta) = \frac{1}{2}[f(\alpha, \beta) + f(\beta, \alpha)]$. If follows immediately that $f_2(\alpha, \beta) = \frac{1}{2}[f(\alpha, \beta) - f(\beta, \alpha)]$. $\square$

We shall, for the rest of this book, assume that $1 + 1 \neq 0$ even where such an assumption is not explicitly mentioned.

EXERCISES

1. Let $\alpha = (x_1, x_2) \in R^2$ and let $\beta = (y_1, y_2, y_3) \in R^3$. Then consider the bilinear form

$$f(\alpha, \beta) = x_1y_1 + 2x_1y_2 - x_2y_1 - x_2y_2 + 6x_1y_3.$$

Determine the 2×3 matrix representing this bilinear form.

2. Express the matrix

$$\begin{bmatrix} 1 & 2 & 3 \\ 4 & 5 & 6 \\ 7 & 8 & 9 \end{bmatrix}$$

as the sum of a symmetric matrix and a skew-symmetric matrix.

3. Show that if B is symmetric, then $P^T BP$ is symmetric for each P, singular or non-singular. Show that if B is skew-symmetric, then $P^T BP$ is skew-symmetric for each P.

4. Show that if A is any $m \times n$ matrix, then $A^T A$ and AA^T are symmetric.

5. Show that a skew-symmetric matrix of odd order must be singular.

6. Let f be a bilinear form defined on U and V. Show that, for each $\alpha \in U$, $f(\alpha, \beta)$ defines a linear functional ϕ_α on V; that is,

$$\phi_\alpha(\beta) = f(\alpha, \beta).$$

With this fixed f show that the mapping of $\alpha \in U$ onto $\phi_\alpha \in \hat{V}$ is a linear transformation of U into $\hat{V}$.

7. (Continuation) Let the linear transformation of U into $\hat{V}$ defined in Exercise 6 be denoted by σ_f. Show that there is an $\alpha \in U$, $\alpha \neq 0$, such that $f(\alpha, \beta) = 0$ for all β if and only if the nullity of σ_f is positive.

8. (Continuation) Show that for each $\beta \in V$, $f(\alpha, \beta)$ defines a linear function ψ_β on U. The mapping of $\beta \in V$ onto $\psi_\beta \in \hat{U}$ is a linear transformation τ_f of V into $\hat{U}$.

9. (Continuation) Show that σ_f and τ_f have the same rank.

10. (Continuation) Show that, if U and V are of different dimensions, there must be either an $\alpha \in U$, $\alpha \neq 0$, such that $f(\alpha, \beta) = 0$ for all $\beta \in V$ or a $\beta \in V$, $\beta \neq 0$, such that $f(\alpha, \beta) = 0$ for all $\alpha \in U$. Show that the same conclusion follows if the matrix representing f is square but singular.

11. Let U_0 be the set of all $\alpha \in U$ such that $f(\alpha, \beta) = 0$ for all $\beta \in V$. Similarly, let V_0 be the set of all $\beta \in V$ such that $f(\alpha, \beta) = 0$ for all $\alpha \in U$. Show that U_0 is a subspace of U and that V_0 is a subspace of V.

12. (Continuation) Show that $m - \dim U_0 = n - \dim V_0$.

13. Show that if f is a skew-symmetric bilinear form, then $f(\alpha, \beta) = -f(\beta, \alpha)$ for all $\alpha, \beta \in V$.

14. Show by an example that, if A and B are symmetric, it is not necessarily true that AB is symmetric. What can be concluded if A and B are symmetric and $AB = BA$?

15. Under what conditions on B does it follow that $X^T B X = 0$ for all X?

16. Show the following: If A is skew-symmetric, then A^2 is symmetric. If A is skew-symmetric and B is symmetric, then $AB - BA$ is symmetric. If A is skew-symmetric and B is symmetric, then AB is skew-symmetric if and only if $AB = BA$.

9 | Quadratic Forms

Definition. A *quadratic form* is a function q on a vector space defined by setting $q(\alpha) = f(\alpha, \alpha)$, where f is a bilinear form on that vector space.

If f is represented as a sum of a symmetric and a skew-symmetric bilinear form, $f(\alpha, \beta) = f_s(\alpha, \beta) + f_{ss}(\alpha, \beta)$ where f_s is symmetric and f_{ss} is skew-symmetric, then $q(\alpha) = f_s(\alpha, \alpha) + f_{ss}(\alpha, \alpha) = f_s(\alpha, \alpha)$. Thus q is completely determined by the symmetric part of f alone. In addition, two different bilinear forms with the same symmetric part must generate the same quadratic form.

We see, therefore, that if a quadratic form is given we should not expect to be able to specify the bilinear form from which it is obtained. At best we can expect to specify the symmetric part of the underlying bilinear form. This symmetric part is itself a bilinear form from which q can be obtained. Each other possible underlying bilinear form will differ from this symmetric bilinear form by a skew-symmetric term.

What is the symmetric part of the underlying bilinear from expressed in terms of the given quadratic form? We can obtain a hint of what it should be by regarding the simple quadratic function x^2 as obtained from the bilinear function xy. Now $(x + y)^2 = x^2 + xy + yx + y^2$. Thus if $xy = yx$ (symmetry), we can express xy as a sum of squares, $xy = \frac{1}{2}[(x + y^2) - x^2 - y^2]$. In general, we see that the symmetric part of the underlying bilinear form can be recovered from the quadratic form by means of the formula

$$\frac{1}{2}[q(\alpha + \beta) - q(\alpha) - q(\beta)]$$
$$= \frac{1}{2}[f(\alpha + \beta, \alpha + \beta) - f(\alpha, \alpha) = f(\beta, \beta)]$$
$$= \frac{1}{2}[f(\alpha, \alpha) + f(\alpha, \beta) + f(\beta, \alpha) + f(\beta, \beta) - f(\alpha, \alpha) - f(\beta, \beta)]$$
$$= \frac{1}{2}[f(\alpha, \beta) + f(\beta, \alpha)]$$
$$= f_s(\alpha, \beta). \tag{9.1}$$

f_s is the symmetric part of f. Thus it is readily seen that

Theorem 9.1. *Every symmetric bilinear form f_s determines a unique quadratic form by the rule $q(\alpha) = f_s(\alpha, \alpha)$, and if $1 + 1 \neq 0$, every quadratic form determines a unique symmetric bilinear form $f_s(\alpha, \beta) = \frac{1}{2}[q(\alpha + \beta) - q(\alpha) - q(\beta)]$ from which it is in turn determined by the given rule. There is a one-to-one correspondence between symmetric bilinear forms and quadratic forms.* $\square$

The significance of Theorem 9.1 is that, to treat quadratic forms adequately, it is sufficient to consider symmetric bilinear forms. It is fortunate that symmetric bilinear forms and symmetric matrices are very easy to handle. Among many possible bilinear forms corresponding to a given quadratic form a symmetric bilinear form can always be selected. Hence, among many possible matrices that could be chosen to represent a given quadratic form, a symmetric matrix can always be selected.

The unique symmetric bilinear form f_s obtainable from a given quadratic form q is called the *polar form* of q.

It is desirable at this point to give a geometric interpretation of quadratic forms and their corresponding polar forms. This application of quadratic forms is by no means the most important, but it the source of much of the terminology. In a Euclidean plane with Cartesian coordinate system, let $(x) = (x_1, x_2)$ be the coordinates of a general point. Then

$$q((x)) = x_1{}^2 - 4x_1x_2 + 2x_2{}^2$$

is a quadratic function of the coordinates and it is a particular quadratic form. The set of all points (x) for which $q((x)) = 1$ is a conic section (in this case a hyperbola).

Now, let $(y) = (y_1, y_2)$ be the coordinates of another point. Then

$$f_s((x), (y)) = x_1y_1 - 2x_1y_2 - 2x_2y_1 + 2x_2y_2$$

is a function of both (x) and (y) and it is linear in the coordinates of each point separately. It is a bilinear form, the polar form of q. For a fixed (x), the set of all (y) for which $f_s((x), (y)) = 1$ is a straight line. This straight line is called the *polar* of (x) and (x) is called the *pole* of the straight line.

The relations between poles and polars are quite interesting and are explored in great depth in projective geometry. One of the simplest relations is that if (x) is on the conic section defined by $q((x)) = 1$, then the polar of (x) is tangent to the conic at (x). This is often shown in courses in analytic geometry and it is an elementary exercise in calculus.

We see that the matrix representing $f_s((x), (y))$, and therefore also $q((x))$, is

$$\begin{bmatrix} 1 & -2 \\ -2 & 2 \end{bmatrix}.$$

EXERCISES

1. Find the symmetric matrix representing each of the following quadratic forms:

    (a) $2x^2 + 3xy + 6y^2$
    (b) $8xy + 4y^2$
    (c) $x^2 + 2xy + 4xz + 3y^2 + yz + 7z^2$
    (d) $4xy$
    (e) $x^2 + 4xy + 4y^2 + 2xz + z^2 + 4yz$
    (f) $x^2 + 4xy - 2y^2$
    (g) $x^2 + 6xy - 2y^2 - 2yz + z^2$.

2. Write down the polar form for each of the quadratic forms of Exercise 1.

3. Show that the polar form f_s of the quadratic form q can be recovered from the quadratic form by the formula

$$f_s(\alpha, \beta) = \tfrac{1}{4}\{q(\alpha + \beta) - q(\alpha - \beta)\}.$$

10 | The Normal Form

Since the symmetry of the polar form f_s is independent of any coordinate system, the matrix representing f_s with respect to any coordinate system will be symmetric. The simplest of all symmetric matrices are those for which the elements not on the main diagonal are all zeros, the diagonal matrices. A great deal of the usefulness and importance of symmetric bilinear forms lies in the fact that for each symmetric bilinear form, over a field in which $1 + 1 \neq 0$, there exists a coordinate system in which the matrix representing the symmetric bilinear form is a diagonal matrix. Neither the coordinate system nor the diagonal matrix is unique.

Theorem 10.1. *For a given symmetric matrix B over a field F (in which $1 + 1 \neq 0$), there is a non-singular matrix P such that $P^T BP$ is a diagonal matrix. In other words, if f_s is the underlying symmetric bilinear (polar) form, there is a basis $A' = \{\alpha_1', \ldots, \alpha_n'\}$ of V such that $f_s(\alpha_i', \alpha_j') = 0$ whenever $i \neq j$.*

PROOF. The proof is by induction on n, the order of B. If $n = 1$, the theorem is obviously true (every 1×1 matrix is diagonal). Suppose the assertion of the theorem has already been established for a symmetric bilinear form in a space of dimension $n - 1$. If $B = 0$, then it is already diagonal. Thus we may as well assume that $B \neq 0$. Let f_s and q be the corresponding symmetric bilinear and quadratic forms. We have already shown that

$$f_s(\alpha, \beta) = \tfrac{1}{2}[q(\alpha + \beta) - q(\alpha) - q(\beta)]. \tag{10.1}$$

The significance of this equation at this point is that if $q(\alpha) = 0$ for all α, then $f_s(\alpha, \beta) = 0$ for all α and β. Hence, there is an $\alpha_1' \in V$ such that $q(\alpha_1') = d_1 \neq 0$.

With this α_1' held fixed, the bilinear form $f_s(\alpha_1', \alpha)$ defines a linear functional ϕ_1' on V. This linear functional is not zero since $\phi_1' \alpha_1' = d_1 \neq 0$. Thus the subspace W_1 annihilated by this linear functional is of dimension $n - 1$.

Consider f_s restricted to W_1. This is a symmetric bilinear form on W_1 and, by assumption, there is a basis $\{\alpha_2', \ldots, \alpha_n'\}$ of W_1 such that $f_s(\alpha_i', \alpha_j') = 0$ if $i \neq j$ and $2 \leq i, j \leq n$. However, $f_s(\alpha_i', \alpha_1') = f_s(\alpha_1', \alpha_i') = 0$ because of symmetry and the fact that $\alpha_i' \in W_1$ for $i \geq 2$. Thus $f_s(\alpha_i', \alpha_j') = 0$ if $i \neq j$ for $1 \leq i, j \leq n$. $\square$

Let P be the matrix of transition from the original basis $A = \{\alpha_1, \ldots, \alpha_n\}$ to the new basis $A' = \{\alpha_1', \ldots, \alpha_n'\}$. Then $P^T BP = B'$ is of the form

$$B' = \begin{bmatrix} d_1 & 0 & \cdots & 0 & \cdot & 0 \\ 0 & d_2 & \cdots & 0 & \cdot & 0 \\ \cdot & \cdot & & & \cdot & \cdot \\ \cdot & \cdot & \cdot & & \cdot & \cdot \\ \cdot & \cdot & & \cdot & & \cdot \\ 0 & 0 & \cdots & d_r & \cdot & 0 \\ \cdot & \cdot & \cdots & & \cdot & \cdot \\ 0 & 0 & \cdots & 0 & \cdot & 0 \end{bmatrix}.$$

In this display of B' the first r elements of the main diagonal are non-zero and all other elements of B' are zero. r is the rank of B' and B, and it is also called the *rank* of the corresponding bilinear or quadratic form.

The d_i's along the main diagonal are not uniquely determined. We can introduce a third basis $A'' = \{\alpha''_1, \ldots, \alpha''_n\}$ such that $\alpha''_i = x_i\alpha'_i$ where $x_i \neq 0$. Then the matrix of transition Q from the basis A' to the basis A'' is a diagonal matrix with $x_1, \ldots, x_n$ down the main diagonal. The matrix B'' representing the symmetric bilinear form with respect to the basis A'' is

$$B'' = Q^T B' Q = \begin{bmatrix} d_1 x_1{}^2 & 0 & \cdots & 0 & \cdot & 0 \\ 0 & d_2 x_2{}^2 & \cdots & 0 & \cdot & 0 \\ \cdot & & \cdot & & \cdot & \cdot \\ \cdot & & & \cdot & & \cdot \\ \cdot & & & & \cdot & \cdot \\ 0 & 0 & \cdots & d_r x_r{}^2 & \cdot & 0 \\ \cdot & & \cdots & & \cdot & \cdot \\ 0 & 0 & \cdots & 0 & \cdot & 0 \end{bmatrix}.$$

Thus the elements in the main diagonal may be multiplied by arbitrary non-zero squares from F.

By taking $B' = \begin{bmatrix} 1 & 0 \\ 0 & -1 \end{bmatrix}$ and $P = \begin{bmatrix} 2 & 1 \\ 1 & 2 \end{bmatrix}$ we get $B'' = P^T B' P = \begin{bmatrix} 3 & 0 \\ 0 & -3 \end{bmatrix}$. Thus, it is possible to change the elements in the main diagonal by factors which are not squares. However, $|B''| = |B'| \cdot |P|^2$ so that it is not possible to change just one element of the main diagonal by a non-square factor. The question of just what changes in the quadratic form can be effected by P with rational elements is a question which opens the door to the arithmetic theory of quadratic forms, a branch of number theory.

Little more can be said without knowledge of which numbers in the field of scalars can be squares. In the field of complex numbers every number is a square; that is, every complex number has at least one square root. Therefore, for each $d_i \neq 0$ we can choose $x_i = \dfrac{1}{\sqrt{d_i}}$ so that $d_i x_i{}^2 = 1$. In this case the non-zero numbers appearing in the main diagonal of B'' are all 1's. Thus we have proved

Theorem 10.2. *If F is the field of complex numbers, then every symmetric matrix B is congruent to a diagonal matrix in which all the non-zero elements are 1's. The number of 1's appearing in the main diagonal is equal to the rank of B.* □

The proof of Theorem 10.1 provides a thoroughly practical method for finding a non-singular P such that $P^T BP$ is a diagonal matrix. The first problem is to find an α'_1 such that $q(\alpha'_1) \neq 0$. The range of choices for such an α'_1 is generally so great that there is no difficulty in finding a suitable choice by trial and error. For the same reason, any systematic method for finding an α'_1 must be a matter of personal preference.

Among other possibilities, an efficient system for finding an α'_1 is the following: First try $\alpha'_1 = \alpha_1$. If $q(\alpha_1) = b_{11} = 0$, try $\alpha'_1 = \alpha_2$. If $q(\alpha_2) = b_{22} = 0$, then $q(\alpha_1 + \alpha_2) = q(\alpha_1) + 2f_s(\alpha_1, \alpha_2) + q(\alpha_2) = 2f_s(\alpha_1 \alpha_2) = 2b_{12}$ so that it is convenient to try $\alpha'_1 = \alpha_1 + \alpha_2$. The point of making this sequence of trials is that the outcome of each is determined by the value of a single element of B. If all three of these fail, then we can pass our attention to α_3, $\alpha_1 + \alpha_3$, and $\alpha_2 + \alpha_3$ with similar ease and proceed in this fashion.

Now, with the chosen α'_1, $f_s(\alpha'_1, \alpha)$ defines a linear functional ϕ'_1 on V. If α'_1 is represented by $(p_{11}, \ldots, p_{n1})$ and α by $(x_1, \ldots, x_n)$, then

$$f_s(\alpha'_1, \alpha) = \sum_{i=1}^{n} \sum_{j=1}^{n} p_{i1} b_{ij} x_j = \sum_{j=1}^{n} \left(\sum_{i=1}^{n} p_{i1} b_{ij} \right) x_j. \tag{10.2}$$

This means that the linear functional ϕ'_1 is represented by $[p_{11} \cdots p_{n1}]B$.

The next step described in the proof is to determine the subspace W_1 annihilated by ϕ'_1. However, it is not necessary to find all of W_1. It is sufficient to find an $\alpha'_2 \in W_1$ such that $q(\alpha'_2) \neq 0$. With this α'_2, $f_s(\alpha'_2, \alpha)$ defines a linear functional ϕ'_2 on V. If α'_2 is represented by $(p_{12}, \ldots, p_{n2})$, then ϕ'_2 is represented by $[p_{12} \cdots p_{n2}]B$.

The next subspace we need is the subspace W_2 of W_1 annihilated by ϕ'_2. Thus W_2 is the subspace annihilated by both ϕ'_1 and ϕ'_2. We then select an α'_2 from W_2 and proceed as before.

Let us illustrate the entire procedure with an example. Consider

$$B = \begin{bmatrix} 0 & 1 & 2 \\ 1 & 0 & 1 \\ 2 & 1 & 0 \end{bmatrix}.$$

Since $b_{11} = b_{22} = 0$, we take $\alpha'_1 = \alpha_1 + \alpha_2 = (1, 1, 0)$. Then the linear functional ϕ'_1 is represented by

$$[1 \quad 1 \quad 0]B = [1 \quad 1 \quad 3].$$

A possible choice for an α'_2 annihilated by this linear functional is $(1, -1, 0)$. The linear functional ϕ'_2 determined by $(1, -1, 0)$ is represented by

$$[1 \quad -1 \quad 0]B = [-1 \quad 1 \quad 1].$$

We should have checked to see that $q(\alpha'_2) \neq 0$, but it is easier to make that check after determining the linear functional ϕ'_2 since $q(\alpha'_2) = \phi'_2 \alpha'_2 = -2 \neq 0$ and the arithmetic of evaluating the quadratic form includes all the steps involved in determining ϕ'_2.

We must now find an α'_3 annihilated by ϕ'_1 and ϕ'_2. This amounts to solving the system of homogeneous linear equations represented by

$$\begin{bmatrix} 1 & 1 & 3 \\ -1 & 1 & 1 \end{bmatrix}.$$

A possible choice is $\alpha'_3 = (-1, -2, 1)$. The corresponding linear functional ϕ'_3 is represented by

$$[-1 \quad -2 \quad 1]B = [0 \quad 0 \quad -4].$$

The desired matrix of transition is

$$P = \begin{bmatrix} 1 & 1 & -1 \\ 1 & -1 & -2 \\ 0 & 0 & 1 \end{bmatrix}.$$

Since the linear functionals we have calculated along the way are the rows of $P^T B$, the calculation of $P^T BP$ is half completed. Thus,

$$P^T BP = \begin{bmatrix} 1 & 1 & 3 \\ -1 & 1 & 1 \\ 0 & 0 & -4 \end{bmatrix} \begin{bmatrix} 1 & 1 & -1 \\ 1 & -1 & -2 \\ 0 & 0 & 1 \end{bmatrix} = \begin{bmatrix} 2 & 0 & 0 \\ 0 & -2 & 0 \\ 0 & 0 & -4 \end{bmatrix}.$$

It is possible to modify the diagonal form by multiplying the elements in the main diagonal by squares from F. Thus, if F is the field of rational numbers we can obtain the diagonal $\{2, -2, -1\}$. If F is the field of real numbers we can get the diagonal $\{1, -1, -1\}$. If F is the field of complex numbers we can get the diagonal $\{1, 1, 1\}$.

Since the matrix of transition P is a product of elementary matrices the diagonal from $P^T BP$ can also be obtained by a sequence of elementary row and column operations, provided the sequence of column operations is exactly the same as the sequence of row operations. This method is commonly used to obtain the diagonal form under the congruence. If an element b_{ii} in the main diagonal is non-zero, it can be used to reduce all other elements in row i and column i to zero. If every element in the main diagonal is zero and $b_{ij} \neq 0$, then adding row j to row i and column j to column i will yield a matrix with $2b_{ij}$ in the ith place of the main diagonal. The method is a little fussy because the same row and column operations must be used, and in the same order.

Another good method for quadratic forms of low order is called *completing the square*. If $X^T BX = \sum_{i,j=1}^n x_i b_{ij} x_j$ and $b_{ii} \neq 0$, then

$$X^T BX - \frac{1}{b_{ii}} (b_{i1}x_1 + \cdots + b_{in}x_n)^2 \tag{10.3}$$

is a quadratic form in which x_i does not appear. Make the substitution

$$x_i' = b_{i1}x_1 + \cdots + b_{in}x_n. \tag{10.4}$$

Continue in this manner if possible. The steps must be modified if at any stage every element in the main diagonal is zero. If $b_{ij} \neq 0$, then the substitution $x_i' = x_i + x_j$ and $x_j' = x_i - x_j$ will yield a quadratic form represented by a matrix with $2b_{ij}$ in the ith place of the main diagonal and $-2b_{ij}$ in the jth place. Then we can proceed as before. In the end we will have

$$X^T B X = \frac{1}{b_{ii}}(x_i')^2 + \cdots \tag{10.5}$$

expressed as a sum of squares; that is, the quadratic form will be in diagonal form.

The method of elementary row and column operations and the method of completing the square have the advantage of being based on concepts much less sophisticated than the linear functional. However, the computational method based on the proof of the theorem is shorter, faster, and more compact. It has the additional advantage of giving the matrix of transition without special effort.

EXERCISES

1. Reduce each of the following symmetric matrices to diagonal form. Use the method of linear functionals, the method of elementary row and column operations, and the method of completing the square,

(a) $\begin{bmatrix} 1 & 2 & 2 \\ 2 & 1 & -2 \\ 2 & -2 & 1 \end{bmatrix}$
(b) $\begin{bmatrix} 1 & 2 & 3 \\ 2 & 0 & -1 \\ 3 & -1 & 1 \end{bmatrix}$

(c) $\begin{bmatrix} 0 & 1 & -1 & 2 \\ 1 & 1 & 0 & -1 \\ -1 & 0 & -1 & 1 \\ 2 & -1 & 1 & 0 \end{bmatrix}$
(d) $\begin{bmatrix} 0 & 1 & 2 & 3 \\ 1 & 0 & 1 & 2 \\ 2 & 1 & 0 & 1 \\ 3 & 2 & 1 & 0 \end{bmatrix}$

2. Using the methods of this section, reduce the quadratic forms of Exercise 1, Section 9, to diagonal form.

3. Each of the quadratic forms considered in Exercise 2 has integral coefficients. Obtain for each a diagonal form in which each coefficient in the main diagonal is a square-free integer.

11 | Real Quadratic Forms

A quadratic form over the complex numbers is not really very interesting. From Theorem 10.2 we see that two different quadratic forms would be distinguishable if and only if they had different ranks. Two quadratic forms of the same rank each have coordinate systems (very likely a different coordinate system for each) in which their representations are the same. Hence, any properties they might have which would be independent of the coordinate system would be indistinguishable.

In this section let us restrict our attention to quadratic forms over the field of real numbers. In this case, not every number is a square; for example, -1 is not a square. Therefore, having obtained a diagonalized representation of a quadratic form, we cannot effect a further transformation, as we did in the proof of Theorem 10.2 to obtain all 1's for the non-zero elements of the main diagonal. ·The best we can do is to change the positive elements to $+1$'s and the negative elements to -1's. There are many choices for a basis with respect to which the representation of the quadratic form has only $+1$'s and -1's along the main diagonal. We wish to show that the number of $+1$'s and the number of -1's are independent of the choice of the basis; that is, these numbers are basic properties of the underlying quadratic form and not peculiarities of the representing matrix.

Theorem 11.1. *Let q be a quadratic form over the real numbers. Let P be the number of positive terms in a diagonalized representation of q and let N be the number of negative terms. In any other diagonalized representation of q the number of positive terms is P and the number of negative terms is N.*

PROOF. Let $A = \{\alpha_1, \ldots, \alpha_n\}$ be a basis which yields a diagonalized representation of q with P positive terms and N negative terms in the main diagonal. Without loss of generality we can assume that the first P elements of the main diagonal are positive. Let $B = \{\beta_1, \ldots, \beta_n\}$ be another basis yielding a diagonalized representation of q with the first P' elements of the main diagonal positive.

Let $U = \langle \alpha_1, \ldots, \alpha_P \rangle$ and let $W = \langle \beta_{P'+1}, \ldots, \beta_n \rangle$. Because of the form of the representation using the basis A, for any *non-zero* $\alpha \in U$ we have $q(\alpha) > 0$. Similarly, for any $\beta \in W$ we have $q(\beta) \le 0$. This shows that $U \cap W = \{0\}$. Now dim $U = P$, dim $W = n - P'$, and dim $(U + W) \le n$. Thus $P + n - P' = $ dim $U + $ dim $W = $ dim $(U + W) + $ dim $(U \cap W) = $ dim $(U + W) \le n$. Hence, $P - P' \le 0$. In the same way it can be shown that $P' - P \le 0$. Thus $P = P'$ and $N = r - P = -P' = N'$. $\square$

Definition. The number $S = P - N$ is called the *signature* of the quadratic form q. Theorem 11.1 shows that S is well defined. A quadratic form is called *non-negative semi-definite* if $S = r$. It is called *positive definite* if $S = n$.

It is clear that a quadratic form is non-negative semi-definite if and only if $q(\alpha) \geq 0$ for all $\alpha \in V$. It is positive definite if and only if $q(\alpha) > 0$ for non-zero $\alpha \in V$. These are the properties of non-negative semi-definite and positive definite forms that make them of interest. We use them extensively in Chapter V.

If the field of constants is a subfield of the real numbers, but not the real numbers, we may not always be able to obtain $+1$'s and -1's along the main diagonal of a diagonalized representation of a quadratic form. However, the statement of Theorem 11.1 and its proof referred only to the diagonal terms as being positive or negative, not necessarily $+1$ or -1. Thus the theorem is equally valid in a subfield of the real numbers, and the definitions of the signature, non-negative semi-definiteness, and positive definiteness have meaning.

In calculus it is shown that

$$\int_{-\infty}^{\infty} e^{-x^2}\, dx = \pi^{1/2}.$$

It happens that analogous integrals of the form

$$I = \int_{-\infty}^{\infty} \cdots \int_{-\infty}^{\infty} e^{-\Sigma x_i a_{ij} x_j}\, dx_1 \cdots dx_n$$

appear in a number of applications. The term $\sum x_i a_{ij} x_j = X^T A X$ appearing in the exponent is a quadratic form, and we can assume it to be symmetric. In order that the integrals converge it is necessary and sufficient that the quadratic form be positive definite. There is a non-singular matrix P such that $P^T A P = L$ is a diagonal matrix. Let $\{\lambda_1, \ldots, \lambda_n\}$ be the main diagonal of L. If $X = (x_1, \ldots, x_n)$ are the old coordinates of a point, then $Y = (y_1, \ldots, y_n)$ are the new coordinates where $x_i = \sum_j p_{ij} y_j$. Since $\dfrac{\partial x_i}{\partial y_j} = p_{ij}$, the Jacobian of the coordinate transformation is $\det P$. Thus,

$$I = \int_{-\infty}^{\infty} \cdots \int_{-\infty}^{\infty} e^{-\Sigma \lambda_i y_i^2} \det P\, dy_1 \cdots dy_n$$

$$= \det P \int_{-\infty}^{\infty} e^{-\lambda_1 y_1^2}\, dy_1 \cdots \int_{-\infty}^{\infty} e^{-\lambda_n y_n^2}\, dy_n$$

$$= \det P \frac{\pi^{1/2}}{\lambda_1^{1/2}} \cdots \frac{\pi^{1/2}}{\lambda_n^{1/2}}$$

$$= \pi^{n/2} \frac{\det P}{(\lambda_1 \cdots \lambda_n)^{1/2}}.$$

Since $\lambda_1 \cdots \lambda_n = \det L = \det P \det A \det P = \det P^2 \det A$, we have

$$I = \frac{\pi^{n/2}}{\det A^{1/2}} \cdot$$

EXERCISES

1. Determine the rank and signature of each of the quadratic forms of Exercise 1, Section 9.

2. Show that the quadratic form $Q(x, y) = ax^2 + bxy + cy^2 (a, b, c$ real) is positive definite if and only if $a > 0$ and $b^2 - 4ac < 0$.

3. Show that if A is a real symmetric positive definite matrix, then there exists a real non-singular matrix P such that $A = P^T P$.

4. Show that if A is a real non-singular matrix, then $A^T A$ is positive definite.

5. Show that if A is a real symmetric non-negative semi-definite matrix—that is, A represents a non-negative semi-definite quadratic form—then there exists a real matrix R such that $A = R^T R$.

6. Show that if A is real, then $A^T A$ is non-negative semi-definite.

7. Show that if A is real and $A^T A = 0$, then $A = 0$.

8. Show that if A is real symmetric and $A^2 = 0$, then $A = 0$.

9. If $A_1, \ldots, A_r$ are real symmetric matrices, show that

$$A_1^2 + \cdots + A_r^2 = 0$$

implies $A_1 = A_2 = \cdots = A_r = 0$.

12 | Hermitian Forms

For the applications of forms to many problems, it turns out that a quadratic form obtained from a bilinear form over the complex numbers is not the most useful generalization of the concept of a quadratic form over the real numbers. As we see later, the property that a quadratic form over the real numbers be positive-definite is a very useful property. While x^2 is positive-definite for real x, it is not positive-definite for complex x. When dealing with complex numbers we need a function like $|x|^2 = \bar{x}x$, where $\bar{x}$ is the conjugate complex of x. $\bar{x}x$ is non-negative for all complex (and real) x, and it is zero only when $x = 0$. Thus $\bar{x}x$ is a form which has the property of being positive definite. In the spirit of these considerations, the following definition is appropriate.

Definition. Let F be the field of complex numbers, or a subfield of the complex numbers, and let V be a vector space over F. A scalar valued function f of two vectors, $\alpha, \beta \in V$ is called a *Hermitian form* if

(1) $\qquad\qquad \overline{f(\alpha, \beta)} = f(\beta, \alpha).$ $\qquad\qquad\qquad\qquad\qquad$ (12.1)

(2) $\qquad\qquad f(\alpha, b_1\beta_1 + b_2\beta_2) = b_1 f(\alpha, \beta_1) + b_2 f(\alpha, \beta_2).$

A Hermitian form differs from a symmetric bilinear form in the taking of the conjugate complex when the roles of the vectors α and β are interchanged. But the appearance of the conjugate complex also affects the bilinearity of the form. Namely,

$$
\begin{aligned}
f(a_1\alpha_1 + a_2\alpha_2, \beta) &= \overline{f(\beta, a_1\alpha_1 + a_2\alpha_2)} \\
&= \overline{a_1 f(\beta, \alpha_1) + a_2 f(\beta, \alpha_2)} \\
&= \overline{a_1 f(\beta, \alpha_1)} + \overline{a_2 f(\beta, \alpha_2)} \\
&= \bar{a}_1 f(\alpha_1, \beta) + \bar{a}_2 f(\alpha_2, \beta).
\end{aligned}
$$

We describe this situation by saying that a Hermitian form is linear in the second variable and *conjugate linear* in the first variable.

Accordingly, it is also convenient to define a more appropriate generalization to vector spaces over the complex numbers of the concept of a bilinear form on vector spaces over the real numbers. A function of two vectors on a vector space over the complex numbers is said to be *conjugate bilinear* if it is conjugate linear in the first variable and linear in the second. We say that a function of two vectors is *Hermitian symmetric* if $f(\alpha, \beta) = \overline{f(\beta, \alpha)}$. This is the most useful generalization to vector spaces over the complex numbers of the concept of symmetry for vector spaces over the real numbers. In this terminology a Hermitian form is a Hermitian symmetric conjugate bilinear form.

For a given Hermitian form f, we define $q(\alpha) = f(\alpha, \alpha)$ and obtain what we call a *Hermitian quadratic form*. In dealing with vector spaces over the field of complex numbers we almost never meet a quadratic form obtained from a bilinear form. The useful quadratic forms are the Hermitian quadratic forms.

Let $A = \{\alpha_1, \ldots, \alpha_n\}$ be any basis of V. Then we can let $f(\alpha_i, \alpha_j) = h_{ij}$ and obtain the matrix $H = [h_{ij}]$ representing the Hermitian form f with respect to A. H has the property that $h_{ij} = f(\alpha_i, \alpha_j) = \overline{f(\alpha_j, \alpha_i)} = \bar{h}_{ji}$, and any matrix which has this property can be used to define a Hermitian form. Any matrix with this property is called a *Hermitian matrix*.

If A is any matrix, we denote by $\bar{A}$ the matrix obtained by taking the conjugate complex of every element of A; that is, if $A = [a_{ij}]$ then $\bar{A} = [\bar{a}_{ij}]$. We denote $\bar{A}^T = \overline{A^T}$ by A^*. In this notation a matrix H is Hermitian if and only if $H^* = H$.

If a new basis $B = \{\beta_1, \ldots, \beta_n\}$ is selected, we obtain the representation

$H' = [h'_{ij}]$ where $h'_{ij} = f(\beta_i, \beta_j)$. Let P be the matrix of transition; that is, $\beta_j = \sum_{i=1}^n p_{ij}\alpha_i$. Then

$$h'_{ij} = f(\beta_i, \beta_j)$$

$$= f\left(\sum_{k=1}^n p_{ki}\alpha_k, \sum_{s=1}^n p_{sj}\alpha_s\right)$$

$$= \sum_{s=1}^n p_{sj} f\left(\sum_{k=1}^n p_{ki}\alpha_k, \alpha_s\right)$$

$$= \sum_{s=1}^n p_{sj} \sum_{k=1}^n \bar{p}_{ki} f(\alpha_k, \alpha_s)$$

$$= \sum_{s=1}^n \sum_{k=1}^n \bar{p}_{ki} h_{ks} p_{sj}. \tag{12.3}$$

In matrix form this equation becomes $H' = P^*HP$.

Definition. If a non-singular matrix P exists such that $H' = P^*HP$, we say that H and H' are *Hermitian congruent*.

Theorem 12.1. *For a given Hermitian matrix H there is a non-singular matrix P such that $H' = P^*HP$ is a diagonal matrix. In other words, if f is the underlying Hermitian form, there is basis* $A' = \{\alpha'_1, \ldots, \alpha'_n\}$ *such that* $f(\alpha'_i, \alpha'_j) = 0$ *whenever $i \neq j$.*

PROOF. The proof is almost identical with the proof of Theorem 10.1, the corresponding theorem for bilinear forms. There is but one place where a modification must be made. In the proof of Theorem 10.1 we made use of a formula for recovering the symmetric part of a bilinear form from the associated quadratic form. For Hermitian forms the corresponding formula is

$$\tfrac{1}{4}[q(\alpha + \beta) - q(\alpha - \beta) - iq(\alpha + i\beta) + iq(\alpha - i\beta)] = f(\alpha, \beta). \tag{12.4}$$

Hence, if f is not identically zero, there is an $\alpha_1 \in V$ such that $q(\alpha_1) \neq 0$. The rest of the proof of Theorem 10.1 then applies without change. $\square$

Again, the elements of the diagonal matrix thus obtained are not unique. We can transform H' into still another diagonal matrix by means of a diagonal matrix Q with $x_1, \ldots, x_n, x_i \neq 0$, along the main diagonal. In this fashion we obtain

$$H'' = Q^*H'Q = \begin{bmatrix} d_1|x_1|^2 & 0 & \cdots & 0 & \cdot & 0 \\ \cdot & d_2|x_2|^2 & \cdots & 0 & \cdot & 0 \\ \cdot & \cdot & \cdot & & \cdot & \cdot \\ \cdot & \cdot & \cdot & & \cdot & \cdot \\ \cdot & \cdot & \cdot & & \cdot & \cdot \\ 0 & 0 & \cdots & d_r|x_r|^2 & \cdot & 0 \\ \cdot & \cdot & & \cdot & \cdot & \cdot \\ 0 & 0 & \cdots & 0 & \cdot & 0 \end{bmatrix}. \tag{12.5}$$

We see that, even though we are dealing with complex numbers, this transformation multiplies the elements along the main diagonal of H' by positive real numbers.

Since $q(\alpha) = f(\alpha, \alpha) = \overline{f(\alpha, \alpha)}$, $q(\alpha)$ is always real. We can, in fact, apply without change the discussion we gave for the real quadratic forms. Let P denote the number of positive terms in the diagonal representation of q, and let N denote the number of negative terms in the main diagonal. The number $S = P - N$ is called the *signature* of the Hermitian quadratic form q. Again, $P + N = r$, the *rank* of q.

The proof that the signature of a Hermitian quadratic form is independent of the particular diagonalized representation is identical with the proof given for real quadratic forms.

A Hermitian quadratic form is called *non-negative semi-definite* if $S = r$. It is called *positive definite* if $S = n$. If f is a Hermitian form whose associated Hermitian quadratic form q is positive-definite (non-negative semi-definite), we say that the Hermitian form f is *positive-definite* (*non-negative semi-definite*).

A Hermitian matrix can be reduced to diagonal form by a method analogous to the method described in Section 10, as is shown by the proof of Theorem 12.1. A modification must be made because the associated Hermitian form is not bilinear, but complex bilinear.

Let α'_1 be a vector for which $q(\alpha'_1) \neq 0$. With this fixed α'_1, $f(\alpha'_1, \alpha)$ defines a linear functional ϕ'_1 on V. If α'_1 is represented by

$$(p_{11}, \ldots, p_{n1}) = P \text{ and } \alpha \text{ by } (x_1, \ldots, x_n) = X,$$

then

$$f(\alpha'_1, \alpha) = \sum_{i=1}^{n} \sum_{j=1}^{n} \overline{p_{i1}} h_{ij} x_j$$

$$= \sum_{j=1}^{n} \left(\sum_{i=1}^{n} \overline{p_{i1}} h_{ij} \right) x_j. \tag{12.6}$$

This means the linear functional ϕ'_1 is represented by P^*H.

EXERCISES

1. Reduce the following Hermitian matrices to diagonal form.

(a) $\begin{bmatrix} 1 & -i \\ i & 1 \end{bmatrix}$

(b) $\begin{bmatrix} 1 & 1-i \\ 1+i & 1 \end{bmatrix}$

2. Let f be an arbitrary complex bilinear form. Define f^* by the rule, $f^*(\alpha, \beta) = \overline{f(\beta, \alpha)}$. Show that f^* is complex bilinear.

3. Show that if H is a positive definite Hermitian matrix—that is, H represents a positive definite Hermitian form—then there exists a non-singular matrix P such that $H = P^*P$.

4. Show that if A is a complex non-singular matrix, then A^*A is a positive definite Hermitian matrix.

5. Show that if H is a Hermitian non-negative semi-definite matrix—that is, H represents a non-negative semi-definite Hermitian quadratic form—then there exists a complex matrix R such that $H = R^*R$.

6. Show that if A is complex, then A^*A is Hermitian non-negative semi-definite.

7. Show that if A is complex and $A^*A = 0$, then $A = 0$.

8. Show that if A is hermitian and $A^2 = 0$, then $A = 0$.

9. If $A_1, \ldots, A_r$ are Hermitian matrices, show that $A_1^2 + \cdots + A_r^2 = 0$ implies $A_1 = \cdots = A_r = 0$.

10. Show by an example that, if A and B are Hermitian, it is not necessarily true that AB is Hermitian. What is true if A and B are Hermitian and $AB = BA$?

chapter V | Orthogonal and unitary transformations, normal matrices

In this chapter we introduce an inner product based on an arbitrary positive definite symmetric bilinear form, or Hermitian form. On this basis the length of a vector and the concept of orthogonality can be defined. From this point on, we concentrate our attention on bases in which the vectors are mutually orthogonal and each is of length 1, the orthonormal bases. The Gram-Schmidt process for obtaining an orthonormal basis from an arbitrary basis is described.

Isometries are linear transformations which preserve length. They also preserve the inner product and therefore map orthonormal bases onto orthonormal bases. It is shown that a matrix representing an isometry has exactly the same properties as a matrix of transition representing a change of bases from one orthonormal basis to another. If the field of scalars is real, these matrices are said to be orthogonal; and if the field of scalars is complex, they are said to be unitary.

If A is an orthogonal matrix, we show that $A^T = A^{-1}$; and if A is unitary, we show that $A^* = A^{-1}$. Because of this fact a matrix representing a linear transformation and a matrix representing a bilinear form are transformed by exactly the same formula under a change of coordinates provided that the change is from one orthonormal basis to another. This observation unifies the discussions of Chapter III and IV.

The penalty for restricting our attention to orthonormal bases is that there is a corresponding restriction in the linear transformations and bilinear forms that can be represented by diagonal matrices. The necessary and sufficient condition that this be possible, expressed in terms of matrices, is that $A^*A = AA^*$. Matrices with this property are called *normal* matrices. Fortunately, the normal matrices constitute a large class of matrices and they happen to include as special cases most of the types that arise in physical problems.

Up to a certain point we can consider matrices with real coefficients to be special cases of matrices with complex coefficients. However, if we wish to restrict our attention to real vector spaces, then the matrices of transition must also be real. This restriction means that the situation for real vector spaces is not a special case of the situation for complex vector spaces. In particular, there are real normal matrices that are unitary similar to diagonal matrices but not orthogonal similar to diagonal matrices. The necessary and sufficient condition that a real matrix be orthogonal similar to a diagonal matrix is that it be symmetric.

The techniques for finding the diagonal normal form of a normal matrix and the unitary or orthogonal matrix of transition are, for the most part, not new. The eigenvalues and eigenvectors are found as in Chapter III. We show that eigenvectors corresponding to different eigenvalues are automatically orthogonal so all that needs to be done is to make sure that they are of length 1. However, something more must be done in the case of multiple eigenvalues. We are assured that there are enough eigenvectors, but we must make sure they are orthogonal. The Gram-Schmidt process provides the method for finding the necessary orthonormal eigenvectors.

1 | Inner Products and Orthogonal Bases

Even when speaking in abstract terms we have tried to draw an analogy between vector spaces and the geometric spaces we have encountered in 2- and 3-dimensional analytic geometry. For example, we have referred to lines and planes through the origin as subspaces; however, we have nowhere used the concept of distance. Some of the most interesting properties of vector spaces and matrices deal with the concept of distance. So in this chapter we introduce the concept of distance and explore the related properties.

For aesthetic reasons, and to show as clearly as possible that we need not have an a priori concept of distance, we use an approach which will emphasize the arbitrary nature of the concept of distance.

It is customary to restrict attention to the field of real numbers or the field of complex numbers when discussing vector space concepts related to distance. However, we need not be quite that restrictive. The scalar field F must be a subfield of the complex numbers with the property that, if $a \in F$, the conjugate complex $\bar{a}$ is also in F. Such a field is said to be *normal* over its real subfield. The real field and the complex field have this property, but so do many other fields. For most of the important applications of the material to follow the field of scalars is taken to be the real numbers or the field of complex numbers. Although most of the proofs given will be valid for any field normal over its real subfield, it will suffice to think in terms of the two most important cases.

In a vector space V of dimension n over the complex numbers (or a subfield of the complex numbers normal over its real subfield), let f be any fixed positive definite Hermitian form. For the purpose of the following development it does not matter which positive definite Hermitian form is chosen, but it will remain fixed for all the remaining discussion. Since this particular Hermitian form is now fixed, we write (α, β) instead of $f(\alpha, \beta)$. (α, β) is called the *inner product*, or *scalar product*, of α and β.

Since we have chosen a positive definite Hermitian form, $(\alpha, \alpha) \geq 0$ and $(\alpha, \alpha) > 0$ unless $\alpha = 0$. Thus $\sqrt{(\alpha, \alpha)} = \|\alpha\|$ is a well-defined nonnegative real number which we call the *length* or *norm* of α. Observe that $\|a\alpha\| = \sqrt{(a\alpha, a\alpha)} = \sqrt{\bar{a}a(\alpha, \alpha)} = |a| \cdot \|\alpha\|$, so that multiplying a vector by a scalar a multiplies its length by $|a|$. We say that the *distance* between two vectors is the norm of their difference; that is, $d(\alpha, \beta) = \|\beta - \alpha\|$. We should like to show that this distance function has the properties we might reasonably expect a distance function to have. But first we have to prove a theorem that has interest of its own and many applications.

Theorem 1.1. *For any vectors $\alpha, \beta \in V$, $|(\alpha, \beta)| \leq \|\alpha\| \cdot \|\beta\|$. This inequality is known as Schwarz's inequality.*

PROOF. For t a real number consider the inequality

$$0 \leq \|(\alpha, \beta)t\alpha - \beta\|^2 = |(\alpha, \beta)|^2 \|\alpha\|^2 t^2 - 2t |(\alpha, \beta)|^2 + \|\beta\|^2. \quad (1.1)$$

If $\|\alpha\| = 0$, the fact that this inequality must hold for arbitrarily large t implies that $|(\alpha, \beta)| = 0$ so that Schwarz's inequality is satisfied. If $\|\alpha\| \neq 0$, take $t = 1/\|\alpha\|^2$. Then (1.1) is equivalent to Schwarz's inequality,

$$|(\alpha, \beta)| \leq \|\alpha\| \cdot \|\beta\|. \square \quad (1.2)$$

This proof of Schwarz's inequality does not make use of the assumption that the inner product is positive definite and would remain valid if the inner product were merely semi-definite. Using the assumption that the inner product is positive definite, however, an examination of this proof of Schwarz's inequality would reveal that equality can hold if and only if

$$\beta - \frac{(\alpha, \beta)}{(\alpha, \alpha)}\alpha = 0; \quad (1.3)$$

that is, if and only if β is a multiple of α.

If $\alpha \neq 0$ and $\beta \neq 0$, Schwarz's inequality can be written in the form

$$\frac{|(\alpha, \beta)|}{\|\alpha\| \cdot \|\beta\|} \leq 1. \quad (1.4)$$

In vector analysis the scalar product of two vectors is equal to the product of the lengths of the vectors times the cosine of the angle between them. The inequality (1.4) says, in effect, that in a vector space over the real numbers the ratio $\dfrac{(\alpha, \beta)}{\|\alpha\| \cdot \|\beta\|}$ can be considered to be a cosine. It would be a diversion for us to push this point much further. We do, however, wish to show that $d(\alpha, \beta)$ behaves like a distance function.

Theorem 1.2.   *For $d(\alpha, \beta) = \|\beta - \alpha\|$, we have,*
(1) $d(\alpha, \beta) = d(\beta, \alpha)$,
(2) $d(\alpha, \beta) \geq 0$ *and* $d(\alpha, \beta) = 0$ *if and only if* $\alpha = \beta$,
(3) $d(\alpha, \beta) \leq d(\alpha, \gamma) + d(\gamma, \beta)$.
PROOF.   (1) and (2) are obvious.   (3) follows from Schwarz's inequality. To see this, observe that

$$
\begin{aligned}
\|\alpha + \beta\|^2 &= (\alpha + \beta, \alpha + \beta) \\
&= (\alpha, \alpha) + (\alpha, \beta) + (\beta, \alpha) + (\beta, \beta) \\
&= \|\alpha\|^2 + (\alpha, \beta) + \overline{(\alpha, \beta)} + \|\beta\|^2 \\
&\leq \|\alpha\|^2 + 2\,|(\alpha, \beta)| + \|\beta\|^2 \\
&\leq \|\alpha\|^2 + 2\,\|\alpha\| \cdot \|\beta\| + \|\beta\|^2 = (\|\alpha\| + \|\beta\|)^2. \qquad (1.5)
\end{aligned}
$$

Replacing α by $\gamma - \alpha$ and β by $\beta - \gamma$, we have

$$
\|\beta - \alpha\| \leq \|\gamma - \alpha\| + \|\beta - \gamma\|. \;\square \qquad (1.6)
$$

(3) is the familiar triangular inequality.   It implies that the sum of two small vectors is also small.   Schwarz's inequality tells us that the inner product of two small vectors is small.   Both of these inequalities are very useful for these reasons.

According to Theorem 12.1 of Chapter IV and the definition of a positive definite Hermitian form, there exists a basis $A = \{\alpha_1, \ldots, \alpha_n\}$ with respect to which the representing matrix is the unit matrix.   Thus,

$$
(\alpha_i, \alpha_j) = \delta_{ij}. \qquad (1.7)
$$

Relative to this fixed positive definite Hermitian form, the inner product, every set of vectors that has this property is called an *orthonormal* set. The word "orthonormal" is a combination of the words "orthogonal" and "normal."   Two vectors α and β are said to be *orthogonal* if $(\alpha, \beta) = (\beta, \alpha) = 0$.   A vector α is *normalized* if it is of length 1; that is, if $(\alpha, \alpha) = 1$.   Thus the vectors of an orthonormal set are mutually orthogonal and normalized.   The basis A chosen above is an *orthonormal basis*.   We shall see that orthonormal bases possess particular advantages for dealing with the properties of a vector space with an inner product.   A vector space over the complex numbers with an inner product such as we have defined is called a *unitary space*. A vector space over the real numbers with an inner product is called a *Euclidean space*.

For $\alpha, \beta \in V$, let $\alpha = \sum_{i=1}^{n} x_i \alpha_i$ and $\beta = \sum_{i=1}^{n} y_i \alpha_i$. Then

$$(\alpha, \beta) = \left(\sum_{i=1}^{n} x_i \alpha_i, \sum_{j=1}^{n} y_j \alpha_j \right)$$

$$= \sum_{i=1}^{n} \bar{x}_i \left[\sum_{j=1}^{n} y_j (\alpha_i, \alpha_j) \right]$$

$$= \sum_{i=1}^{n} \bar{x}_i y_i. \tag{1.8}$$

If we represent α by the n-tuple $(x_1, \ldots, x_n) = X$, and β by the n-tuple $(y_1, \ldots, y_n) = Y$, the inner product can be written in the form

$$(\alpha, \beta) = \sum_{i=1}^{n} \bar{x}_i y_i = X^* Y. \tag{1.9}$$

This is a familiar formula in vector analysis where it is also known as the inner product, scalar, or dot product.

Theorem 1.3. *An orthonormal set is linearly independent.*

PROOF. Suppose that $\{\xi_1, \xi_2, \ldots\}$ is an orthonormal set and that $\sum_i x_i \xi_i = 0$. Then $0 = (\xi_j, 0) = (\xi_j, \sum_i x_i \xi_i) = \sum_i x_i (\xi_j, \xi_i) = x_j$. Thus the set is linearly independent. $\square$

It is an immediate consequence of Theorem 1.3 that an orthonormal set cannot contain more than n elements.

Since V has at least one orthonormal basis and orthonormal sets are linearly independent, some questions naturally arise. Are there other orthonormal bases? Can an orthonormal set be extended to an orthonormal basis? Can a linearly independent set be modified to form an orthonormal set? For infinite dimensional vector spaces the question of the existence of even one orthonormal basis is a non-trivial question. For finite dimensional vector spaces all these questions have nice answers, and the technique employed in giving these answers is of importance in infinite dimensional vector spaces as well.

Theorem 1.4. *If* A $= \{\alpha_1, \ldots, \alpha_s\}$ *is any linearly independent set whatever in* V, *there exists an orthonormal set* X $= \{\xi_1, \ldots, \xi_s\}$ *such that* $\xi_k = \sum_{i=1}^{k} a_{ik} \alpha_i$.

PROOF. (The Gram-Schmidt orthonormalization process). Since α_1 is an element of a linearly independent set $\alpha_1 \neq 0$, and therefore $\|\alpha_1\| > 0$.

$$\text{Let } \xi_1 = \frac{1}{\|\alpha_1\|} \alpha_1. \text{ Clearly, } \|\xi_1\| = 1.$$

Suppose, then, $\{\xi_1, \ldots, \xi_r\}$ has been found so that it is an orthonormal set and such that each ξ_k is a linear combination of $\{\alpha_1, \ldots, \alpha_k\}$. Let

$$\alpha'_{r=1} = \alpha_{r+1} - (\xi_1, \alpha_{r+1})\xi_1 - \cdots - (\xi_r, \alpha_{r+1})\xi_r. \tag{1.10}$$

Then for any ξ_i, $1 \leq i \leq r$, we have

$$(\xi_i, \alpha'_{r+1}) = (\xi_i, \alpha_{r+1}) - (\xi_i, \alpha_{r+1}) = 0. \tag{1.11}$$

Furthermore, since each ξ_k is a linear combination of the $\{\alpha_1, \ldots, \alpha_k\}$, α'_{r+1} is a linear combination of the $\{\alpha_1, \ldots, \alpha_{r+1}\}$. Also, α'_{r+1} is not zero since $\{\alpha_1, \ldots, \alpha_{r+1}\}$ is a linearly independent set and the coefficient of α_{r+1} in the representation of α'_{r+1} is 1. Thus we can define

$$\xi_{r+1} = \frac{1}{\|\alpha'_{r+1}\|} \alpha'_{r+1}. \tag{1.12}$$

Clearly, $\{\xi_1, \ldots, \xi_{r+1}\}$ is an orthonormal set with the desired properties. We can continue in this fashion until we exhaust the elements of A. The set $X = \{\xi_1, \ldots, \xi_s\}$ has the required properties. $\square$

The Gram-Schmidt process is completely effective and the computations can be carried out exactly as they are given in the proof of Theorem 1.4. For example, let $A = \{\alpha_1 = (1, 1, 0, 1), \alpha_2 = (3, 1, 1, -1), \alpha_3 = (0, 1, -1, 1)\}$. Then

$$\xi_1 = \frac{1}{\sqrt{3}} (1, 1, 0, 1),$$

$$\alpha'_2 = (3, 1, 1, -1) - \frac{3}{\sqrt{3}} \frac{1}{\sqrt{3}} (1, 1, 0, 1) = (2, 0, 1, -2),$$

$$\xi_2 = \tfrac{1}{3}(2, 0, 1, -2),$$

$$\alpha'_3 = (0, 1, -1, 1) - \frac{2}{\sqrt{3}} \frac{1}{\sqrt{3}} (1, 1, 0, 1) - \frac{-3}{3} \frac{1}{3} (2, 0, 1, -2)$$

$$= \tfrac{1}{3}(0, 1, -2, -1),$$

$$\xi_3 = \frac{1}{\sqrt{6}} (0, 1, -2, -1).$$

It is easily verified that $\{\xi_1, \xi_2, \xi_3\}$ is an orthonormal set.

Corollary 1.5. *If $A = \{\alpha_1, \ldots, \alpha_n\}$ is a basis of V, the orthonormal set $X = \{\xi_1, \ldots, \xi_n\}$, obtained from A by the application of the Gram-Schmidt process, is an orthonormal basis of V.*

PROOF. Since X is orthonormal it is linearly independent. Since it contains n vectors it also spans V and is a basis. $\square$

Theorem 1.4 and its corollary are used in much the same fashion in which we used Theorem 3.6 of Chapter I to obtain a basis (in this case an orthonormal basis) such that a subset spans a given subspace.

Theorem 1.6. *Given any vector α_1 of length 1, there is an orthonormal basis with α_1 as the first element.*

PROOF.   Since the set $\{\alpha_1\}$ is linearly independent it can be extended to a basis with α_1 as the first element. Now, when the Gram-Schmidt process is applied, the first vector, being of length 1, is unchanged and becomes the first vector of an orthonormal basis. □

EXERCISES

In the following problems we assume that all n-tuples are representations of their vectors with respect to orthonormal bases.

1. Let $A = \{\alpha_1, \ldots, \alpha_4\}$ be an orthonormal basis of R^4 and let $\alpha,\ \beta \in V$ be represented by $(1, 2, 3, -1)$ and $(2, 4, -1, 1)$, respectively. Compute (α, β).

2. Let $\alpha = (1, i, 1 + i)$ and $\beta = (i, 1, i - 1)$ be vectors in C^3, where C is the field of complex numbers. Compute (α, β).

3. Show that the set $\{(1, i, 2), (1, i, -1), (1, -i, 0)\}$ is orthogonal in C^3.

4. Show that $(\alpha, 0) = (0, \alpha) = 0$ for all $\alpha \in V$.

5. Show that $\|\alpha + \beta\|^2 + \|\alpha - \beta\|^2 = 2\|\alpha\|^2 + 2\|\beta\|^2$.

6. Show that if the field of scalars is real and $\|\alpha\| = \|\beta\|$, then $\alpha - \beta$ and $\alpha + \beta$ are orthogonal, and conversely.

7. Show that if the field of scalars is real and $\|\alpha + \beta\|^2 = \|\alpha\|^2 + \|\beta\|^2$, then α and β are orthogonal, and conversely.

8. Verify Schwarz's inequality for the vectors α and β in Exercises 1 and 2.

9. The set $\{(1, -1, 1), (2, 0, 1), (0, 1, 1)\}$ is linearly independent, and hence a basis for F^3. Apply the Gram-Schmidt process to obtain an orthonormal basis.

10. Given the basis $\{(1, 0, 1, 0), (1, 1, 0, 0), (0, 1, 1, 1,), (0, 1, 1, 0)\}$ apply the Gram-Schmidt process to obtain an orthonormal basis.

11. Let W be a subspace of V spanned by $\{(0, 1, 1, 0), (0, 5, -3, -2), (-3, -3, 5, -7)\}$. Find an orthonormal basis for W.

12. In the space of real integrable functions let the inner product be defined by

$$\int_{-1}^{1} f(x)g(x)\, dx.$$

Find a polynomial of degree 2 orthogonal to 1 and x. Find a polynomial of degree 3 orthogonal to 1, x, and x^2. Are these two polynomials orthogonal?

13. Let $X = \{\xi_1, \ldots, \xi_m\}$ be a set of vectors in the n-dimensional space V. Consider the matrix $G = [g_{ij}]$ where

$$g_{ij} = (\xi_i, \xi_j).$$

Show that if X is linearly dependent, then the columns of G are also linearly dependent. Show that if X is linearly independent, then the columns of G are also linearly independent. Det G is known as the *Gramian* of the set X. Show that X is linearly dependent if and only if det $G = 0$. Choose an orthonormal basis in V and represent the vectors in X with respect to that basis. Show that G can be represented as the product of an $m \times n$ matrix and an $n \times m$ matrix. Show that det $G \geq 0$.

*2 | Complete Orthonormal Sets

We now develop some properties of orthonormal sets that hold in both finite and infinite dimensional vector spaces. These properties are deep and important in infinite dimensional vector spaces, but in finite dimensional vector spaces they could easily be developed in passing and without special terminology. It is of some interest, however, to borrow the terminology of infinite dimensional vector spaces and to give proofs, where possible, which are valid in infinite as well as finite dimensional vector spaces.

Let $X = \{\xi_1, \xi_2, \dots\}$ be an orthonormal set and let α be any vector in V. The numbers $\{\alpha_i = (\xi_i, \alpha)\}$ are called the *Fourier coefficients* of α.

There is, first, the question of whether an expression like $\sum_i x_i \xi_i$ has any meaning in cases where infinitely many of the x_i are non-zero. This is a question of the convergence of an infinite series and the problem varies from case to case so that we cannot hope to deal with it in all generality. We have to assume for this discussion that all expressions like $\sum_i x_i \xi_i$ that we write down have meaning.

Theorem 2.1. *The minimum of* $\|\alpha - \sum_i x_i \xi_i\|$ *is attained if and only if all*

$$x_i = (\xi_i, x) = a_i.$$

PROOF.

$$\begin{aligned}
\|\alpha - \sum_i x_i \xi_i\|^2 &= (\alpha - \sum_i x_i \xi_i, \alpha - \sum_i x_i \xi_i) \\
&= (\alpha, \alpha) - \sum_i x_i \bar{a}_i - \sum_i \bar{x}_i a_i + \sum_i \bar{x}_i x_i \\
&= \sum_i \bar{a}_i a_i - \sum_i x_i \bar{a}_i - \sum_i \bar{x}_i a_i + \sum_i \bar{x}_i x_i + (\alpha, \alpha) - \sum_i \bar{a}_i a_i \\
&= \sum_i (\bar{a}_i - \bar{x}_i)(a_i - x_i) + \|\alpha\|^2 - \sum_i \bar{a}_i a_i \\
&= \sum_i |a_i - x_i|^2 + \|\alpha\|^2 - \sum_i |a_i|^2.
\end{aligned} \tag{2.1}$$

Only the term $\sum_i |a_i - x_i|^2$ depends on the x_i and, being a sum of real squares, it takes on its minimum value of zero if and only if all $x_i = a_i$. □

Theorem 2.1 is valid for any orthonormal set X, whether it is a basis or not. If the norm is used as a criterion of smallness, then the theorem says that the best approximation of α in the form $\sum_i x_i \xi_i$ (using only the $\xi_i \in X$) is obtained if and only if all x_i are the Fourier coefficients.

Theorem 2.2 $\sum_i |a_i|^2 \leq \|\alpha\|^2$. *This inequality is known as Bessel's inequality.*

PROOF. Setting $x_i = a_i$ in equation (2.1) we have

$$\|\alpha\|^2 - \sum_i |a_i|^2 = \|\alpha - \sum_i a_i\xi_i\|^2 \geq 0. \quad \square \qquad (2.2)$$

It is desirable to know conditions under which the Fourier coefficients will represent the vector α. This means we would like to have $\alpha = \sum_i a_i\xi_i$. In a finite dimensional vector space the most convenient sufficient condition is that X be an orthonormal basis. In the theory of Fourier series and other orthogonal functions it is generally not possible to establish the validity of an equation like $\alpha = \sum_i a_i\xi_i$ without some modification of what is meant by convergence or a restriction on the set of functions under consideration. Instead, we usually establish a condition known as completeness. An orthonormal set is said to be *complete* if and only if it is not a subset of a larger orthonormal set.

Theorem 2.3. *Let* $X = \{\xi_i\}$ *be an orthonormal set. The following three conditions are equivalent:*

(1) *For each* $\alpha, \beta \in V$, $(\alpha, \beta) = \sum_i \overline{(\xi_i, \alpha)}(\xi_i, \beta)$. $\qquad (2.3)$

(2) *For each* $\alpha \in V$, $\|\alpha\|^2 = \sum_i |(\xi_i, \alpha)|^2$. $\qquad (2.4)$

(3) X *is complete.*

Equations (2.3) *and* (2.4) *are both known as Parseval's identities.*

PROOF. Assume (1). Then $\|\alpha\|^2 = (\alpha, \alpha) = \sum_i \overline{(\xi_i, \alpha)}(\xi_i, \alpha)$

$$= \sum_i |(\xi_i, \alpha)|^2.$$

Assume (2). If X were not complete, it would be contained in a larger orthonormal set Y. But for any $\alpha_0 \in Y$, $\alpha_0 \notin X$, we would have

$$1 = \|\alpha_0\|^2 = \sum_i |(\xi_i, \alpha_0)|^2 = 0$$

because of (2) and the assumption that Y is orthonormal. Thus X is complete.

Now, assume (3). Let β be any vector in V and consider $\beta' = \beta - \sum_i (\xi_i, \beta)\xi_i$. Then

$$(\xi_i, \beta') = \left(\xi_i, \beta - \sum_j (\xi_j, \beta)\xi_j\right)$$

$$= (\xi_i, \beta) - \sum_j (\xi_j, \beta)(\xi_i, \xi_j)$$

$$= (\xi_i, \beta) - (\xi_i, \beta) = 0;$$

that is, β' is orthogonal to all $\xi_i \in X$. If $\|\beta'\| \neq 0$, then $X \cup \left\{\dfrac{1}{\|\beta'\|} \beta'\right\}$

would be a larger orthonormal set. Hence, $\|\beta'\| = 0$. Using the assumption that the inner product is positive definite we can now conclude that $\beta' = 0$. However, it is not necessary to use this assumption and we prefer to avoid using it. What we really need to conclude is that if α is any vector in V then $(\alpha, \beta') = 0$, and this follows from Schwarz's inequality. Thus we have

$$0 = (\alpha, \beta') = (\alpha, \beta - \sum_i (\xi_i, \beta)\xi_i)$$
$$= (\alpha, \beta) - \sum_i (\xi_i, \beta)(\alpha, \xi_i)$$
$$= (\alpha, \beta) - \sum_i \overline{(\xi_i, \alpha)}(\xi_i, \beta),$$

or

$$(\alpha, \beta) = \sum_i \overline{(\xi_i, \alpha)}(\xi_i, \beta).$$

This completes the cycle of implications and proves that conditions (1), (2), and (3) are equivalent. $\square$

Theorem 2.4. *The following two conditions are equivalent:*
(4) *The only vector orthogonal to all vectors in X is the zero vector.*
(5) *For each $\alpha \in V$, $\alpha = \sum_i (\xi_i, \alpha)\xi_i$.* (2.5)

PROOF. Assume (4). Let α be any vector in V and consider $\alpha' = \alpha - \sum_i (\xi_i, \alpha)\xi_i$. Then

$$(\xi_i, \alpha') = (\xi_i, \alpha - \sum_j (\xi_j, \alpha)\xi_j)$$
$$= (\xi_i, \alpha) - \sum_j (\xi_j, \alpha)(\xi_i, \xi_j)$$
$$= (\xi_i, \alpha) - (\xi_i, \alpha) = 0;$$

that is, α' is orthogonal to all $\xi_i \in X$. Thus $\alpha' = 0$ and $\alpha = \sum_i (\xi_i, \alpha)\xi_i$.

Now, assume (5) and let α be orthogonal to all $\xi_i \in X$. Then $\alpha = \sum_i (\xi_i, \alpha)\xi_i = 0$. $\square$

Theorem 2.5. *The conditions (4) or (5) imply the conditions (1), (2), and (3).*

PROOF. Assume (5). Then

$$(\alpha, \beta) = (\sum_i (\xi_i, \alpha)\xi_i, \sum_j (\xi_j, \beta)\xi_j)$$
$$= \sum_i \overline{(\xi_i, \alpha)} \sum_j (\xi_j, \beta)(\xi_i, \xi_j)$$
$$= \sum_i \overline{(\xi_i, \alpha)}(\xi_i, \beta). \square$$

Theorem 2.6. *If the inner product is positive definite, the conditions (1), (2), or (3) imply the conditions (4) and (5).*

PROOF. In the proof that (3) implies (1) we showed that if $\alpha' = \alpha - \sum_i (\xi_i, \alpha)\xi_i$, then $\|\alpha'\| = 0$. If the inner product is positive definite, then $\alpha' = 0$ and, hence,

$$\alpha = \sum_i (\xi_i, \alpha)\xi_i. \quad \square$$

The proofs of Theorems 2.3, 2.4, and 2.5 did not make use of the positive definiteness of the inner product and they remain valid if the inner product is merely non-negative semi-definite. Theorem 2.6 depends critically on the fact that the inner product is positive definite.

For finite dimensional vector spaces we always assume that the inner product is positive definite so that the three conditions of Theorem 2.3 and the two conditions of Theorem 2.4 are equivalent. The point of our making a distinction between these two sets of conditions is that there are a number of important inner products in infinite dimensional vector spaces that are not positive definite. For example, the inner product that occurs in the theory of Fourier series is of the form

$$(\alpha, \beta) = \frac{1}{2\pi} \int_{-\pi}^{\pi} \overline{\alpha(x)}\beta(x) \, dx. \tag{2.6}$$

This inner product is non-negative semi-definite, but not positive definite if V is the set of integrable functions. Hence, we cannot pass from the completeness of the set of orthogonal functions to a theorem about the convergence of a Fourier series to the function from which the Fourier coefficients were obtained.

In using theorems of this type in infinite dimensional vector spaces in general and Fourier series in particular, we proceed in the following manner. We show that any $\alpha \in V$ can be approximated arbitrarily closely by finite sums of the form $\sum_i x_i\xi_i$. For the theory of Fourier series this theorem is known as the Weierstrass approximation theorem. A similar theorem must be proved for other sets of orthogonal functions. This implies that the minimum mentioned in Theorem 2.1 must be zero. This in turn implies that condition (2) of Theorem 2.3 holds. Thus Parseval's equation, which is equivalent to the completeness of an orthonormal set, is one of the principal theorems of any theory of orthogonal functions. Condition (5), which is the convergence of a Fourier series to the function which it represents, would follow if the inner product were positive definite. Unfortunately, this is usually not the case. To get the validity of condition (5) we must either add further conditions or introduce a different type of convergence.

EXERCISES

1. Show that if X is an orthonormal basis of a finite dimensional vector space, then condition (5) holds.

2. Let X be a finite set of mutually orthogonal vectors in V. Suppose that the only vector orthogonal to each vector in X is the zero vector. Show that X is a basis of V.

3 | The Representation of a Linear Functional by an Inner Product

For a fixed vector $\beta \in V$, (β, α) is a linear function of α. Thus there is a linear functional $\phi \in \hat{V}$ such that $\phi(\alpha) = (\beta, \alpha)$ for all α. We denote the linear functional defined in this way by ϕ_β. The following theorem is a converse of this observation.

Theorem 3.1. *Given a linear functional $\phi \in \hat{V}$, there exists a unique $\eta \in V$ such that $\phi(\alpha) = (\eta, \alpha)$ for all $\alpha \in V$.*

PROOF. Let $X = \{\xi_1, \ldots, \xi_n\}$ be an orthonormal basis of V, and let $\hat{X} = \{\phi_1, \ldots, \phi_n\}$ be the dual basis. Let $\phi \in \hat{V}$ have the representation $\phi = \sum_{i=1}^n y_i \phi_i$. Define $\eta = \sum_{i=1}^n \bar{y}_i \xi_i$. Then for each ξ_j, $(\eta, \xi_j) = (\sum_{i=1}^n \bar{y}_i \xi_i, \xi_j) = \sum_{i=1}^n y_i(\xi_i, \xi_j) = y_j = \sum_{i=1}^n y_i \phi_i(\xi_j) = \phi(\xi_j)$. But then $\phi(\alpha)$ and (η, α) are both linear functionals on V that coincide on the basis, and hence coincide on all of V.

If η_1 and η_2 are two choices such that $(\eta_1, \alpha) = (\eta_2, \alpha) = \phi(\alpha)$ for all $\alpha \in V$, then $(\eta_1 - \eta_2, \alpha) = 0$ for all $\alpha \in V$. For $\alpha = \eta_1 - \eta_2$ this means $(\eta_1 - \eta_2, \eta_1 - \eta_2) = 0$. Hence, $\eta_1 - \eta_2 = 0$ and the choice for η is unique. $\square$

Call the mapping defined by this theorem η; that is, for each $\phi \in \hat{V}$, $\eta(\phi) \in V$ has the property that $\phi(\alpha) = (\eta(\phi), \alpha)$ for all $\alpha \in V$.

Theorem 3.2. *The correspondence between $\phi \in \hat{V}$ and $\eta(\phi) \in V$ is one-to-one and onto V.*

PROOF. In Theorem 3.1 we have already shown that $\eta(\phi)$ is well defined. Let β be any vector in V and let ϕ_β be the linear functional in $\hat{V}$ such that $\phi_\beta(\alpha) = (\beta, \alpha)$ for all α. Then $\beta = \eta(\phi_\beta)$ and the mapping is onto. Since (β, α), as a function of α, determines a unique linear functional ϕ_β the correspondence is one-to-one. $\square$

Theorem 3.3. *If the inner product is symmetric, η is an isomorphism of $\hat{V}$ onto V.*

PROOF. We have already shown in Theorem 3.2 that η is one-to-one and onto. Let $\phi = \sum_i b_i \phi_i$ and consider $\beta = \sum_i b_i \eta(\phi_i)$. Then $(\beta, \alpha) = (\sum_i b_i \eta(\phi_i), \alpha) = (\alpha, \sum_i b_i \eta(\phi_i)) = \sum_i b_i(\alpha, \eta(\phi_i)) = \sum_i b_i(\eta(\phi_i), \alpha) = \sum_i b_i \phi_i(\alpha) = \phi(\alpha)$. Thus $\eta(\phi) = \beta = \sum_i b_i \eta(\phi_i)$ and η is linear. $\square$

Notice that η is not linear if the scalar field is complex and the inner product is Hermitian. Then for $\phi = \sum_i b_i \phi_i$ we consider $\gamma = \sum_i \bar{b}_i \eta(\phi_i)$. We see that $(\gamma, \alpha) = (\sum_i \bar{b}_i \eta(\phi_i), \alpha) = \sum_i b_i(\eta(\phi_i), \alpha) = \sum_i b_i \phi_i(\alpha) = \phi(\alpha)$.

Thus $\eta(\phi) = \gamma = \sum_i \bar{b}_i \eta(\phi_i)$ and η is conjugate linear. It should be observed that even when η is conjugate linear it maps subspaces of $\hat{V}$ onto subspaces of V.

We describe this situation by saying that we can "represent a linear functional by an inner product." Notice that although we made use of a particular basis to specify the η corresponding to ϕ, the uniqueness shows that this choice is independent of the basis used. If V is a vector space over the real numbers, ϕ and η happen to have the same coordinates. This happy coincidence allows us to represent $\hat{V}$ in V and make V do double duty. This fact is exploited in courses in vector analysis. In fact, it is customary to start immediately with inner products in real vector spaces with orthonormal bases and not to mention $\hat{V}$ at all. All is well as long as things remain simple. As soon as things get a little more complicated, it is necessary to separate the structure of $\hat{V}$ superimposed on V. The vectors representing themselves in V are said to be *contravariant* and the vectors representing linear functionals in $\hat{V}$ are said to be *covariant*.

We can see from the proof of Theorem 3.1 that, if V is a vector space over the complex numbers, ϕ and the corresponding η will not necessarily have the same coordinates. In fact, there is no choice of a basis for which each ϕ and its corresponding η will have the same coordinates.

Let us examine the situation when the basis chosen in V is not orthonormal. Let $A = \{\alpha_1, \ldots, \alpha_n\}$ be any basis of V, and let $\hat{A} = \{\psi_1, \ldots, \psi_n\}$ be the corresponding dual basis of $\hat{V}$. Let $b_{ij} = (\alpha_i, \alpha_j)$. Since the inner product is Hermitian, $b_{ij} = \bar{b}_{ji}$, or $[b_{ij}] = B = B^*$. Since the inner product is positive definite, B has rank n. That is, B is non-singular. Let $\phi = \sum_{i=1}^n c_i \psi_i$ be an arbitrary linear functional in V. What are the coordinates of the corresponding η? Let $\eta = \sum_{i=1}^n y_i \alpha_i$. Then

$$(\eta, \alpha_j) = \left(\sum_{i=1}^n y_i \alpha_i, \alpha_j \right)$$

$$= \sum_{i=1}^n \bar{y}_i (\alpha_i, \alpha_j)$$

$$= \sum_{i=1}^n \bar{y}_i b_{ij}$$

$$= \phi(\alpha_j)$$

$$= \sum_{k=1}^n c_k \psi_k(\alpha_j)$$

$$= c_j. \tag{3.1}$$

Thus, we have to solve the equations

$$\sum_{i=1}^n \bar{y}_i b_{ij} = \sum_{i=1}^n \bar{b}_{ji} \bar{y}_i = c_j, \, j = 1, \ldots, n. \tag{3.2}$$

In matrix form this becomes

$$\bar{B}\bar{Y} = C^T,$$

where

$$C = [c_1 \quad \cdots \quad c_n],$$

or

$$Y = B^{-1}C^* = (CB^{-1})^*. \tag{3.3}$$

Of course this means that it is rather complicated to obtain the coordinate representation of η from the coordinate representation of ϕ. But that is not the cause for all the fuss about covarient and contravariant vectors. After all, we have shown that η corresponds to ϕ independently of the basis used and the coordinates of η transform according to the same rules that apply to any other vector in V. The real difficulty stems from the insistence upon using (1.9) as the definition of the inner product, instead of using a definition not based upon coordinates.

If $\eta = \sum_{i=1}^n y_i\alpha_i$, and $\xi = \sum_{i=1}^n x_i\alpha_i$, we see that

$$(\eta, \xi) = \left(\sum_{i=1}^n y_i\alpha_i, \sum_{j=1}^n x_j\alpha_j \right)$$

$$= \sum_{i=1}^n \sum_{j=1}^n \bar{y}_i b_{ij} x_j$$

$$= Y^*BX. \tag{3.4}$$

Thus, if η represents the linear functional ϕ, we have

$$(\eta, \xi) = Y^*BX$$

$$= (CB^{-1})BX$$

$$= CX$$

$$= (C^*)^*X. \tag{3.5}$$

Elementary treatments of vector analysis prefer to use C^* as the representation of η. This preference is based on the desire to use (1.9) as the definition of the inner product so that (3.5) is the representation of (η, ξ), rather than to use a coordinate-free definition which would lead to (η, ξ) being represented by (3.4). The elements of C^* are called the covariant components of η. We obtained C by representing ϕ in $\hat{V}$. Since the dual space is not available in such an elementary treatment, some kind of artifice must be used. It is then customary to introduce a *reciprocal basis* $A^* = \{\alpha_1^*, \ldots, \alpha_n^*\}$, where α_i^* has the property $(\alpha_i^*, \alpha_j) = \delta_{ij} = \phi_i(\alpha_j)$. A^* is the representation of the dual basis A in V. But C was the original representation of ϕ in terms of the dual basis. Thus, the insistence upon representing linear functionals by the inner product does not result in a single computational advantage. The confusion that it introduces is a severe price to pay to avoid introducing linear functionals and the dual space at the beginning.

4 | The Adjoint Transformation

Theorem 4.1. *For a given linear transformation σ on V, there is a unique linear transformation σ^* on V such that $(\sigma^*(\alpha), \beta) = (\alpha, \sigma(\beta))$ for all $\alpha, \beta \in V$.*

PROOF. Let σ be given. Then for a fixed α, $(\alpha, \sigma(\beta))$ is a linear function of β, that is, a linear functional on V. By Theorem 3.1 there is a unique $\eta \in V$ such that $(\alpha, \sigma(\beta)) = (\eta, \beta)$. Define $\sigma^*(\alpha)$ to be this η.

Now, $(a_1\alpha_1 + a_2\alpha_2, \sigma(\beta)) = \bar{a}_1(\alpha_1, \sigma(\beta)) + \bar{a}_2(\alpha_2, \sigma(\beta)) = \bar{a}_1(\sigma^*(\alpha_1), \beta) + \bar{a}_2(\sigma^*(\alpha_2), \beta) = (a_1\sigma^*(\alpha_1) + a_2\sigma^*(\alpha_2), \beta)$ so that $a_1\sigma^*(\alpha_1) + a_2\sigma^*(\alpha_2) = \sigma^*(a_1\alpha_1 + a_2\alpha_2)$ and σ^* is linear. $\square$

Since for each α the choice for $\sigma^*(\alpha)$ is unique, σ^* is uniquely defined by σ. σ^* is called the *adjoint* of σ.

Theorem 4.2. *The relation between σ and σ^* is symmetric, that is, $(\sigma^*)^* = \sigma$.*

PROOF. Let σ be given. Then σ^* is defined uniquely by $(\sigma^*(\alpha), \beta) = (\alpha, \sigma(\beta))$ for all $\alpha, \beta \in V$. Then $(\sigma^*)^*$, which we denote by σ^{**}, is defined by $(\sigma^{**}(\alpha), \beta) = (\alpha, \sigma^*(\beta))$ for all $\alpha, \beta \in V$. Now the inner product is Hermitian so that $(\sigma^{**}(\alpha), \beta) = (\alpha, \sigma^*(\beta)) = \overline{(\sigma^*(\beta), \alpha)} = \overline{(\beta, \sigma(\alpha))} = (\sigma(\alpha), \beta)$. Thus $\sigma^{**}(\alpha) = \sigma(\alpha)$ for all $\alpha \in V$; that is, $\sigma^{**} = \sigma$. It then follows also that $(\sigma(\alpha), \beta) = (\alpha, \sigma^*(\beta))$. $\square$

Let $A = [a_{ij}]$ be the matrix representing σ with respect to an orthonormal basis $X = \{\xi_i, \ldots, \xi_n\}$ and let us find the matrix representing σ^*.

$$(\sigma^*(\xi_j), \xi_k) = (\xi_j, \sigma(\xi_k))$$
$$= \left(\xi_j, \sum_{i=1}^{n} a_{ik}\xi_i\right)$$
$$= \sum_{i=1}^{n} a_{ik}(\xi_j, \xi_i)$$
$$= a_{jk}$$
$$= \sum_{i=1}^{n} a_{ji}(\xi_i, \xi_k)$$
$$= \left(\sum_{i=1}^{n} \bar{a}_{ji}\xi_i, \xi_k\right). \tag{4.1}$$

Since this equation holds for all ξ_k, $\sigma^*(\xi_j) = \sum_{i=1}^{n} \bar{a}_{ji}\xi_i$. Thus σ^* is represented by the conjugate transpose of A; that is, σ^* is represented by A^*.

The adjoint σ^* is closely related to the dual $\hat{\sigma}$ defined on page 142. σ is a linear transformation of V into itself, so the dual $\hat{\sigma}$ is a linear transformation of $\hat{V}$ into itself. Since η establishes a one-to-one correspondence between $\hat{V}$ and V, we can define a mapping of V into itself corresponding to $\hat{\sigma}$ on

$\hat{V}$. For any $\alpha \in V$ we can map α onto $\eta\{\hat{\sigma}[\eta^{-1}(\alpha)]\}$ and denote this mapping by $\eta(\hat{\sigma})$. Then for any α, $\beta \in V$ we have

$$
\begin{aligned}
(\eta(\hat{\sigma})(\alpha), \beta) &= (\eta\{\hat{\sigma}[\eta^{-1}(\alpha)]\}, \beta) \\
&= \hat{\sigma}[\eta^{-1}(\alpha)](\beta) \\
&= \eta^{-1}(\alpha)[\sigma(\beta)] \\
&= (\alpha, \sigma(\beta)) \\
&= (\sigma^*(\alpha), \beta).
\end{aligned} \tag{4.2}
$$

Hence, $\eta(\hat{\sigma})(\alpha) = \sigma^*(\alpha)$ for all $\alpha \in V$, that is, $\eta(\hat{\sigma}) = \sigma^*$. The adjoint is a representation of the dual. Because the mapping η of $\hat{V}$ onto V is conjugate linear instead of linear, and because the vectors in $\hat{V}$ are represented by row matrices while those in V are represented by columns, the matrix representing σ^* is the transpose of the complex conjugate of the matrix representing $\hat{\sigma}$. Thus σ^* is represented by A^*.

We shall maintain the distinction between the dual $\hat{\sigma}$ defined on $\hat{V}$ and the adjoint σ^* defined on V. This distinction is not always made and quite often both terms are used for both purposes. Actually, this confusion seldom causes any trouble. However, it can cause trouble when discussing the matrix representation of $\hat{\sigma}$ or σ^*. If σ is represented by A, we have chosen also to represent $\hat{\sigma}$ by A with respect to the dual basis. If we had chosen to represent linear functionals by columns instead of rows, $\hat{\sigma}$ would have been represented by A^T. It would have been represented by A^T in either the real or the complex case. But the adjoint σ^* is represented by A^*. No convention will allow $\hat{\sigma}$ and σ^* to be represented by the same matrix in the complex case because the mapping η is conjugate linear. Because of this we have chosen to make clear the distinction between $\hat{\sigma}$ and σ^*, even to the extent of having the matrix representations look different. Furthermore, the use of rows to represent linear functionals has the advantage of making some of the formulas look simpler. However, this is purely a matter of choice and taste, and other conventions, used consistently, would serve as well.

Since we now have a model of $\hat{V}$ in V, we can carry over into V all the terminology and theorems on linear functionals in Chapter IV. In particular, we see that an orthonormal basis can also be considered to be its own dual basis since $(\xi_i, \xi_j) = \delta_{ij}$.

Recall that, when a basis is changed in V and P is the matrix of transition, $(P^T)^{-1}$ is the matrix of transition for the dual bases in $\hat{V}$. In mapping $\hat{V}$ onto V, $\overline{(P^T)^{-1}} = (P^*)^{-1}$ becomes the matrix of transition for the representation of dual basis in V. Since an orthonormal basis is dual to itself, if P is the matrix of transition from one orthonormal basis to another, then P must also be the matrix of transition for the dual basis; that is, $(P^*)^{-1} = P$. This important property of the matrices of transition from one orthonormal basis to another will be established independently in Section 6.

Let W be a subset of V. In Chapter IV-4, we defined $W^\perp$ to be the annihilator of W in $\hat{V}$. The mapping η of $\hat{V}$ onto V maps $W^\perp$ onto a subspace of V. It is easily seen that $\eta(W^\perp)$ is precisely the set of all vectors orthogonal to every vector in W. Since we are in the process of dropping $\hat{V}$ as a separate space and identifying it with V, we denote the set of all vectors in orthogonal to all vectors in W by $W^\perp$ and call it the *annihilator* of W.

Theorem 4.3. *If W is a subspace of dimension p, $W^\perp$ is of dimension $n - p$. $W \cap W^\perp = \{0\}$. $W \oplus W^\perp = V$.*

PROOF. That $W^\perp$ is of dimension $n - p$ follows from Theorem 4.1 of Chapter IV. The other two assertions had no meaning in the context of Chapter IV. If $\alpha \in W \cap W^\perp$, then $\|\alpha\|^2 = (\alpha, \alpha) = 0$ so that $\alpha = 0$. Since dim $\{W + W^\perp\} = $ dim $W + $ dim $W^\perp - $ dim $\{W \cap W^\perp\} = p + (n - p) = n$, $W \oplus W^\perp = V$. $\square$

When W_1 and W_2 are subspaces of V such that their sum is direct and W_1 and W_2 are also orthogonal, we use the notation $W_1 \perp W_2$ to denote this sum. Actually, the fact that the sum is direct is a consequence of the fact that the subspaces are orthogonal. In this notation, the direct sum in the conclusion of Theorem 4.3 takes the form $V = W \perp W^\perp$.

Theorem 4.4. *Let W be a subspace invariant under σ. $W^\perp$ is then invariant under σ^*.*

PROOF. Let $\alpha \in W^\perp$. Then, for any $\beta \in W$, $(\sigma^*(\alpha), \beta) = (\alpha, \sigma(\beta)) = 0$ since $\sigma(\beta) \in W$. Thus $\sigma^*(\alpha) \in W^\perp$. $\square$

Theorem 4.5. $K(\sigma^*) = \text{Im}(\sigma)^\perp$.

PROOF. By definition $(\alpha, \sigma(\beta)) = (\sigma^*(\alpha), \beta)$. $(\alpha, \alpha(\beta)) = 0$ for all $\beta \in V$ if and only if $\alpha \in \text{Im}(\sigma)^\perp$ and $(\sigma^*(\alpha), \beta) = 0$ for all $\beta \in V$ if and only if $\alpha \in K(\sigma^*)$. Thus $K(\sigma^*) = \text{Im}(\sigma)^\perp$. $\square$

Theorem 4.5 here is equivalent to Theorem 5.3 of Chapter IV.

Theorem 4.6. *If S and T are subspaces of V, then $(S + T)^\perp = S^\perp \cap T^\perp$ and $(S \cap T)^\perp = S^\perp + T^\perp$.*

PROOF. This theorem is equivalent to Theorem 4.4 of Chapter IV. $\square$

Theorem 4.7. *For each conjugate bilinear form f, there is a linear transformation σ such that $f(\alpha, \beta) = (\alpha, \sigma(\beta))$ for all $\alpha, \beta \in V$.*

PROOF. For a fixed $\alpha \in V$, $f(\alpha, \beta)$ is linear in β. Thus by Theorem 3.1 there is a unique $\eta \in V$ such that $f(\alpha, \beta) = (\eta, \beta)$ for all $\beta \in V$. Define $\sigma^*(\alpha) = \eta$. σ^* is linear since $(\sigma^*(a_1\alpha_1 + a_2\alpha_2), \beta) = f(a_1\alpha_1 + a_2\alpha_2, \beta) = \bar{a}_1 f(\alpha_1, \beta) + \bar{a}_2 f(\alpha_2, \beta) = \bar{a}_1(\sigma^*(\alpha_1), \beta) + \bar{a}_2(\sigma^*(\alpha_2), \beta) = (a_1\sigma^*(\alpha_1) + a_2\sigma^*(\alpha_2), \beta)$. Let $\sigma^{**} = \sigma$ be the linear transformation of which σ^* is the adjoint. Then $f(\alpha, \beta) = (\sigma^*(\alpha), \beta) = (\alpha, \sigma(\beta))$. $\square$

We shall call σ the *linear transformation associated* with the conjugate bilinear form f. The *eigenvalues* and *eigenvectors* of a conjugate bilinear form are defined to be the eigenvalues and eigenvectors of the associated linear transformation. Conversely, to each linear transformation σ there is associated a conjugate bilinear form $(\alpha, \sigma(\beta))$, and we shall also freely transfer terminology in the other direction. Thus a linear transformation will be called *symmetric*, or *skew-symmetric*, etc., if it is associated with a symmetric, or skew-symmetric bilinear form.

Theorem 4.8. *The conjugate bilinear form f and the linear transformation σ for which $f(\alpha, \beta) = (\alpha, \sigma(\beta))$ are represented by the same matrix with respect to an orthonormal basis.*

PROOF. Let $X = \{\xi_1, \ldots, \xi_n\}$ be an orthonormal basis and let $A = [a_{ij}]$ be the matrix representing σ with respect to this basis. Then $f(\xi_i, \xi_j) = (\xi_i, \sigma(\xi_j)) = (\xi_i, \sum_{k=1}^{n} a_{kj}\xi_k) = \sum_{k=k}^{n} a_{kj}(\xi_i, \xi_k) = a_{ij}$. $\square$

A linear transformation is called *self-adjoint* if $\sigma^* = \sigma$. Clearly, a linear transformation is self-adjoint if and only if the matrix representing it (with respect to an orthonormal basis) is Hermitian. However, by means of Theorem 4.7 self-adjointness of a linear transformation can be related to the Hermitian character of a conjugate bilinear form without the intervention of matrices. Namely, if f is a Hermitian form then $(\sigma^*(\alpha), \beta) = (\alpha, \sigma(\beta)) = f(\alpha, \beta) = \overline{f(\beta, \alpha)} = \overline{(\beta, \sigma(\alpha))} = (\sigma(\alpha), \beta)$.

Theorem 4.9. *If σ and τ are linear transformations on V such that $(\sigma(\alpha), \beta) = (\tau(\alpha), \beta)$ for all $\alpha, \beta \in V$, then $\sigma = \tau$.*

PROOF. If $(\sigma(\alpha), \beta) - (\tau(\alpha), \beta) = ((\sigma - \tau)(\alpha), \beta) = 0$ for all α, β, then for each α and $\beta = (\sigma - \tau)(\alpha)$ we have $\|(\sigma - \tau)(\alpha)\|^2 = 0$. Hence, $(\sigma - \tau)(\alpha) = 0$ for all α and $\sigma = \tau$. $\square$

Corollary 4.10. *If σ and τ are linear transformations on V such that $(\alpha, \sigma(\beta)) = (\alpha, \tau(\beta))$ for all $\alpha, \beta \in V$, then $\sigma = \tau$.* $\square$

Theorem 4.9 provides an independent proof that the adjoint operator σ^* is unique. Corollary 4.10 shows that the linear transformation σ corresponding to the bilinear form f such that $f(\alpha, \beta) = (\alpha, \sigma(\beta))$ is also unique. Since, in turn, each linear transformation σ defines a bilinear form f by the formula $f(\alpha, \beta) = (\alpha, \sigma(\beta))$, this establishes a one-to-one correspondence between conjugate bilinear forms and linear transformations.

Theorem 4.11. *Let V be a unitary vector space. If σ and τ are linear transformations on V such that $(\sigma(\alpha), \alpha) = (\tau(\alpha), \alpha)$ for all $\alpha \in V$, then $\sigma = \tau$.*

PROOF. It can be checked that

$$(\sigma(\alpha), \beta) = \tfrac{1}{4}\{(\sigma(\alpha + \beta), \alpha + \beta) - (\sigma(\alpha - \beta), \alpha - \beta)$$
$$- i(\sigma(\alpha + i\beta), \alpha + i\beta) + i(\sigma(\alpha - i\beta), \alpha - i\beta)\}. \quad (4.3)$$

It follows from the hypothesis that $(\sigma(\alpha), \beta) = (\tau(\alpha), \beta)$ for all $\alpha, \beta \in V$. Hence, by Theorem 4.9, $\sigma = \tau$. $\square$

It is curious to note that this theorem can be proved because of the relation (4.3), which is analogous to formula (12.4) of Chapter IV. But the analogue of formula (10.1) in the real case does not yield the same conclusion. In fact, if V is a vector space over the real numbers and σ is skew-symmetric, then $(\sigma(\alpha), \alpha) = (\alpha, \sigma^*(\alpha)) = (\alpha, -\sigma(\alpha)) = -(\alpha, \sigma(\alpha)) = -(\sigma(\alpha), \alpha)$ for all α. Thus $(\sigma(\alpha), \alpha) = 0$ for all α. In the real case the best analogue of this theorem is that if $(\sigma(\alpha), \alpha) = (\tau(\alpha), \alpha)$ for all $\alpha \in V$, then $\sigma + \sigma^* = \tau + \tau^*$, or σ and τ have the same symmetric part.

EXERCISES

1. Show that $(\sigma\tau)^* = \tau^*\sigma^*$.

2. Show that if $\sigma^*\sigma = 0$, then $\sigma = 0$.

3. Let σ be a skew-symmetric linear transformation on a vector space over the real numbers. Show that $\sigma^* = -\sigma$.

4. Let f be a skew-Hermitian form—that is, $f(\alpha, \beta) = -\overline{f(\beta, \alpha)}$—and let σ be the associated skew-Hermitian linear transformation. Show that $\sigma^* = -\sigma$.

5. Show that eigenvalues of a real skew-symmetric linear transformation are either 0 or pure imaginary. Show that the same is true for a skew-Hermitian linear transformation.

6. For what kind of linear transformation σ is it true that $(\xi, \sigma(\xi)) = 0$ for all $\xi \in V$?

7. For what kind of linear transformation σ is it true that $\sigma(\xi) \in \xi^{\perp}$ for all $\xi \in V$?

8. Show that if W is an invariant subspace under σ, then $W^{\perp}$ is an invariant subspace under σ^*.

9. Show that if σ is self-adjoint and W is invariant under σ, then $W^{\perp}$ is also invariant under σ.

10. Let π be the projection of V onto S along T. Let π^* be the adjoint of π. Show that π^* is the projection of V onto $T^{\perp}$ along $S^{\perp}$.

11. Let $W = \sigma(V)$. Show that $W^{\perp}$ is the kernel of σ^*.

12. Show that σ and σ^* have the same rank.

13. Let $W = \sigma(V)$. Show that $\sigma^*(V) = \sigma^*(W)$.

14. Show that $\sigma^*(V) = \sigma^*\sigma(V)$. Show that $\sigma(V) = \sigma\sigma^*(V)$.

15. Show that if $\sigma^*\sigma = \sigma^*\sigma$, then $\sigma^*(V) = \sigma(V)$.

16. Show that if $\sigma^*\sigma = \sigma\sigma^*$, then σ and σ^* have the same kernel.

17. Show that $\sigma + \sigma^*$ is self-adjoint.

18. Show that if $\sigma + \sigma^* = 0$, then σ is skew-symmetric, or skew-Hermitian.

19. Show that $\sigma - \sigma^*$ is skew-symmetric, or skew-Hermitian.

20. Show that every linear transformation is the sum of a self-adjoint transformation and a skew-Hermitian transformation.

21. Show that if $\sigma\sigma^* = \sigma^*\sigma$, then $\text{Im}(\sigma)$ is an invariant subspace under σ. In fact, show that $\sigma^n(V) = \sigma(V)$ for all $n \geq 1$.

22. Show that if σ is a scalar transformation, that is $\sigma(\alpha) = a\alpha$, then $\sigma^*(\alpha) = \bar{a}\alpha$.

5 | Orthogonal and Unitary Transformations

Definition. A linear transformation of V into itself is called an *isometry* if it preserves length; that is, σ is an isometry if and only if $\|\sigma(\alpha)\| = \|\alpha\|$ for all $\alpha \in V$. An isometry in a vector space over the real numbers is called an *orthogonal transformation*. An isometry in a vector space over the complex numbers is called a *unitary transformation*. We try to save duplication and repetition by treating the real and complex cases together whenever possible.

Theorem 5.1. *A linear transformation σ of V into itself is an isometry if and only if it preserves the inner product; that is, if and only if $(\alpha, \beta) = (\sigma(\alpha), \sigma(\beta))$ for all $\alpha, \beta \in V$.*

PROOF. Certainly, if σ preserves the inner product then it preserves length since $\|\sigma(\alpha)\|^2 = (\sigma(\alpha), \sigma(\alpha)) = (\alpha, \alpha) = \|\alpha\|^2$.

The converse requires the separation of the real and complex cases. For an inner product over the real numbers we have

$$(\alpha, \beta) = \tfrac{1}{2}\{(\alpha + \beta, \alpha + \beta) - (\alpha, \alpha) - (\beta, \beta)\}$$
$$= \tfrac{1}{2}\{\|\alpha + \beta\|^2 - \|\alpha\|^2 - \|\beta\|^2\}. \tag{5.1}$$

For an inner product over the complex numbers we have

$$(\alpha, \beta) = \tfrac{1}{4}\{(\alpha + \beta, \alpha + \beta) - (\alpha - \beta, \alpha - \beta)$$
$$- i(\alpha + i\beta, \alpha + i\beta) + i(\alpha - i\beta, \alpha - i\beta)\}$$
$$= \tfrac{1}{4}\{\|\alpha + \beta\|^2 - \|\alpha - \beta\|^2 - i\|\alpha + i\beta\|^2 + i\|\alpha - i\beta\|^2\}. \tag{5.2}$$

In either case, any linear transformation which preserves length will preserve the inner product. □

Theorem 5.2. *A linear transformation σ of V into itself is an isometry if and only if it maps an orthonormal basis onto an orthonormal basis.*

PROOF. It follows immediately from Theorem 5.1 that if σ is an isometry, then σ maps every orthonormal set onto an orthonormal set and, therefore, an orthonormal basis onto an orthonormal basis.

On the other hand, let $X = \{\xi_1, \ldots, \xi_n\}$ be any orthonormal basis which is mapped by σ onto an orthonormal basis $\{\sigma(\xi_1), \ldots, \sigma(\xi_n)\}$. For an arbitrary vector $\alpha \in V$, $\alpha = \sum_{i=1}^{n} x_i \xi_i$, we have

$$\|\sigma(\alpha)\|^2 = (\sigma(\alpha), \sigma(\alpha))$$

$$= \left(\sum_{i=1}^{n} x_i \sigma(\xi_i), \sum_{j=1}^{n} x_j \sigma(\xi_j) \right)$$

$$= \sum_{i=1}^{n} \bar{x}_i \sum_{j=1}^{n} x_j (\sigma(\xi_i), \sigma(\xi_j))$$

$$= \sum_{i=1}^{n} \bar{x}_i \sum_{j=1}^{n} x_j \delta_{ij}$$

$$= \sum_{i=1}^{n} \bar{x}_i x_i = \|\alpha\|^2. \tag{5.3}$$

Thus σ preserves length and it is an isometry. $\square$

Theorem 5.3. *σ is an isometry if and only if $\sigma^* = \sigma^{-1}$.*

PROOF. If σ is an isometry, then $(\sigma(\alpha), \sigma(\beta)) = (\alpha, \beta)$ for all $\alpha, \beta \in V$. By the definition of σ^*, $(\alpha, \beta) = (\sigma^*[\sigma(\alpha)], \beta) = (\sigma^*\sigma(\alpha), \beta)$. Since this equation holds for all $\beta \in V$, $\sigma^*\sigma(\alpha)$ is uniquely defined and $\sigma^*\sigma(\alpha) = \alpha$. Thus $\sigma^*\sigma$ is the identity transformation, that is, $\sigma^* = \sigma^{-1}$.

Conversely, suppose that $\sigma^* = \sigma^{-1}$. Then $(\sigma(\alpha), \sigma(\beta)) = (\sigma^*[\sigma(\alpha)], \beta) = (\sigma^*\sigma(\alpha), \beta) = (\alpha, \beta)$ for all $\alpha, \beta \in V$, and σ is an isometry. $\square$

EXERCISES

1. Let σ be an isometry and let λ be an eigenvalue of σ. Show that $|\lambda| = 1$.

2. Show that the real eigenvalues of an isometry are ± 1.

3. Let $X = \{\xi_1, \xi_2\}$ be an orthonormal basis of V. Find an isometry that maps ξ_1 onto $\dfrac{1}{\sqrt{2}} (\xi_1 + \xi_2)$.

4. Let $X = \{\xi_1, \xi_2, \xi_3\}$ be an orthonormal basis of V. Find an isometry that maps ξ_1 onto $\frac{1}{3}(\xi_1 + 2\xi_2 + 2\xi_3)$.

6 | Orthogonal and Unitary Matrices

Let σ be an isometry and let $U = [u_{ij}]$ be a matrix representing σ with respect to an orthonormal basis $X = \{\xi_i, \ldots, \xi_n\}$. Since σ is an isometry, the set $X' = \{\sigma(\xi_1), \ldots, \sigma(\xi_n)\}$ must also be orthonormal. Thus

$$\delta_{ij} = (\sigma(\xi_i), \sigma(\xi_j))$$

$$= \left(\sum_{k=1}^{n} u_{ki} \xi_k, \sum_{l=1}^{n} u_{lj} \xi_l \right)$$

$$= \sum_{k=1}^{n} \overline{u_{ki}} \left(\sum_{l=1}^{n} u_{lj} (\xi_k, \xi_l) \right)$$

$$= \sum_{k=1}^{n} \overline{u_{ki}} u_{kj}. \tag{6.1}$$

This is equivalent to the matrix equation $U*U = I$, which also follows from Theorem 5.3.

It is also easily seen that if $U*U = I$, then σ must map an orthonormal basis onto an orthonormal basis. By Theorem 5.2 σ is then an isometry. Thus,

Theorem 6.1.   *A matrix U whose elements are complex numbers represents a unitary transformation (with respect to an orthonormal basis) if and only if $U* = U^{-1}$. A matrix with this property is called a unitary matrix.* $\square$

If the underlying field of scalars is the real numbers instead of the complex numbers, then U is real and $U* = U^T$. Nothing else is really changed and we have the corresponding theorem for vector spaces over the real numbers.

Theorem 6.2.   *A matrix U whose elements are real numbers represents an orthogonal transformation (with respect to an orthonormal basis) if and only if $U^T = U^{-1}$. A real matrix with this property is called an orthogonal matrix.* $\square$

As is the case in Theorems 6.1 and 6.2, quite a bit of the discussion of unitary and orthogonal transformations and matrices is entirely parallel. To avoid unnecessary duplication we discuss unitary transformations and matrices and leave the parallel discussion for orthogonal transformations and matrices implicit. Up to a certain point, an orthogonal matrix can be considered to be a unitary matrix that happens to have real entries. This viewpoint is not quite valid because a unitary matrix with real coefficients represents a unitary transformation, an isometry on a vector space over the complex numbers. This viewpoint, however, leads to no trouble until we make use of the algebraic closure of the complex numbers, the property of complex numbers that every polynomial equation with complex coefficients possesses at least one complex solution.

It is customary to read equations (6.1) as saying that the columns of U are orthonormal. Conversely, if the columns of U are orthonormal, then $U* = U^{-1}$ and U is unitary. Also, $U*$ as a left inverse is also a right inverse; that is, $UU* = I$. Thus,

$$\sum_{k=1}^{n} u_{ik}\overline{u_{jk}} = \delta_{ij} = \sum_{k=1}^{n} \overline{u_{jk}}u_{ik}. \tag{6.2}$$

Thus U is unitary if and only if the rows of U are orthonormal. Hence,

Theorem 6.3.   *Unitary and orthogonal matrixes are characterized by the property that their columns are orthonormal. They are equally characterized by the property that their rows are orthonormal.* $\square$

Theorem 6.4.   *The product of unitary matrices is unitary. The product of orthogonal matrices is orthogonal.*

PROOF. This follows immediately from the observation that unitary and orthogonal matrices represent isometries, and one isometry followed by another results in an isometry. □

A proof of Theorem 6.4 based on the characterizing property $U^* = U^{-1}$ (or $U^T = U^{-1}$ for orthogonal matrices) is just as brief. Namely, $(U_1U_2)^* = U_2^*U_1^* = U_2^{-1}U_1^{-1} = (U_1U_2)^{-1}$.

Now suppose that $X = \{\xi_1, \ldots, \xi_n\}$ and $X' = \{\xi_1', \ldots, \xi_n'\}$ are two orthonormal bases, and that $P = [p_{ij}]$ is the matrix of transition from the basis X to the basis X'. By definition,

$$\xi_j' = \sum_{i=1}^{n} p_{ij}\xi_i. \tag{6.3}$$

Thus,

$$(\xi_j', \xi_k') = \left(\sum_{i=1}^{n} p_{ij}\xi_i, \sum_{s=1}^{n} p_{sk}\xi_s \right)$$

$$= \sum_{i=1}^{n} \bar{p}_{ij} \sum_{s=1}^{n} p_{sk}(\xi_i, \xi_s)$$

$$= \sum_{i=1}^{n} \bar{p}_{ij}p_{ik} = \delta_{jk}. \tag{6.4}$$

This means the columns of P are orthonormal and P is unitary (or orthogonal). Thus we have

Theorem 6.5. *The matrix of transition from one orthonormal basis to another is unitary (or orthogonal if the underlying field is real).* □

We have seen that two matrices representing the same linear transformation with respect to different bases are similar. If the two bases are both ortho-normal, then the matrix of transition is unitary (or orthogonal). In this case we say that the two matrices are *unitary similar* (or *orthogonal similar*). The matrices A and A' are unitary (orthogonal) similar if and only if there exists a unitary (orthogonal) matrix P such that $A' = P^{-1}AP = P^*AP$ ($A' = P^{-1}AP = P^TAP$).

If H and H' are matrices representing the same conjugate bilinear form with respect to different bases, they are Hermitian congruent and there exists a non-singular matrix P such that $H' = P^*HP$. P is the matrix of transition and, if the two bases are orthonormal, P is unitary. Then $H' = P^*HP = P^{-1}HP$. Hence, if we restrict our attention to orthonormal bases in vector spaces over the complex numbers, we see that matrices representing linear transformations and matrices representing conjugate bilinear forms transform according to the same rules; they are unitary similar.

If B and B' are matrices representing the same real bilinear form with respect to different bases, they are congruent and there exists a non-singular matrix P such that $B' = P^T BP$. P is the matrix of transition and, if the two bases are orthonormal, P is orthogonal. Then $B' = P^T BP = P^{-1}BP$. Hence, if we restrict our attention to orthonormal bases in vector spaces over the real numbers, we see that matrices representing linear transformations and matrices representing bilinear forms transform according to the same rules; they are orthogonal similar.

In our earlier discussions of similarity we sought bases with respect to which the representing matrix had a simple form, usually a diagonal form. We were not always successful in obtaining a diagonal form. Now we restrict the set of possible bases even further by demanding that they be orthonormal. But we can also restrict our attention to the set of matrices which are unitary (or orthogonal) similar to diagonal matrices. It is fortunate that this restricted class of matrices includes a rather wide range of cases occurring in some of the most important applications of matrices. The main goal of this chapter is to define and characterize the class of matrices unitary similar to diagonal matrices and to organize computational procedures by means of which these diagonal matrices and the necessary matrices of transition can be obtained. We also discuss the special cases so important in the applications of the theory of matrices.

EXERCISES

1. Test the following matrices for orthogonality. If a matrix is orthogonal, find its inverse.

(a)
$$\begin{bmatrix} \dfrac{1}{2} & \dfrac{\sqrt{3}}{2} \\[2mm] \dfrac{-\sqrt{3}}{2} & \dfrac{1}{2} \end{bmatrix}$$

(b)
$$\begin{bmatrix} \dfrac{1}{2} & \dfrac{\sqrt{3}}{2} \\[2mm] \dfrac{\sqrt{3}}{2} & \dfrac{1}{2} \end{bmatrix}$$

(c)
$$\begin{bmatrix} 0.6 & 0.8 \\ 0.8 & -0.6 \end{bmatrix}$$

2. Which of the following matrices are unitary?

(a)
$$\begin{bmatrix} \dfrac{1+i}{2} & \dfrac{1-i}{2} \\[2mm] \dfrac{1-i}{2} & \dfrac{1+i}{2} \end{bmatrix}$$

(b)
$$\begin{bmatrix} 1 & i \\ i & 1 \end{bmatrix}$$

(c)
$$\begin{bmatrix} 1 & -i \\ i & 1 \end{bmatrix}$$

3. Find an orthogonal matrix with $(1/\sqrt{2}, 1/\sqrt{2})$ in the first column. Find an orthogonal matrix with $(\tfrac{1}{3}, \tfrac{2}{3}, \tfrac{2}{3})$ in the first column.

4. Find a symmetric orthogonal matrix with $(\tfrac{1}{3}, \tfrac{2}{3}, \tfrac{2}{3})$ in the first column. Compute its square.

5. The following matrices are all orthogonal. Describe the geometric effects in real Euclidean 3-space of the linear transformations they represent.

$$(a) \begin{bmatrix} 1 & 0 & 0 \\ 0 & 1 & 0 \\ 0 & 0 & -1 \end{bmatrix} \quad (b) \begin{bmatrix} -1 & 0 & 0 \\ 0 & -1 & 0 \\ 0 & 0 & 1 \end{bmatrix} \quad (c) \begin{bmatrix} -1 & 0 & 0 \\ 0 & -1 & 0 \\ 0 & 0 & -1 \end{bmatrix}$$

$$(d) \begin{bmatrix} \cos\theta & -\sin\theta & 0 \\ \sin\theta & \cos\theta & 0 \\ 0 & 0 & 1 \end{bmatrix} \quad (e) \begin{bmatrix} \cos\theta & -\sin\theta & 0 \\ \sin\theta & \cos\theta & 0 \\ 0 & 0 & -1 \end{bmatrix}.$$

Show that these five matrices, together with the identity matrix, each have different eigenvalues (provided θ is not $0°$ or $180°$), and that the eigenvalues of any third-order orthogonal matrix must be one of these six cases.

6. If a matrix represents a rotation of R^2 around the origin through an angle of θ, then it has the form

$$A(\theta) = \begin{bmatrix} \cos\theta & -\sin\theta \\ \sin\theta & \cos\theta \end{bmatrix}.$$

Show that $A(\theta)$ is orthogonal. Knowing that $A(\theta) \cdot A(\psi) = A(\theta + \psi)$, prove that $\sin(\theta + \psi) = \sin\theta\cos\psi + \cos\theta\sin\psi$. Show that if U is an orthogonal 2×2 matrix, then $U^{-1}A(\theta)U = A(\pm\theta)$.

7. Find the matrix B representing the real quadratic form $q(x, y) = ax^2 + 2bxy + cy^2$. Show that the discriminant $D = ac - b^2$ is the determinant of B. Show that the discriminant is invariant under orthogonal coordinate changes, that is, changes of coordinates for which the matrix of transition is orthogonal.

7 | Superdiagonal Form

In this section we restrict our attention to vector spaces (and to matrices) over the field of complex numbers. We have already observed that not every matrix is similar to a diagonal matrix. Thus, it is also true that not every matrix is unitary similar to a diagonal matrix. We later restrict our attention to a class of matrices which are unitary similar to diagonal matrices. As an intermediate step we obtain a relatively simple form to which every matrix can be reduced by unitary similar transformations.

Theorem 2.1. Let σ be any linear transformation of V, a finite dimensional vector space over the complex numbers, into itself. There exists an orthonormal basis of V with respect to which the matrix representing σ is in superdiagonal form; that is, every element below the main diagonal is zero.

PROOF. The proof is by induction on n, the dimension of V. The theorem says there is an orthonormal basis $Y = \{\eta_1, \ldots, \eta_n\}$ such that $\sigma(\eta_k) = \sum_{i=1}^{k} a_{ik}\eta_i$, the important property being that the summation ends with the kth term. The theorem is certainly true for $n = 1$.

Assume the theorem is true for vector spaces of dimensions $<n$. Since V is a vector space over the complex numbers, σ has at least one eigenvalue. Let λ_1 be an eigenvalue for σ and let $\xi_1' \neq 0$, $\|\xi_1'\| = 1$, be a corresponding eigenvector. There exists a basis, and hence an orthonormal basis, with ξ_1' as the first element. Let the basis be $X' = \{\xi_1', \ldots, \xi_n'\}$ and let W be the subspace spanned by $\{\xi_2', \ldots, \xi_n'\}$. W is the subspace consisting of all vectors orthogonal to ξ_1'. For each $\alpha = \sum_{i=1}^n a_i \xi_i'$ define $\tau(\alpha) = \sum_{i=2}^n a_i \xi_i' \in W$. Then $\tau\sigma$ restricted to W is a linear transformation of W into itself. According to the induction assumption, there is an orthonormal basis $\{\eta_2, \ldots, \eta_n\}$ of W such that for each η_k, $\tau\sigma(\eta_k)$ is expressible in terms of $\{\eta_2, \ldots, \eta_k\}$ alone. We see from the way τ is defined that $\sigma(\eta_k)$ is expressible in terms of $\{\xi_1', \eta_2, \ldots, \eta_k\}$ alone. Let $\eta_1 = \xi_1'$. Then $Y = \{\eta_1, \eta_2, \ldots, \eta_n\}$ is the required basis.

ALTERNATE PROOF. The proof just given was designed to avoid use of the concept of adjoint introduced in Section 4. Using that concept, a very much simpler proof can be given. This proof also proceeds by induction on n. The assertion for $n = 1$ is established in the same way as in the first proof given. Assume the theorem is true for vector spaces of dimension $<n$. Since V is a vector space over the complex numbers, σ^* has at least one eigenvalue. Let λ_n be an eigenvalue for σ^* and let η_n, $\|\eta_n\| = 1$, be a corresponding eigenvector. Then by Theorem 4.4, $W = \langle \eta_n \rangle^\perp$ is an invariant subspace under σ. Since $\eta_n \neq 0$, W is of dimension $n - 1$. According to the induction assumption, there is an orthonormal basis $\{\eta_1, \ldots, \eta_{n-1}\}$ of W such that $\sigma(\eta_k) = \sum_{i=1}^k a_{ik}\eta_i$, for $k = 1, 2, \ldots, n - 1$. However, $\{\eta_1, \ldots, \eta_n\}$ is also an orthonormal basis of U and $\sigma(\eta_n) = \sum_{i=1}^k a_{ik}\eta_i$, for $k = 1, \ldots, n$. $\square$

Corollary 7.2. *Over the field of complex numbers, every matrix is unitary similar to a superdiagonal matrix.* $\square$

Theorem 7.1 and Corollary 7.2 depend critically on the assumption that the field of scalars is the field of complex numbers. The essential feature of this condition is that it guarantees the existence of eigenvalues and eigenvectors. If the field of scalars is not algebraically closed, the theorem is simply not true.

Corollary 7.3. *The diagonal terms of the superdiagonal matrix representing σ are the eigenvalues of σ.*

PROOF. If $A = [a_{ij}]$ is in superdiagonal form, then the characteristic polynomial is $(a_{11} - x)(a_{22} - x) \cdots (a_{nn} - x)$. $\square$

EXERCISES

1. Let σ be a linear transformation mapping U into V. Let A be any basis of U whatever. Show that there is an orthonormal basis B of V such that the matrix

representing σ with respect to A and B is in superdiagonal form. (In this case where U and V need not be of the same dimension so that the matrix representing σ need not be square, by superdiagonal form we mean that all elements below the main diagonal are zeros.)

2. Let σ be a linear transformation on V and let $Y = \{\eta_1, \ldots, \eta_n\}$ be an orthonormal basis such that the matrix representing σ with respect to Y is in superdiagonal form. Show that the matrix representing σ^* with respect to Y is in subdiagonal form; that is, all elements above the main diagonal are zeros.

3. Let σ be a linear transformation on V. Show that there is an orthonormal basis Y of V such that the matrix representing σ with respect to Y is in subdiagonal form.

8 | Normal Matrices

It is possible to give a necessary and sufficient condition that a matrix be unitary similar to a diagonal matrix. The real value in establishing this condition is that several important types of matrices do satisfy this condition.

Theorem 8.1. *A matrix A in superdiagonal form is a diagonal matrix if and only if $A^*A = AA^*$.*

PROOF. Let $A = [a_{ij}]$ where $a_{ij} = 0$ if $i > j$. Suppose that $A^*A = AA^*$. This means, in particular, that

$$\sum_{j=1}^{n} \bar{a}_{ji} a_{ji} = \sum_{k=1}^{n} a_{ik} \bar{a}_{ik}. \tag{8.1}$$

But since $a_{ij} = 0$ for $i > j$, this reduces to

$$\sum_{j=1}^{i} |a_{ji}|^2 = \sum_{k=i}^{n} |a_{ik}|^2. \tag{8.2}$$

Now, if A were not a diagonal matrix, there would be a first index i for which there exists an index $k > i$ such that $a_{ik} \neq 0$. For this choice of the index i the sum on the left in (8.2) reduces to one term while the sum on the right contains at least two non-zero terms. Thus,

$$\sum_{j=1}^{i} |a_{ji}|^2 = |a_{ji}|^2 = \sum_{k=i}^{n} |a_{ik}|^2, \tag{8.3}$$

which is a contradiction. Thus A must be a diagonal matrix.

Conversely, if A is a diagonal matrix, then clearly $A^*A = AA^*$. $\square$

A matrix A for which $A^*A = AA^*$ is called a *normal matrix*.

Theorem 8.2. *A matrix is unitary similar to a diagonal matrix if and only if it is normal.*

PROOF.   If A is a normal matrix, then any matrix unitary similar to A is also normal. Namely, if U is unitary, then

$$(U^*AU)^*(U^*AU) = U^*A^*UU^*AU$$
$$= U^*A^*AU$$
$$= U^*AA^*U$$
$$= U^*AUU^*A^*U.$$
$$= (U^*AU)(U^*AU)^*.   \tag{8.4}$$

Thus, if A is normal, the superdiagonal form to which it is unitary similar is also normal and, hence, diagonal. Conversely, if A is unitary similar to a diagonal matrix, it is unitary similar to a normal matrix and it is therefore normal itself. □

Theorem 8.3.   *Unitary matrices and Hermitian matrices are normal.*
PROOF.   If U is unitary then $U^*U = U^{-1}U = UU^{-1} = UU^*$. If H is Hermitian then $H^*H = HH = HH^*$. □

EXERCISES

1. Determine which of the following matrices are orthogonal, unitary, symmetric Hermitian, skew-symmetric, skew-Hermitian, or normal.

(a) $\begin{bmatrix} 1 & -2 \\ 2 & 1 \end{bmatrix}$ (b) $\begin{bmatrix} 1 & i \\ i & 1 \end{bmatrix}$ (c) $\begin{bmatrix} 1 & i \\ i & 2 \end{bmatrix}$

(d) $\begin{bmatrix} 0 & 1 \\ 1 & 0 \end{bmatrix}$ (e) $\begin{bmatrix} 0 & -1 \\ 1 & 0 \end{bmatrix}$ (f) $\begin{bmatrix} 1 & 1-i \\ 1+i & 3 \end{bmatrix}$

(g) $\frac{1}{3}\begin{bmatrix} 1 & -2 & 2 \\ 2 & 2 & 1 \\ 2 & -1 & -2 \end{bmatrix}$ (h) $\frac{1}{3}\begin{bmatrix} 1 & 2 & 2 \\ 2 & -2 & 1 \\ 2 & 1 & -2 \end{bmatrix}$ (i) $\begin{bmatrix} 0 & -1 & -2 \\ 1 & 0 & -3 \\ 2 & 3 & 0 \end{bmatrix}$

(j) $\begin{bmatrix} 1 & 1 \\ 0 & 1 \end{bmatrix}$ (k) $\begin{bmatrix} 0 & -1 & -1 \\ 1 & 0 & -1 \\ 1 & 1 & 0 \end{bmatrix}$

2. Which of the matrices of Exercise 1 are unitary similar to diagonal matrices?

3. Show that a real skew-symmetric matrix is normal.

4. Show that a skew-Hermitian matrix is normal.

5. Show by example that there is a skew-symmetric complex matrix which is not normal.

6. Show by example that there is a symmetric complex matrix which is not normal.

7. Find an example of a normal matrix which is not Hermitian or unitary.

8. Show that if $M = A + Bi$ where A and B are real and symmetric, then M is normal if and only if A and B commute.

9 | Normal Linear Transformations

Theorem 9.1. *If there exists an orthonormal basis consisting of eigenvectors of a linear transformation σ, then $\sigma^*\sigma = \sigma\sigma^*$.*

PROOF.   Let $X = \{\xi_1, \ldots, \xi_n\}$ be an orthonormal basis consisting of eigenvectors of σ. Let λ_i be the eigenvalue corresponding to ξ_i. Then.
$(\sigma^*(\xi_i), \xi_j) = (\xi_i, \sigma(\xi_j)) = (\xi_i, \lambda_j\xi_j) = \lambda_j\delta_{ij} = \lambda_i\delta_{ij} = \lambda_i(\xi_i, \xi_j) = (\bar{\lambda}_i\xi_i, \xi_j)$.
For a fixed ξ_i this equation holds for all ξ_j and, hence, $(\sigma^*(\xi_i), \alpha) = (\bar{\lambda}_i\xi_i, \alpha)$ for all $\alpha \in V$. This means $\sigma^*(\xi_i) = \bar{\lambda}_i\xi_i$ and ξ_i is an eigenvector of σ^* with eigenvalue $\bar{\lambda}_i$. Then $\sigma\sigma^*(\xi_i) = \sigma(\bar{\lambda}_i\xi_i) = \bar{\lambda}_i\lambda_i\xi_i = \sigma^*\sigma(\xi_i)$. Since $\sigma\sigma^* = \sigma^*\sigma$ on a basis of V, $\sigma\sigma^* = \sigma^*\sigma$ on all of V. $\square$

A linear transformation σ for which $\sigma^*\sigma = \sigma\sigma^*$ is called a *normal linear transformation.* Clearly, a linear transformation is normal if and only if the matrix representing it (with respect to an orthonormal basis) is normal.

In the proof of Theorem 9.1 the critical step is showing that an eigenvector of σ is also an eigenvector of σ^*. The converse is also true.

Theorem 9.2. *If ξ is an eigenvector of a normal linear transformation σ corresponding to the eigenvalue λ, then ξ is an eigenvector of σ^* corresponding to $\bar{\lambda}$.*

PROOF.   Since σ is normal $(\sigma(\xi), \sigma(\xi)) = (\sigma^*\sigma(\xi), \xi) = (\sigma\sigma^*(\xi), \xi) = (\sigma^*(\xi), \sigma^*(\xi))$. Since ξ is an eigenvector of σ corresponding to λ, $\sigma(\xi) = \lambda\xi$ so that

$$
\begin{aligned}
0 = \|\sigma(\xi) - \lambda\xi\|^2 &= (\sigma(\xi) - \lambda\xi, \sigma(\xi) - \lambda\xi) \\
&= (\sigma(\xi), \sigma(\xi)) - \bar{\lambda}(\xi, \sigma(\xi)) - \lambda(\sigma(\xi), \xi) + \bar{\lambda}\lambda(\xi, \xi) \\
&= (\sigma^*(\xi), \sigma^*(\xi)) - \bar{\lambda}(\sigma^*(\xi), \xi) - \lambda(\xi, \sigma^*(\xi)) + \bar{\lambda}\lambda(\xi, \xi) \\
&= (\sigma^*(\xi) - \bar{\lambda}\xi, \sigma^*(\xi) - \bar{\lambda}\xi) \\
&= \|\sigma^*(\xi) - \bar{\lambda}\xi\|^2. \qquad (9.1)
\end{aligned}
$$

Thus $\sigma^*(\xi) - \bar{\lambda}\xi = 0$, or $\sigma^*(\xi) = \bar{\lambda}\xi$. $\square$

Theorem 9.3. *For a normal linear transformation, eigenvectors corresponding to different eigenvalues are orthogonal.*

PROOF.   Suppose $\sigma(\xi_1) = \lambda_1\xi_1$ and $\sigma(\xi_2) = \lambda_2\xi_2$ where $\lambda_1 \neq \lambda_2$. Then $\lambda_2(\xi_1, \xi_2) = (\xi_1, \lambda_2\xi_2) = (\xi_1, \sigma(\xi_2)) = (\sigma^*(\xi_1), \xi_2) = (\bar{\lambda}_1\xi_1, \xi_2) = \lambda_1(\xi_1, \xi_2)$. Thus $(\lambda_1 - \lambda_2)(\xi_1, \xi_2) = 0$. Since $\lambda_1 - \lambda_2 \neq 0$ we see that $(\xi_1, \xi_2) = 0$; that is, ξ_1 and ξ_2 are orthogonal. $\square$

Theorem 9.4. *If σ is normal, then $(\sigma(\alpha), \sigma(\beta)) = (\sigma^*(\alpha), \sigma^*(\beta))$ for all $\alpha, \beta \in V$.*

PROOF. $(\sigma(\alpha), \sigma(\beta)) = (\sigma^*\sigma(\alpha), \beta) = (\sigma\sigma^*(\alpha), \beta) = (\sigma^*(\alpha), \sigma^*(\beta))$. $\square$

Corollary 9.5. *If σ is normal, $\|\sigma(\alpha)\| = \|\sigma^*(\alpha))\|$ for all $\alpha \in V$.* $\square$

Theorem 9.6. *If $(\sigma(\alpha), \sigma(\beta)) = (\sigma^*(\alpha), \sigma^*(\beta))$ for all α, $\beta \in V$, then σ is normal.*

PROOF. $(\alpha, \sigma\sigma^*(\beta)) = (\sigma^*(\alpha), \sigma^*(\beta)) = (\sigma(\alpha), \sigma(\beta)) = (\alpha, \sigma^*\sigma(\beta))$ for all α, $\beta \in V$. By Corollary 4.10, $\sigma\sigma^* = \sigma^*\sigma$ and σ is normal. $\square$

Theorem 9.7. *If $\|\sigma(\alpha)\| = \|\sigma^*(\alpha)\|$ for all $\alpha \in V$, then σ is normal.*

PROOF. We must divide this proof into two cases:

1. V is a vector space over F, a subfield of the real numbers. Then

$$(\sigma(\alpha), \sigma(\beta)) = \tfrac{1}{4}\{\|\sigma(\alpha + \beta)\|^2 - \|\sigma(\alpha - \beta)\|^2\}.$$

It then follows from the hypothesis that $(\sigma(\alpha), \sigma(\beta)) = (\sigma^*(\alpha), \sigma^*(\beta))$ for all α, $\beta \in V$, and σ is normal.

2. V is a vector space over F, a non-real normal subfield of the complex numbers. Let $a \in F$ be chosen so that $a \neq \bar{a}$. Then

$$(\sigma(\alpha), \sigma(\beta)) = \frac{1}{2(\bar{a} - a)}$$

$$\times \{\bar{a} \, \|\sigma(\alpha + \beta)\|^2 - \bar{a} \, \|\sigma(\alpha - \beta)\|^2 - \|\sigma(\alpha + a\beta)\|^2 + \|\sigma(\alpha - a\beta)\|^2\}.$$

Again, it follows that σ is normal. $\square$

Theorem 9.8. *If σ is normal then $K(\sigma) = K(\sigma^*)$.*

PROOF. Since $\|\sigma(\alpha)\| = \|\sigma^*(\alpha)\|$, $\sigma(\alpha) = 0$ if and only if $\sigma^*(\alpha) = 0$. $\square$

Theorem 9.9. *If σ is normal, $K(\sigma) = \text{Im}(\sigma)^{\perp}$.*

PROOF. By Theorem 4.5 $K(\sigma^*) = \text{Im}(\sigma)^{\perp}$, and by Theorem 9.8 $K(\sigma) = K(\sigma^*)$. $\square$

Theorem 9.10. *If σ is normal, $\text{Im } \sigma = \text{Im } \sigma^*$.*

PROOF. $\text{Im } \sigma = K(\sigma)^{\perp} = \text{Im } \sigma^*$. $\square$

Theorem 9.11. *If σ is a normal linear transformation and W is a set of eigenvectors of σ, then $W^{\perp}$ is an invariant subspace under σ.*

PROOF. $\alpha \in W^{\perp}$ if and only if $(\xi, \alpha) = 0$ for all $\xi \in W$. But then $(\xi, \sigma(\alpha)) = (\sigma^*(\xi), \alpha) = (\bar{\lambda}\xi, \alpha) = \lambda(\xi, \alpha) = 0$. Hence, $\sigma(\alpha) \in W^{\perp}$ and $W^{\perp}$ is invariant under σ. $\square$

Notice it is not necessary that W be a subspace, it is not necessary that W contain all the eigenvectors corresponding to any particular eigenvalue, and it is not necessary that the eigenvectors in W correspond to the same eigenvalue. In particular, if ξ is an eigenvector of σ, then $\{\xi\}^{\perp}$ is an invariant subspace under σ.

Theorem 9.12. *Let V be a vector space with an inner product, and let σ be a normal linear transformation of V into itself. If W is a subspace which is invariant under both σ and σ^*, then σ is normal on W.*

PROOF. Let $\underline{\sigma}$ denote the linear transformation of W into itself induced by σ. Let $\underline{\sigma}^*$ denote the adjoint of $\underline{\sigma}$ on W. Then for all, $\alpha, \beta \in W$ we have

$$(\underline{\sigma}^*(\alpha), \beta) = (\alpha, \underline{\sigma}(\beta)) = (\alpha, \sigma(\beta)) = (\sigma^*(\alpha), \beta).$$

Since $((\underline{\sigma}^* - \sigma^*)(\alpha), \beta) = 0$ for all $\alpha, \beta \in W$, $\underline{\sigma}^*$ and σ^* coincide on W. Thus $\underline{\sigma}^*\underline{\sigma} = \sigma^*\sigma = \sigma\sigma^* = \underline{\sigma}\underline{\sigma}^*$ on W, and $\underline{\sigma}$ is normal. $\square$

Theorem 9.13. *Let V be a finite dimensional vector space over the complex numbers, and let σ be a normal linear transformation on V. If W is invariant under σ, then W is invariant under σ^* and σ is normal on W.*

PROOF. By Theorem 4.4, $W^\perp$ is invariant under σ^*. Let $\overline{\sigma^*}$ be the restriction of σ^* with $W^\perp$ as domain and codomain. Since $W^\perp$ is also a finite dimensional vector space over the complex numbers, $\overline{\sigma^*}$ has at least one eigenvalue λ and corresponding to it a non-zero eigenvector ξ. Thus $\overline{\sigma^*}(\xi) = \lambda\xi = \sigma^*(\xi)$. Thus, we have found an eigenvector for σ^* in $W^\perp$.

Now proceed by induction. The theorem is certainly true for spaces of dimension 1. Assume the theorem holds for vector spaces of dimension $<n$. By Theorem 9.2, ξ is an eigenvector of σ. By Theorem 9.11, $\langle\xi\rangle^\perp$ is invariant under both σ and σ^*. By Theorem 9.12, σ is normal on $\langle\xi\rangle^\perp$. Since $\langle\xi\rangle \subset W^\perp$, $W \subset \langle\xi\rangle^\perp$. Since dim $\langle\xi\rangle^\perp = n - 1$, the induction assumption applies. Hence, σ is normal on W and W is invariant under σ^*. $\square$

Theorem 9.13 is also true for a vector space over any subfield of the complex numbers, but the proof is not particularly instructive and this more general form of Theorem 9.13 will not be needed later.

We should like to obtain a converse of Theorem 9.1 and show that a normal linear transformation has enough eigenvectors to make up an orthonormal basis. Such a theorem requires some condition to guarantee the existence of eigenvalues or eigenvectors. One of the most important general conditions is to assume we are dealing with vector spaces over the complex numbers.

Theorem 9.14. *If V is a finite dimensional vector space over the complex numbers and σ is a normal linear transformation, then V has an orthonormal basis consisting of eigenvectors of σ.*

PROOF. Let n be the dimension of V. The theorem is certainly true for $n = 1$, for if $\{\xi_1\}$ is a basis $\sigma(\xi_1) = a_1\xi_1$.

Assume the theorem holds for vector spaces of dimension $<n$. Since V is a finite dimensional vector space over the complex numbers, σ has at least one eigenvalue λ_1, and corresponding to it a non-zero eigenvector ξ_1 which we can take to be normalized. By Theorem 9.11, $\{\xi_1\}^\perp$ is an invariant subspace under σ. This means that σ acts like a linear transformation on $\{\xi_1\}^\perp$ when we confine our attention to $\{\xi_1\}^\perp$. But then σ is also normal on

$\{\xi_1\}^\perp$. Since $\{\xi_1\}^\perp$ is of dimension $n - 1$, our induction assumption applies and $\{\xi_1\}^\perp$ has an orthonormal basis $\{\xi_2, \ldots, \xi_n\}$ consisting of eigenvectors of σ. $\{\xi_1, \xi_2, \ldots, \xi_n\}$ is the required orthonormal basis of V consisting of eigenvectors of σ. $\square$

We can observe from examining the proof of Theorem 9.1 that the conclusion that σ and σ^* commute followed immediately after we showed that the eigenvectors of σ were also eigenvectors of σ^*. Thus the following theorem follows immediately.

Theorem 9.15. *If there exists a basis (orthonormal or not) consisting of vectors which are eigenvectors for both σ and τ, then $\sigma\tau = \tau\sigma$.* $\square$

Any possible converse to Theorem 9.15 requires some condition to ensure the existence of the necessary eigenvectors. In the following theorem we accomplish this by assuming that the field of scalars is the field of complex numbers, any set of conditions that would imply the existence of the eigenvectors could be substituted.

Theorem 9.16. *Let V be a finite dimensional vector space over the complex numbers and let σ and τ be normal linear transformations on V. If $\sigma\tau = \tau\sigma$, then there exists an orthonormal basis consisting of vectors which are eigenvectors for both σ and τ.*

PROOF. Suppose $\sigma\tau = \tau\sigma$. Let λ be an eigenvalue of σ and let $S(\lambda)$ be the eigenspace of σ consisting of all eigenvectors of σ corresponding to λ. Then for each $\xi \in S(\lambda)$ we have $\sigma\tau(\xi) = \tau\sigma(\xi) = \tau(\lambda\xi) = \lambda\tau(\xi)$. Hence, $\tau(\xi) \in S(\lambda)$. This shows that $S(\lambda)$ is an invariant subspace under τ; that is, τ confined to $S(\lambda)$ can be considered to be a normal linear transformation of $S(\lambda)$ into itself. By Theorem 9.14 there is an orthonormal basis of $S(\lambda)$ consisting of eigenvectors of τ. Being in $S(\lambda)$ they are also eigenvectors of σ. By Theorem 9.3 the basis vectors obtained in this way in eigenspaces corresponding to different eigenvalues of σ are orthogonal. Again, by Theorem 9.14 there is a basis of V consisting of eigenvectors of σ. This implies that the eigenspaces of σ span V and, hence, the entire orthonormal set obtained in this fashion is an orthonormal basis of V. $\square$

As we have seen, self-adjoint linear transformations and isometries are particular cases of normal linear transformations. They can also be characterized by the nature of their eigenvalues.

Theorem 9.17. `Let V be a finite dimensional vector space over the complex numbers. A normal linear transformation σ on V is self-adjoint if and only if all its eigenvalues are real.*

PROOF. Suppose σ is self-adjoint. Let λ be an eigenvalue for σ and let ξ be an eigenvector corresponding to λ. Then $\|\sigma(\xi)\|^2 = (\sigma(\xi), \sigma(\xi)) = (\sigma^*(\xi), \sigma(\xi)) = \lambda^2 \|\xi\|^2$. Thus λ^2 is real non-negative and λ is real.

On the other hand, suppose σ is a normal linear transformation and that all its eigenvalues are real. Since σ is normal there exists a basis $X = \{\xi_1, \ldots, \xi_n\}$ of eigenvectors of σ. Let λ_i be the eigenvalue corresponding to ξ_i. Then $\sigma^*(\xi_i) = \bar{\lambda}_i\xi_i = \lambda_i\xi_i = \sigma(\xi_i)$. Since σ^* coincides with σ on a basis of V, $\sigma = \sigma^*$ on all of V. $\square$

Theorem 9.18. *Let V be a finite dimensional vector space over the complex numbers. A normal linear transformation σ on V is an isometry if and only if all its eigenvalues are of absolute value 1.*

PROOF. Suppose σ is an isometry. Let λ be an eigenvalue of σ and let ξ be an eigenvector corresponding to λ. Then $\|\xi\|^2 = \|\sigma(\xi)\|^2 = (\sigma(\xi), \sigma(\xi)) = (\lambda\xi, \lambda\xi) = |\lambda|^2 (\xi, \xi)$. Hence $|\lambda|^2 = 1$.

On the other hand suppose σ is a normal linear transformation and that all its eigenvalues are of absolute value 1. Since σ is normal there exists a basis $X = \{\xi_1, \ldots, \xi_n\}$ of eigenvectors of σ. Let λ_i be the eigenvalue corresponding to ξ_i. Then $(\sigma(\xi_i), \sigma(\xi_j)) = (\lambda_i\xi_i, \lambda_j\xi_j) = \bar{\lambda}_i\lambda_j(\xi_i, \xi_j) = \delta_{ij}$. Hence, σ maps an orthonormal basis onto an orthonormal basis and it is an isometry. $\square$

EXERCISES

1. Prove Theorem 9.2 directly from Corollary 9.5.

2. Show that if there exists an orthonormal basis such that σ and τ are both represented by diagonal matrices, then $\sigma\tau = \tau\sigma$.

3. Show that if σ and τ are normal linear transformations such that $\sigma\tau = \tau\sigma$, then there is an orthonormal basis of V such that the matrices representing σ and τ are both diagonal; that is, σ and τ can be diagonalized simultaneously.

4. Show that the linear transformation associated with a Hermitian form is self-adjoint.

5. Let f be a Hermitian form and let σ be the associated linear transformation. Let $X = \{\xi_1, \ldots, \xi_n\}$ be a basis of eigenvectors of σ (show that such a basis exists) and let $\{\lambda_1, \ldots, \lambda_n\}$ be the corresponding eigenvalues. Let $\alpha = \sum_{i=1}^{n} a_i\xi_i$ and $\beta = \sum_{i=1}^{n} b_i\xi_i$ be arbitrary vectors in V. Show that $f(\alpha, \beta) = \sum_{i=1}^{n} \bar{a}_ib_i\lambda_i$.

6. (Continuation) Let q be the Hermitian quadratic form associated with the Hermitian form f. Let S be the set of all unit vectors in V; that is, $\alpha \in S$ if and only if $\|\alpha\| = 1$. Show that the maximum value of $q(\alpha)$ for $\alpha \in S$ is the maximum eigenvalue, and the minimum value of $q(\alpha)$ for $\alpha \in S$ is the minimum eigenvalue. Show that $q(\alpha) \neq 0$ for all non-zero $\alpha \in V$ if all the eigenvalues of f are non-zero and of the same sign.

7. Let σ be a normal linear transformation and let $\{\lambda_1, \ldots, \lambda_k\}$ be the distinct eigenvalues of σ. Let M_i be the subspace of eigenvectors of σ corresponding to λ_i. Show that $V = M_1 \perp \cdots \perp M_k$.

8. (Continuation) Let π_i be the projection of V onto M_i along $M_i^\perp$. Show that $1 = \pi_1 + \cdots + \pi_k$. Show that $\sigma = \lambda_1 \pi_1 + \cdots + \lambda_k \pi_k$. Show that $\sigma^r = \lambda_1{}^r \pi_1 + \cdots + \lambda_k{}^r \pi_k$. Show that if $p(x)$ is a polynomial, then $p(\sigma) = \sum_{i=1}^k p(\lambda_i) \pi_i$.

10 | Hermitian and Unitary Matrices

Although all the results we state in this section have already been obtained, they are sufficiently useful to deserve being summarized separately. In this section we are considering matrices whose entries are complex numbers.

Theorem 10.1.   *If H is Hermitian, then*
(1) *H is unitary similar to a diagonal matrix D.*
(2) *The elements along the main diagonal of D are the eigenvalues of H.*
(3) *The eigenvalues of H are real.*
Conversely, if H is normal and all its eigenvalues are real, then H is Hermitian.

PROOF.   We have already observed that a Hermitian matrix is normal so that (1) and (2) follow immediately. Since D is diagonal and Hermitian, $\bar{D} = D^* = D$ and the eigenvalues are real.

Conversely, if H is a normal matrix with real eigenvalues, then the diagonal form to which it is unitary similar must be real and hence Hermitian. Thus H itself must be Hermitian. $\square$

Theorem 10.2.   *If A is unitary, then*
(1) *A is unitary similar to a diagonal matrix D.*
(2) *The elements along the main diagonal of D are the eigenvalues of A.*
(3) *The eigenvalues of A are of absolute value 1.*
Conversely, if A is normal and all its eigenvalues are of absolute value 1, then A is unitary.

PROOF.   We have already observed that a unitary matrix is normal so that (1) and (2) follow immediately. Since D is also unitary, $\bar{D}D = D^*D = I$ so that $|\lambda_i|^2 = \bar{\lambda_i}\lambda_i = 1$ for each eigenvalue λ_i.

Conversely, if A is a normal matrix with eigenvalues of absolute value 1, then from the diagonal form D we have $D^*D = \bar{D}D = I$ so that D and A are unitary. $\square$

Corollary 10.3.   *If A is orthogonal, then*
(1) *A is unitary similar to a diagonal matrix D.*
(2) *The elements along the main diagonal of D are the eigenvalues of A.*
(3) *The eigenvalues of A are of absolute value 1.* $\square$

This is a conventional statement of this corollary and in this form it is somewhat misleading. If A is a unitary matrix that happens to be real, then this corollary says nothing that is not contained in Theorem 10.2. A little more information about A and its eigenvalues is readily available. For example, the characteristic equation is real so that the eigenvalues occur in conjugate pairs. An orthogonal matrix of odd order has at least one real eigenvalue, etc. If A is really an orthogonal matrix, representing an isometry in a vector space over the real numbers, then the unitary matrix mentioned in the corollary does not necessarily represent a permissible change of basis. An orthogonal matrix is not always orthogonal similar to a diagonal matrix. As an example, consider the matrix representing a 90° rotation in the Euclidean plane. However, properly interpreted, the corollary is useful.

EXERCISES

1. Find the diagonal matrices to which the following matrices are unitary similar. Classify each as to whether it is Hermitian, unitary, or orthogonal.

(a) $\begin{bmatrix} \dfrac{1+i}{2} & \dfrac{1-i}{2} \\[2mm] \dfrac{1-i}{2} & \dfrac{1+i}{2} \end{bmatrix}$ (b) $\begin{bmatrix} 1 & -i \\ i & 1 \end{bmatrix}$ (c) $\begin{bmatrix} 0.6 & -0.8 \\ 0.8 & 0.6 \end{bmatrix}$

(d) $\begin{bmatrix} 3 & 1+i \\ 1-i & 2 \end{bmatrix}$ (e) $\begin{bmatrix} 1 & i & 0 \\ -i & 1 & i \\ 0 & -i & 1 \end{bmatrix}.$

2. Let A be an arbitrary square complex matrix. Since $A*A$ is Hermitian, there is a unitary matrix P such that $P*A*AP$ is a diagonal matrix D. Let $F = P*AP$. Show that $F*F = D$. Show that D is real and the elements of D are non-negative.

3. Show that every complex matrix can be written as the sum of a real matrix and an imaginary matrix; that is, if M is complex, then $M = A + Bi$ where A and B are real. Show that M is Hermitian if and only if A is symmetric and B is skew-symmetric. Show that M is skew-Hermitian if and only if A is skew-symmetric and B is symmetric.

11 | Real Vector Spaces

We now wish to consider linear transformation and matrices in vector spaces over the real numbers. Much of what has been done for complex vector spaces can be carried over to real vectors spaces without any difficulty. We must be careful, however, when it comes to theorems depending on the existence of eigenvalues and eigenvectors. In particular, Theorems 7.1 and 7.2 do not carry over as stated. Those parts of Section 8 and 9 which depend on these theorems must be reexamined carefully before their implications for real vector spaces can be established.

An examination of the proof of Theorem 7.1 will reveal that the only use made of any special properties of the complex numbers not shared by the real numbers was at the point where it was asserted that each linear transformation has at least one eigenvalue. In stating a corresponding theorem for real vector spaces we have to add an assumption concerning the existence of eigenvalues. Thus we have the following modification of Theorem 7.1 for real vector spaces.

Theorem 11.1   *Let* V *be a finite dimensional vector space over the real numbers, and let* σ *be a linear transformation on* V *whose characteristic polynomial factors into real linear factors. Then there exists an orthonormal basis of* V *with respect to which the matrix representing* σ *is in superdiagonal form.*

PROOF.   Let n be the dimension of V. The theorem is certainly true for $n = 1$.

Assume the theorem is true for real vector spaces of dimensions $<n$. Let λ_1 be an eigenvalue for σ and let $\xi_1' \neq 0$, $\|\xi_1'\| = 1$, be a corresponding eigenvector. There exists an orthonormal basis with ξ_1' as the first element. Let the basis be $X' = \{\xi_1', \ldots, \xi_n'\}$ and let W be the subspace spanned by $\{\xi_2', \ldots, \xi_n'\}$. For each $\alpha = \sum_{i=1}^{n} a_i \xi_i'$ define $\tau(\alpha) = \sum_{i=2}^{n} a_i \xi_i' \in W$. Then $\tau\sigma$ restricted to W is a linear transformation of W into itself.

In the proof of Theorem 7.1 we could apply the induction hypothesis to $\tau\sigma$ without any difficulty since the assumptions of Theorem 7.1 applied to all linear transformations on V. Now we are dealing with a set of linear transformations, however, whose characteristic polynomials factor into real linear factors. Thus we must show that the characteristic polynomial for $\tau\sigma$ factors into real linear factors.

First, consider $\tau\sigma$ as defined on all of V. Since $\tau\sigma(\xi_1') = \tau(\lambda_1 \xi_1') = 0$, $\tau\sigma(\alpha) = \tau[\tau(\alpha)] = \tau\sigma\tau(\alpha)$ for all $\alpha \in V$. This implies that $(\tau\sigma)^k(\alpha) = \tau\sigma^k(\alpha)$ since any τ to the right of a σ can be omitted if there is a τ to the left of that σ.

Let $f(x)$ be the characteristic polynomial for σ. It follows from the observations of the previous paragraph that $\tau f(\tau\sigma) = \tau f(\sigma) = 0$ on V. But on W, τ acts like the identity transformation, so that $f(\tau\sigma) = 0$ when restricted to W. Hence, the minimum polynomial for $\tau\sigma$ on W divides $f(x)$. By assumption, $f(x)$ factors into real linear factors so that the minimum polynomial for $\tau\sigma$ on W must also factor into real linear factors. This means that the hypotheses of the theorem are satisfied for $\tau\sigma$ on W. By induction, there is an orthogonal basis $\{\eta_2, \ldots, \eta_n\}$ of W such that for each η_k, $\tau\sigma(\eta_k)$ is expressible in terms of $\{\eta_2, \ldots, \eta_k\}$ alone. We see from the way τ is defined that $\sigma(\eta_k)$ is expressible in terms of $\{\xi_1^r, \eta_2, \ldots, \eta_k\}$ alone. Let $\eta_1 = \xi_1'$. Then $Y = \{\eta_1, \eta_2, \ldots, \eta_n\}$ is the required basis. $\square$

Since any $n \times n$ matrix with real entries represents some linear transformation with respect to any orthonormal basis, we have

Theorem 11.2. *Let A be a real matrix with real characteristic values. Then A is orthogonal similar to a superdiagonal matrix.* $\square$

Now let us examine the extent to which Sections 8 and 9 apply to real vector spaces. Theorem 8.1 applies to matrices with coefficients in any subfield of the complex numbers and we can use it for real matrices without reservation. Theorem 8.2 does not hold for real matrices, however. To obtain the corresponding theorem over the real numbers we must add the assumption that the characteristic values are real. A normal matrix with real characteristic values is Hermitian and, being real, it must then be symmetric. On the other hand a real symmetric matrix has all real characteristic values. Hence, we have

Theorem 11.3. *A real matrix is orthogonal similar to a diagonal matrix if and only if it is symmetric.* $\square$

Because of the importance of real quadratic forms, in many applications this is a very useful and important theorem, one of the most important of this chapter. We describe some of the applications in Chapter VI and show how this theorem is used.

Of the theorems in Section 9 only Theorems 9.14 and 9.16 fail to hold as stated for real vector spaces. As before, adding the assumption that all the characteristic values of the linear transformation σ are real to the condition that σ is normal amounts to assuming that σ is self-adjoint. Hence, the theorems corresponding to Theorems 9.14 and 9.16 are

Theorem 11.4. *If V is a finite dimensional vector space over the real numbers and σ is a self-adjoint linear transformation on V, then V has an orthonormal basis consisting of eigenvectors of σ.* $\square$

Theorem 11.5. *Let V be a finite dimensional vector space over the real numbers and let σ and τ be self-adjoint linear transformations on V. If $\sigma\tau = \tau\sigma$, then there exists an orthonormal basis of V consisting of vectors which are eigenvectors for both σ and τ.* $\square$

Theorem 9.18 must be modified by substituting the words "characteristic values" for "eigenvalues." Thus,

Theorem 11.6. *A normal linear transformation σ defined on a real vector space V is an isometry if and only if all its characteristic values are of absolute value 1.* $\square$

EXERCISES

1. For those of the following matrices which are orthogonal similar to diagonal matrices, find the diagonal form.

(a) $\begin{bmatrix} 13 & 6 \\ 6 & -3 \end{bmatrix}$ (b) $\begin{bmatrix} 1 & 2 \\ 5 & 4 \end{bmatrix}$ (c) $\begin{bmatrix} 1 & -i \\ i & 1 \end{bmatrix}$

(d) $\begin{bmatrix} 7 & 4 & -4 \\ 4 & -8 & -1 \\ -4 & -1 & -8 \end{bmatrix}$ (e) $\begin{bmatrix} 13 & -4 & 2 \\ -4 & 13 & -2 \\ 2 & -2 & 10 \end{bmatrix}$

(f) $\begin{bmatrix} 3 & -4 & 2 \\ -4 & -1 & 6 \\ 2 & 6 & -2 \end{bmatrix}$ (g) $\begin{bmatrix} -4 & 4 & -2 \\ 4 & -4 & 2 \\ -2 & 2 & -1 \end{bmatrix}$

(h) $\begin{bmatrix} 1 & -1 & 0 \\ -1 & 1 & 0 \\ 0 & 0 & 1 \end{bmatrix}$ (i) $\begin{bmatrix} 0 & 1 & 2 \\ 1 & -\frac{4}{3} & -1 \\ 2 & -1 & -\frac{15}{4} \end{bmatrix}$

(j) $\begin{bmatrix} 1 & 2 & 2 \\ 2 & 1 & -2 \\ 2 & -2 & 1 \end{bmatrix}$ (k) $\begin{bmatrix} 5 & 2 & 2 \\ 2 & 2 & -4 \\ 2 & -4 & 2 \end{bmatrix}.$

2. Which of the matrices of Exercise 1, Section 8, are orthogonal similar to diagonal matrices?

3. Let A and B be real symmetric matrices with A positive definite. There is a non-singular matrix P such that $P^T A P = I$. Show that $P^T B P$ is symmetric. Show that there exists a non-singular matrix Q such that $Q^T A Q = I$ and $Q^T B Q$ is a diagonal matrix.

4. Show that every real skew-symmetric matrix A has the form $A = P^T B P$ where P is orthogonal and B^2 is diagonal.

5. Show that if A and B are real symmetric matrices, and A is positive definite, then the roots of det $(B - xA) = 0$ are all real.

6. Show that a real skew-symmetric matrix of positive rank is not orthogonal similar to a diagonal matrix.

7. Show that if A is a real 2×2 normal matrix with at least one element equal to zero, then it is symmetric or skew-symmetric.

8. Show that if A is a real 2×2 normal matrix with no zero element, then A is symmetric or a scalar multiple of an orthogonal matrix.

9. Let σ be a skew-symmetric linear transformation on the vector space V over the real numbers. The matrix A representing σ with respect to an orthonormal

basis is skew-symmetric. Show that the real characteristic values of A are zeros. The characteristic equation may have complex solutions. Show that all complex solutions are pure imaginary. Why are these solutions not eigenvalues of σ?

10. (Continuation) Show that σ^2 is symmetric. Show that the characteristic values of A^2 are real. Show that the non-zero eigenvalues of A^2 are negative. Let $-\mu^2$ be a non-zero eigenvalue of σ^2 and let ξ be a corresponding eigenvector. Define η to be $\dfrac{1}{\mu}\sigma(\xi)$. Show that $\sigma(\eta) = -\mu\xi$. Show that ξ and η are orthogonal. Show that η is also an eigenvector of σ^2 corresponding to $-\mu^2$.

11. (Continuation) Let σ be the skew-symmetric linear transformation considered in Exercises 9 and 10. Show that there exists an orthonormal basis of V such that the matrix representing σ has all zero elements except for a sequence of 2×2 matrices down the main diagonal of the form

$$\begin{bmatrix} 0 & -\mu_k \\ \mu_k & 0 \end{bmatrix},$$

where the numbers μ_k are defined as in Exercise 10.

12. Let σ be an orthogonal linear transformation on a vector space V over the real numbers. Show that the real characteristic values of σ are ± 1. Show that any eigenvector of σ corresponding to a real eigenvalue is also an eigenvector of σ^* corresponding to the same eigenvalue. Show that these eigenvectors are also eigenvectors of $\sigma + \sigma^*$ corresponding to the eigenvalues ± 2.

13. (Continuation) Show that $\sigma + \sigma^*$ is self-adjoint. Show that there exists a basis of eigenvectors of $\sigma + \sigma^*$. Show that if an eigenvector of $\sigma + \sigma^*$ is also an eigenvector of σ, then the corresponding eigenvalue is ± 2. Let 2μ be an eigenvalue of $\sigma + \sigma^*$ for which the corresponding eigenvector ξ is not an eigenvector of σ. Show that μ is real and that $|\mu| < 1$. Show that $(\xi, \sigma(\xi)) = \mu(\xi, \xi)$.

14. (Continuation) Define η to be $\dfrac{\sigma(\xi) - \mu\xi}{\sqrt{1 - \mu^2}}$. Show that ξ and η are orthogonal. Show that $\sigma(\xi) = \mu\xi + \sqrt{1 - \mu^2}\,\eta$, and $\sigma(\eta) = -\sqrt{1 - \mu^2}\,\xi + \mu\eta$.

15. (Continuation) Let σ be the orthogonal linear transformation considered in Exercises 12, 13, 14. Show that there exists an orthonormal basis of V such that the matrix representing σ has all zero elements except for a sequence of ± 1's and/or 2×2 matrices down the main diagonal of the form

$$\begin{bmatrix} \cos\theta_k & -\sin\theta_k \\ \sin\theta_k & \cos\theta_k \end{bmatrix},$$

where $\mu_k = \cos\theta_k$ are defined as in Exercise 13.

12 | The Computational Processes

We now summarize a complete set of computational steps which will effectively determine a unitary (or orthogonal) matrix of transition for diagonalizing a given normal matrix. Let A be a given normal matrix.

1. Determine the characteristic matrix $C(x) = A - xI$.
2. Compute the characteristic polynomial $f(x) = \det(A - xI)$.
3. Determine all eigenvalues of A by finding all the solutions of the characteristic equation $f(x) = 0$. In any but very special or contrived examples this step is tedious and lengthy. In an arbitrarily given example we can find at best only approximate solutions. In that case all the following steps are also approximate. In some applications special information derivable from the peculiarities of the application will give information about the eigenvalues or the eigenvectors without our having to solve the characteristic equation.
4. For each eigenvalue λ_i find the corresponding eigenvectors by solving the homogeneous linear equations

$$C(\lambda_i)X = 0. \tag{12.1}$$

Each such system of linear equations is of rank less than n. Thus the technique of Chapter II-7 is the recommended method.
5. Find an orthonormal basis consisting of eigenvectors of A. If the eigenvalues are distinct, Theorem 9.3 assures us that they are mutually orthogonal. Thus all that must be done is to normalize each vector and the required orthonormal basis is obtained immediately.

Even where a multiple eigenvalue λ_i occurs, Theorem 8.2 or Theorem 9.14 assures us that an orthonormal basis of eigenvectors exists. Thus, the nullity of $C(\lambda_i)$ must be equal to the algebraic multiplicity of λ_i. Hence, there is no difficulty in obtaining a basis of eigenvectors. The problem is that the different eigenvectors corresponding to the multiple eigenvalue λ_i are not automatically orthogonal; however, that is easily remedied. All we need to do is to take a basis of eigenvectors and use the Gram-Schmidt orthonormalization process in each eigenspace. The vectors obtained in this way will still be eigenvectors since they are linear combinations of eigenvectors corresponding to the same eigenvalue. Vectors from different eigenspaces will be orthogonal because of Theorem 9.3. Since eigenspaces are seldom of very high dimensions, the amount of work involved in applying the Gram-Schmidt process is usually quite nominal.

We now give several examples to illustrate the computational procedures and the various diagonalization theorems. Remember that these examples are contrived so that the characteristic equation can easily be solved. Randomly given examples of high order are very likely to result in vexingly difficult characteristic equations.

Example 1. A real symmetric matrix with distinct eigenvalues. Let

$$A = \begin{bmatrix} 1 & -2 & 0 \\ -2 & 2 & -2 \\ 0 & -2 & 3 \end{bmatrix}.$$

We first determine the characteristic matrix,

$$C(x) = \begin{bmatrix} 1-x & -2 & 0 \\ -2 & 2-x & -2 \\ 0 & -2 & 3-x \end{bmatrix},$$

and then the characteristic polynomial,

$$f(x) = \det C(x) = -x^3 + 6x^2 - 3x - 10 = -(x+1)(x-2)(x-5).$$

The eigenvalues are $\lambda_1 = -1$, $\lambda_2 = 2$, $\lambda_3 = 5$.

Solving the equations $C(\lambda_i)X = 0$ we obtain the eigenvectors $\alpha_1 = (2, 2, 1)$, $\alpha_2 = (-2, 1, 2)$, $\alpha_3 = (1, -2, 2)$. Theorem 9.3 assures us that these eigenvectors are orthogonal, and upon checking we see that they are. Normalizing them, we obtain the orthonormal basis

$$X = \{\xi_1 = \tfrac{1}{3}(2, 2, 1),\ \xi_2 = \tfrac{1}{3}(-2, 1, 2),\ \xi_3 = \tfrac{1}{3}(1, -2, 2)\}.$$

The orthogonal matrix of transition is

$$P = \tfrac{1}{3}\begin{bmatrix} 2 & -2 & 1 \\ 2 & 1 & -2 \\ 1 & 2 & 2 \end{bmatrix}.$$

Example 2. A real symmetric matrix with repeated eigenvalues. Let

$$A = \begin{bmatrix} 5 & 2 & 2 \\ 2 & 2 & -4 \\ 2 & -4 & 2 \end{bmatrix}.$$

The corresponding characteristic matrix is

$$C(x) = \begin{bmatrix} 5-x & 2 & 2 \\ 2 & 2-x & -4 \\ 2 & -4 & 2-x \end{bmatrix},$$

and the characteristic polynomial is

$$f(x) = -x^3 + 9x^2 - 108 = -(x + 3)(x - 6)^2.$$

The eigenvalues are $\lambda_1 = -3$, $\lambda_2 = \lambda_3 = 6$.

Corresponding to $\lambda_1 = -3$, we obtain the eigenvector $\alpha_1 = (1, -2, -2)$. For $\lambda_2 = \lambda_3 = 6$ we find that the eigenspace $S(6)$ is of dimension 2 and is the set of all solutions of the equation

$$x_1 - 2x_2 - 2x_3 = 0.$$

Thus $S(6)$ has the basis $\{(2, 1, 0), (2, 0, 1)\}$. We can now apply the Gram-Schmidt process to obtain the orthonormal basis

$$\left\{ \frac{1}{\sqrt{5}} (2, 1, 0), \frac{1}{3\sqrt{5}} (2, -4, 5) \right\}.$$

Again, by Theorem 9.3 we are assured that α_1 is orthogonal to all vectors in $S(6)$, and to these vectors in particular. Thus,

$$X = \left\{ \frac{1}{3} (1, -2, -2), \frac{1}{\sqrt{5}} (2, 1, 0), \frac{1}{3\sqrt{5}} (2, -4, 5) \right\}$$

is an orthonormal basis of eigenvectors. The orthogonal matrix of transition is

$$P = \begin{bmatrix} \dfrac{1}{3} & \dfrac{2}{\sqrt{5}} & \dfrac{2}{3\sqrt{5}} \\[2mm] \dfrac{-2}{3} & \dfrac{1}{\sqrt{5}} & \dfrac{-4}{3\sqrt{5}} \\[2mm] \dfrac{-2}{3} & 0 & \dfrac{5}{3\sqrt{5}} \end{bmatrix}.$$

It is worth noting that, whereas the eigenvector corresponding to an eigenvalue of multiplicity 1 is unique up to a factor of absolute value 1, the orthonormal basis of the eigenspace corresponding to a multiple eigenvalue is not unique. In this example, any vector orthogonal to $(1, -2, -2)$ must be in $S(6)$. Thus $\{\frac{1}{3}(2, 2, -1), \frac{1}{3}(2, -1, 2)\}$ would be another choice for an orthonormal basis for $S(6)$. It happens to result in a slightly simpler orthogonal matrix of transition (in this case a matrix over the rational numbers.)

Example 3. A Hermitian matrix. Let

$$A = \begin{bmatrix} 2 & 1 - i \\ 1 + i & 3 \end{bmatrix}.$$

Then

$$C(x) = \begin{bmatrix} 2 - x & 1 - i \\ 1 + i & 3 - x \end{bmatrix},$$

and $f(x) = x^2 - 5x + 4 = (x - 1)(x - 4) = 0$ is the characteristic equation. The eigenvalues are $\lambda_1 = 1$ and $\lambda_2 = 4$. (The example is contrived so that the eigenvalues are rational, but the fact that they are real is assured by Theorem 10.1.) Corresponding to $\lambda_1 = 1$ we obtain the normalized eigenvector $\xi_1 = \frac{1}{\sqrt{3}}(-1 + i, 1)$, and corresponding to $\lambda_2 = 4$ we obtain the normalized eigenvector $\xi_2 = \frac{1}{\sqrt{3}}(1, 1 + i)$. The unitary matrix of transition is

$$U = \frac{1}{\sqrt{3}} \begin{bmatrix} -1 + i & 1 \\ 1 & 1 + i \end{bmatrix}.$$

Example 4. An orthogonal matrix. Let

$$A = \tfrac{1}{3} \begin{bmatrix} 1 & -2 & 2 \\ -2 & 1 & 2 \\ -2 & -2 & -1 \end{bmatrix}.$$

This orthogonal matrix is real but not symmetric. Therefore, it is unitary similar to a diagonal matrix but it is not orthogonal similar to a diagonal matrix. We have

$$C(x) = \begin{bmatrix} \tfrac{1}{3} - x & -\tfrac{2}{3} & \tfrac{2}{3} \\ -\tfrac{2}{3} & \tfrac{1}{3} - x & \tfrac{2}{3} \\ -\tfrac{2}{3} & -\tfrac{2}{3} & -\tfrac{1}{3} - x \end{bmatrix},$$

and, hence, $-x^3 + \tfrac{1}{3}x^2 - \tfrac{1}{3}x + 1 = -(x - 1)(x^2 + \tfrac{2}{3}x + 1) = 0$ is the characteristic equation. Notice that the real eigenvalues of an orthogonal matrix are particularly easy to find since they must be of absolute value 1. The eigenvalues are $\lambda_1 = 1$, $\lambda_2 = \dfrac{-1 + 2\sqrt{2}i}{3}$, and $\lambda_3 = \dfrac{-1 - 2\sqrt{2}i}{3}$. The corresponding normalized eigenvectors are $\xi_1 = \dfrac{1}{\sqrt{2}}(1, -1, 0)$, $\xi_2 = \tfrac{1}{2}(1, 1, \sqrt{2}i)$, and $\xi_3 = \tfrac{1}{2}(1, 1, -\sqrt{2}i)$. Thus, the unitary matrix of transition is

$$U = \begin{bmatrix} 1/\sqrt{2} & \tfrac{1}{2} & \tfrac{1}{2} \\ -1/\sqrt{2} & \tfrac{1}{2} & \tfrac{1}{2} \\ 0 & i/\sqrt{2} & -i/\sqrt{2} \end{bmatrix}.$$

EXERCISES

1. Apply the computational methods outlined in this section to obtain the orthogonal or unitary matrices of transition to diagonalize each of the normal matrices given in Exercises 1 of Sections 8, 10, and 11.

2. Carry out the program outlined in Exercises 12 through 15 of Section 11. Consider the orthogonal linear transformation σ represented by the orthogonal matrix

$$A = \begin{bmatrix} \frac{1}{3} & -\frac{2}{3} & \frac{2}{3} \\ -\frac{2}{3} & \frac{1}{3} & \frac{2}{3} \\ -\frac{2}{3} & -\frac{2}{3} & -\frac{1}{3} \end{bmatrix}.$$

Find an orthonormal basis of eigenvectors of $\sigma + \sigma^*$. Find the representation of σ with respect to this basis. Since $\sigma + \sigma^*$ has one eigenvalue of multiplicity 2, the pairing described in Exercise 14 of Section 11 is not necessary. If $\sigma + \sigma^*$ had an eigenvalue of multiplicity 4 or more, such a pairing would be required to obtain the desired form.

chapter VI | Selected applications of linear algebra

In general, the application of any mathematical theory to any realistic problem requires constructing a model of the problem in mathematical terminology. How each concept in the model corresponds to a concept in the problem requires understanding of both areas on the part of the person making the application. If the problem is physical, he must understand the physical facts that are to be related. He must also understand how the mathematical concepts are related so that he can establish a correspondence between the physical concepts and the mathematical concepts.

If this correspondence has been established in a meaningful way, presumably the conclusions in the mathematical model will also have physical meaning. If it were not for this aspect of the use of mathematical models, mathematics could make little contribution to the problem for it could otherwise not reveal any fact or conclusion not already known. The usefulness of the model depends on how removed from obvious the conclusions are, and how experience verifies the validity of the conclusions.

It must be emphasized that there is no hope of making any meaningful numerical computations until the model has been constructed and understood. Anyone who attempts to apply a mathematical theory to a real problem without understanding of the model faces the danger of making inappropriate applications, or the restriction of doing only what someone who does understand has instructed him to do. Too many students limit their aims to remembering a sequence of steps that "give the answer" instead of understanding the basic principles.

In the applications chosen for illustration here it is not possible to devote more than token attention to the concepts in the field of the application. For more details reference will have to be made to other sources. We do identify the connection between the concept in the application and the concept in the model. Considerable attention is given to the construction of complete and coherent models. In some cases the model already exists in the material that has been developed in the first five chapters. This is true to a large extent of the applications to geometry, communication theory, differential equations, and small oscillations. In other cases, extensive portions of the models must be constructed here. This has been necessary for the applications to linear inequalities, linear programming, and representation theory.

1 | Vector Geometry

This section requires Chapter I, the first eight sections of Chapter II, and the first four sections of Chapter IV for background.

We have already used the geometric interpretation of vectors to give the concepts we were discussing a reality. We now develop this interpretation in more detail. In doing this we find that the vector space concepts are powerful tools in geometry and the geometric concepts are suggestive models for corresponding facts about vector spaces.

We use vector algebra to construct an algebraic model for geometry. In our imagination we identify a point P with a vector α from the origin to P. α is called the *position vector* of P. In this way we establish a one-to-one correspondence between points and vectors. The correspondence will depend on the choice of the origin. If a new origin is chosen, there will result a different one-to-one correspondence between points and vectors. The type of geometry that is described by the model depends on the type of algebraic structure that is given to the model. It is not our purpose here to get involved in the details of various types of geometry. We shall be more concerned with the ways geometric concepts can be identified with algebraic concepts.

Let V be a vector space of dimension n. We call a subspace S of dimension 1 a *straight line through the origin*. In the familiar model we have in mind there are straight lines which do not pass through the origin, so this definition must be generalized. A *straight line* is a set L of the form

$$L = \alpha + S \tag{1.1}$$

where α is a fixed vector and S is a subspace of dimension 1. We describe this situation by saying that S is a line through the origin and α displaces S "parallel" to itself to a new position.

In general, a *linear manifold* or *flat* is a set L of the form

$$L = \alpha + S \tag{1.2}$$

where α is a fixed vector and S is a subspace. If S is of dimension r, we say the linear manifold is of dimension r. *A point* is a linear manifold of dimension 0, a *line* is of dimension 1, a *plane* is of dimension 2, and a *hyperplane* is a linear manifold of dimension $n - 1$.

Let $\hat{V}$ be the dual space of V and let $S^{\perp}$ be the annihilator of S. For every $\phi \in S^{\perp}$ we have

$$\phi(L) = \phi(\alpha) + \phi(S) = \phi(\alpha). \tag{1.3}$$

On the other hand, let β be any point in V for which $\phi(\beta) = \phi(\alpha)$ for all $\phi \in S^{\perp}$. Then $\phi(\beta - \alpha) = \phi(\beta) - \phi(\alpha) = 0$ so that $\beta - \alpha \in S$; that is, $\beta \in \alpha + S = L$. This means that S is identified by β as well as by α; that is, L is determined by S and any vector in it.

Let $L = \alpha + S$ be of dimension r. Then $S^{\perp}$ is of dimension $n - r$. Let $\{\phi_1, \ldots, \phi_{n-r}\}$ be a basis of $S^{\perp}$. Then

$$\phi_i(L) = \phi_i(\alpha) + \phi_i(S) = \phi_i(\alpha) = c_i \tag{1.4}$$

for $i = 1, \ldots, n - r$. Then $\beta \in L$ if and only if $\phi_i(\beta) = c_i$ for $i = 1, \ldots,$ $n - r$. Thus a linear manifold is determined by giving these $n - r$ conditions, known as *linear conditions*. The linear manifold L is of dimension r if and only if the $n - r$ linear conditions are independent.

Two linear manifolds $L_1 = \alpha_1 + S_1$ and $L_2 = \alpha_2 + S_2$ are said to be *parallel* if and only if either $S_1 \subset S_2$ or $S_2 \subset S_1$. If L_1 and L_2 are of the same dimension, they are parallel if and only if $S_1 = S_2$.

Let L_1 and L_2 be parallel and, to be definite, let us take $S_1 \subset S_2$. Suppose L_1 and L_2 have a point $\beta = \alpha_1 + \sigma_1 = \alpha_2 + \sigma_2$ in common. Then $\alpha_1 = \alpha_2 + (\sigma_2 - \sigma_1) \in \alpha_2 + S_2$. Hence, $L_1 \subset L_2$. Thus, if two parallel linear manifolds have a point in common, one is a subset of the other.

Let $L = \alpha + S$ be a linear manifold and let $\{\alpha_1, \ldots, \alpha_r\}$ be a basis for S. Then every vector $\beta \in L$ can be written in the form

$$\beta = \alpha + t_1\alpha_1 + \cdots + t_r\alpha_r. \tag{1.5}$$

As the $t_1, \ldots, t_r$ run through all values in the field F, β runs through the linear manifold L. For this reason (1.5) is called a *parametric representation*

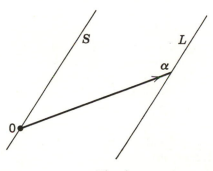

Fig. 3

of the linear manifold L. Since α and the basis vectors are subject to a wide variety of choices, there is no unique parametric representation.

Example. Let $\alpha = (1, 2, 3)$ and let S be a subspace of dimension 1 with basis $\{(2, -1, 1)\}$. Then $\xi = (x_1, x_2, x_3) \in \alpha + S$ must satisfy the conditions

$$(x_1, x_2, x_3) = (1, 2, 3) + t(2, -1, 1).$$

In analytic geometry these conditions are usually written in the form

$$x_1 = 1 + 2t$$
$$x_2 = 2 - t$$
$$x_3 = 3 + t,$$

the conventional or extended form of the parametric equations of a line. The annihilator of S in this case has the basis $\{[1 \quad 2 \quad 0], [0 \quad 1 \quad 1]\}$. The equations of the line $\alpha + S$ are then

$$x_1 + 2x_2 + 0x_3 = 1 \cdot 1 + 2 \cdot 2 + 0 = 5,$$
$$0x_1 + x_2 + x_3 = 0 + 1 \cdot 2 + 1 \cdot 3 = 5.$$

With a little practice, vector methods are more intuitive and easier than the methods of analytic geometry.

Suppose we wish to find out whether two lines are coplanar, or whether they intersect. Let $L_1 = \alpha_1 + S_1$ and $L_2 = \alpha_2 + S_2$ be two given lines. Then $S_1 + S_2$ is the smallest subspace parallel to both L_1 and L_2. $S_1 + S_2$ is of dimension 2 unless $S_1 = S_2$, a special case which is easy to handle. Thus, assume $S_1 + S_2$ is of dimension 2. $\alpha_1 + S_1 + S_2$ is a plane containing L_1 and parallel to L_2. Thus L_1 and L_2 are coplanar and L_1 intersects L_2 if and only if $\alpha_2 \in \alpha_1 + S_1 + S_2$. To determine this find the annihilator of $S_1 + S_2$. As in (1.3) $\alpha_2 \in \alpha_1 + S_1 + S_2$ if and only if $(S_1 + S_2)^\perp$ has the same effect on both α_1 and α_2.

Example. Let $L_1 = (1, 2, 3) + t(2, -1, 1)$ and then let $L_2 = (1, 1, 0) + s(1, 0, 2)$. Then $S_1 + S_2 = \langle (2, -1, 1), (1, 0, 2) \rangle$ and $(S_1 + S_2)^\perp = \langle [2 \quad 3 \quad -1] \rangle$. Since $[2 \quad 3 \quad -1](1, 2, 3) = 5$ and $[2 \quad 3 \quad -1](1, 1, 0) = 5$, the lines L_1 and L_2 both lie in the plane $M = (1, 2, 3) + \langle (2, -1, 1), (1, 0, 2) \rangle$.

We can easily find the intersection of L_1 and L_2. Since S_1 is a proper subspace of $S_1 + S_2$, $(S_1 + S_2)^\perp$ is a proper subspace of $S_1^\perp$. In this case the difference in dimension is 1, so we can find one linearly independent functional in $S_1^\perp$ that is not in $(S_1 + S_2)^\perp$, for example, $[0 \quad 1 \quad 1]$. Then a point of L_2 is in L_1 if and only if $[0 \quad 1 \quad 1]$ has the same effect on it as it has on $(1, 2, 3)$. Since $[0 \quad 1 \quad 1](1, 2, 3) = 5$ and $[0 \quad 1 \quad 1]\{(1, 1, 0) + s(1, 0, 2)\} = 1 + 2s$, we see that $s = 2$. It is easily verified that $(1, 1, 0) + 2(1, 0, 2) = (3, 1, 4) = (1, 2, 3) + (2, -1, 1)$ is in both L_1 and L_2.

An important problem that must be considered is that of finding the smallest linear manifold containing a given set of points. Let $\{P_0, P_1, \ldots, P_n\}$ be a given set of points and let $\{\alpha_0, \alpha_1, \ldots, \alpha_n\}$ be the corresponding set of position vectors. For the sake of providing a geometric interpretation of the algebra, we shall speak as though the linear manifold containing the set of points is the same as the linear manifold containing the set of vectors.

A linear manifold containing these vectors must be of the form $L = \alpha + S$ where S is a subspace. Since α can be any vector in L, we may as well take $L = \alpha_0 + S$. Since $\alpha_i - \alpha_0 \in S$, S contains $\{\alpha_1 - \alpha_0, \ldots, \alpha_r - \alpha_0\}$. On the other hand, if S contains $\{\alpha_1 - \alpha_0, \ldots, \alpha_r - \alpha_0\}$, then $\alpha_0 + S$ will contain $\{\alpha_0, \alpha_1, \ldots, \alpha_r\}$. Thus $\alpha_0 + \langle \alpha_1 - \alpha_0, \ldots, \alpha_r - \alpha_0 \rangle$ is the smallest linear manifold containing $\{\alpha_0, \ldots, \alpha_r\}$. If the $\{\alpha_0, \ldots, \alpha_r\}$ are given arbitrarily, there is no assurance that $\{\alpha_1 - \alpha_0, \ldots, \alpha_r - \alpha_0\}$ will be a linearly independent set. If it is, the linear manifold L will be of dimension r. In general, two points determine a line, three points determine a plane, and n points determine a hyperplane.

An arbitrary vector β will be in L if and only if it can be represented in the form

$$\beta = \alpha_0 + t_1(\alpha_1 - \alpha_0) + t_2(\alpha_2 - \alpha_0) + \cdots + t_r(\alpha_r - \alpha_0). \quad (1.6)$$

By setting $1 - t_1 - \cdots - t_r = t_0$, this expression can be written in the form

$$\beta = t_0\alpha_0 + t_1\alpha_1 + \cdots + t_r\alpha_r, \quad (1.7)$$

where

$$t_0 + t_1 + \cdots + t_r = 1. \quad (1.8)$$

It is easily seen that (1.8) is a necessary and sufficient condition on the t_i in order that a linear combination of the form (1.7) lie in L.

We should also like to be able to determine whether the linear manifold generated by $\{\alpha_0, \alpha_1, \ldots, \alpha_r\}$ is of dimension r or of a lower dimension. For example, when are three points colinear? The dimension of L is less than r if and only if $\{\alpha_1 - \alpha_0, \ldots, \alpha_r - \alpha_0\}$ is a linearly dependent set; that is, if there exists a non-trivial linear relation of the form

$$c_1(\alpha_1 - \alpha_0) + \cdots + c_r(\alpha_r - \alpha_0) = 0. \quad (1.9)$$

This, in turn, can be written in the form

$$c_0\alpha_0 + c_1\alpha_1 + \cdots + c_r\alpha_r = 0, \quad (1.10)$$

where

$$c_0 + c_1 + \cdots + c_r = 0. \quad (1.11)$$

It is an easy computational problem to determine the c_i, if a non-zero solution exists. If α_i is represented by $(a_{1i}, a_{2j}, \ldots, a_{nj})$, then (1.10) becomes equivalent to the system of n equations

$$a_{i0}c_0 + c_{i1}c_1 + \cdots + a_{ir}c_r = 0, \qquad i = 1, \ldots, n. \quad (1.12)$$

These equations together with (1.11) form a system of $n + 1$ homogeneous linear equations to solve. As a system, they are equivalent to determining whether the set $\{(1, a_{1j}, a_{2j}, \ldots, a_{nj}) \mid j = 0, 1, \ldots, r\}$ is linearly dependent.

Geometrically, the pair of conditions (1.7), (1.8) and the pair of conditions (1.10), (1.11) are independent of the choice of an origin. Suppose the points $P_0, P_1, \ldots, P_r$ are identified with position vectors from some other point. For example, let 0 be the origin and let $0'$ be a new choice for an origin. If α' is the position vector of $0'$ with reference to the old origin, then

$$\alpha_i' = \alpha_i - \alpha' \tag{1.13}$$

is the position vector of P_i relative to $0'$. If β in (1.7) is the position vector of B relative to 0 and $\beta' = \beta - \alpha'$ is the position vector of B relative to $0'$, then (1.7) takes the form

$$\beta' = \beta - \alpha' = \sum_{k=0}^{r} t_k \alpha_k - \alpha'$$

$$= \sum_{k=0}^{r} t_k (\alpha_k - \alpha')$$

$$= \sum_{k=0}^{r} t_k \alpha_k'. \tag{1.7}'$$

Also, (1.10) takes the form

$$\sum_{k=0}^{r} c_k \alpha_k' = \sum_{k=0}^{r} c_k (\alpha_k - \alpha')$$

$$= \sum_{k=0}^{r} c_k \alpha_k - \sum_{k=0}^{r} c_k \alpha'$$

$$= \sum_{k=0}^{r} c_k \alpha_k = 0. \tag{1.10}'$$

Since the pair of conditions (1.7), (1.8) is related to a geometric property, it should be expected that if a pair of conditions like (1.7), (1.8) hold with one choice of an origin, then a pair of similar conditions should hold for another choice. The real importance of the observation above is that the new pair of conditions involve the same coefficients.

If (1.7) holds subject to (1.8), we call β an *affine* combination of $\{\alpha_0, \alpha_1, \ldots, \alpha_r\}$. If (1.10) holds with coefficients not all zero subject to (1.11), we say that $\{\alpha_0, \alpha_1, \ldots, \alpha_r\}$ is an *affinely dependent* set. The concepts of affine combinations, affine dependence, and affine independence are related to each other in much the same way that linear combinations, linear dependence, and linear independence are related to each other. For example, the affine combination (1.7) is unique if and only if the set $\{\alpha_0, \alpha_1, \ldots, \alpha_r\}$ is affinely independent. The set $\{\alpha_0, \alpha_1, \ldots, \alpha_r\}$ is affinely dependent if and only if one vector is an affine combination of the preceding vectors.

Affine geometry is important, but its study is much neglected in American high schools and universities. The reason for this is primarily that it is difficult to study affine geometry without using linear algebra as a tool. A good picture of what concepts are involved in affine geometry can be obtained by considering Euclidean geometry in the plane or 3-space in which we "forget" about distance. In Euclidean geometry we study properties that are unchanged by rigid motions. In affine geometry we allow non-rigid motions, provided they preserve intersection, parallelism, and colinearity. If an origin is introduced, a rigid motion can be represented as an orthogonal transformation followed by a translation. An affine transformation can be represented by a non-singular linear transformation followed by a translation. If we accept this assertion, it is easy to see that affine combinations are preserved under affine transformations.

Theorem 1.1. *Let $\{L_\lambda : \lambda \in \Lambda\}$ be any collection of linear manifolds in V. Either $\cap_{\lambda \in \Lambda} L_\lambda$ is empty or it is a linear manifold.*

PROOF. If $\cap_{\lambda \in \Lambda} L_\lambda$ is not empty, let $\alpha_0 \in \cap_{\lambda \in \Lambda} L_\lambda$. Since $\alpha_0 \in L_\lambda$ and L_λ is a linear manifold, $L_\lambda = \alpha_0 + S_\lambda$ where S_λ is a subspace of V. Then $\cap_{\lambda \in \Lambda} L_\lambda = \cap_{\lambda \in \Lambda}(\alpha_0 + S_\lambda) = \alpha_0 + \cap_{\lambda \in \Lambda} S_\lambda$. Since $\cap_{\lambda \in \Lambda} S_\lambda$ is a subspace, $\cap_{\lambda \in \Lambda} L_\lambda$ is a linear manifold. $\square$

Definition. Given any non-empty subset S of V, the *affine closure* of S is the smallest linear manifold containing S. In view of Theorem 1.1, the affine closure of S is the intersection of all linear manifolds containing S, and this shows that a smallest linear manifold containing S actually exists. The affine closure of S is denoted by $A(S)$.

Theorem 1.2. *Let S be any subset of V. Let $\bar{S}$ be the set of all affine combinations of finite subsets of S. Then $\bar{S}$ is a linear manifold.*

PROOF. Let $\{\beta_0, \beta_i, \ldots, \beta_k\}$ be a finite subset of $\bar{S}$. Each β_i is of the form

$$\beta_j = \sum_{i=0}^{r_j} x_{ij} \alpha_{ij}$$

where

$$\sum_{i=0}^{r} x_{ij} = 1$$

and each $\alpha_{ij} \in S$. Then for any β of the form

$$\beta = \sum_{j=0}^{r} t_j \beta_j$$

and

$$\sum_{j=0}^{r} t_j = 1,$$

we have

$$\beta = \sum_{j=0}^{r} t_j \left(\sum_{i=0}^{r_j} x_{ij} \alpha_{ij} \right)$$

$$= \sum_{j=0}^{r} \sum_{i=0}^{r} x_{ij} t_j \alpha_{ij}$$

and

$$\sum_{j=0}^{r} \sum_{i=0}^{r_j} x_{ij} t_j = \sum_{j=0}^{r} t_j \sum_{i=0}^{r_j} x_{ij}$$

$$= \sum_{j=0}^{r} t_j$$

$$= 1.$$

Thus $\beta \in \bar{S}$. This shows that $\bar{S}$ is closed under affine combinations, that is, an affine combination of a finite number of elements in $\bar{S}$ is also in $\bar{S}$.

The observation of the previous paragraph will allow us to conclude that $\bar{S}$ is a linear manifold. Let α_0 be any fixed element in S. Let $\bar{S} - \alpha_0$ denote the set of elements of the form $\alpha - \alpha_0$ where $\alpha \in \bar{S}$. If $\{\beta_1, \beta_2, \ldots, \beta_r\}$ is a finite subset of $\bar{S} - \alpha_0$, where $\beta_i = \alpha_i - \alpha_0$, then $\sum_{i=1}^{k} c_i \beta_i = \sum_{i=1}^{k} c_i \alpha_i - \sum_{i=1}^{k} c_i \alpha_0 = \sum_{i=0}^{k} c_i \alpha_i - \alpha_0$ where $c_0 = 1 - \sum_{i=1}^{k} c_i$. Thus $\sum_{i=0}^{k} c_i \alpha_i \in \bar{S}$ and $\sum_{i=1}^{k} c_i \beta_i \in \bar{S} - \alpha_0$. This shows that $\bar{S} - \alpha_0$ is a subspace. Hence, $\bar{S} = \alpha_0 + (\bar{S} - \alpha_0)$ is a linear manifold. $\square$

Theorem 1.3. *The affine closure of S is the set of all affine combinations of finite subsets of S.*

Since $\bar{S}$ is a linear manifold containing S, $A(S) \subset \bar{S}$. On the other hand, $A(S)$ is a linear manifold containing S and, hence, $A(S)$ contains all affine combinations of elements in S. Thus $\bar{S} \subset A(S)$. This shows that $\bar{S} = A(S)$. $\square$

Theorem 1.4. *Let $L_1 = \alpha_1 + S_1$ and let $L_2 = \alpha_2 + S_2$. Then $A(L_1 \cup L_2) = \alpha_1 + \langle \alpha_2 - \alpha_1 \rangle + S_1 + S_2$.*

PROOF. Clearly, $L_1 \cup L_2 \subset \alpha_1 + \langle \alpha_2 - \alpha_1 \rangle + S_1 + S_2$ and $\alpha_1 + \langle \alpha_2 - \alpha_1 \rangle + S_1 + S_2$ is a linear manifold containing $L_1 \cup L_2$. Since $\alpha_1 \in L_1 \cup L_2 \subset A(L_1 \cup L_2)$, $A(L_1 \cup L_2)$ is of the form $\alpha_1 + S$ where S is a subspace. And since $\alpha_2 \in L_1 \cup L_2 \subset \alpha_1 + S$, $\alpha_1 + S = \alpha_2 + S$. Thus $\alpha_2 - \alpha_1 \in S$. Since $\alpha_1 + S_1 = L_1 \subset L_1 \cup L_2 \subset \alpha_1 + S$, $S_1 \subset S$. Similarly, $S_2 \subset S$. Thus $\langle \alpha_2 - \alpha_1 \rangle + S_1 + S_2 \subset \alpha_1 + S = A(L_1 \cup L_2)$. This shows that $A(L_1 \cup L_2) = \alpha_1 + \langle \alpha_2 - \alpha_1 \rangle + S_1 + S_2$. $\square$

Theorem 1.5. *Let $L_1 = \alpha_1 + S_1$ and let $L_2 = \alpha_2 + S_2$ be linear manifolds. $L_1 \cap L_2 = \emptyset$ if and only if $\alpha_2 - \alpha_1 \notin S_1 + S_2$.*

PROOF. If $L_1 \cap L_2$ is not empty, let $\alpha_0 \in L_1 \cap L_2$. Then $L_1 = \alpha_0 + S_1 = \alpha_1 + S_1$ and $L_2 = \alpha_0 + S_2 = \alpha_2 + S_2$. Thus $\alpha_0 - \alpha_1 \in S_1$ and $\alpha_2 - \alpha_0 \in S_2$.

Hence $\alpha_2 - \alpha_1 = (\alpha_0 - \alpha_1) + (\dot{\alpha}_2 - \alpha_0) \in S_1 + S_2$. Conversely, if $\alpha_2 - \alpha_1 \in S_1 + S_2$ then $\alpha_2 - \alpha_1 = \gamma_1 + \gamma_2$ where $\gamma_1 \in S_1$ and $\gamma_2 \in S_2$. Thus $\alpha_1 + \gamma_1 = \alpha_2 - \gamma_2 \in (\alpha_1 + S_1) \cap (\alpha_2 + S_2)$. $\square$

Corollary 1.6. $L_1 \cap L_2 = \emptyset$ *if and only if* $\dim A(L_1 \cup L_2) = \dim (S_1 + S_2) + 1$. *If* $L_1 \cap L_2 \neq \emptyset$, $\dim A(L_1 \cup L_2) = \dim (S_1 + S_2)$. $\square$

We now wish to introduce the idea of betweenness. This is a concept closely tied to the real numbers, so we assume for the rest of this section that F is the field of real numbers or a subfield of the real numbers.

Let α_1 and α_2 be any two vectors. We have seen that every vector β in the line generated by α_1 and α_2 can be written in the form $\beta = t_1\alpha_1 + t_2\alpha_2$, where $t_1 + t_2 = 1$. This is equivalent to the form

$$\beta = (1 - t)\alpha_1 + t\alpha_2. \tag{1.14}$$

For $t = 0$, $\beta = \alpha_1$, and for $t = 1$, $\beta = \alpha_2$. We say that β is *between* α_1 and α_2 if and only if t is between 0 and 1. The *line segment* joining α_1 and α_2 consists of all points of the form $t_1\alpha_1 + t_2\alpha_2$ where $t_1 + t_2 = 1$ and $t_1 \geq 0$ and $t_2 \geq 0$. A subset C of V is said to be *convex* if, whenever two points are in C, every point of the line segment joining them is also in C. Clearly, the space V itself is convex, and every subspace of V is convex. Exercise 11 of Chapter IV-4 amounts to showing that the two sides of a hyperplane are convex.

Theorem 1.7. *The intersection of any number of convex sets is convex.*

PROOF. Let $\{C_\lambda\}_{\lambda \in \Lambda}$ be a collection of convex sets. If α_1 and α_2 are in $\cap_{\lambda \in \Lambda} C_\lambda$, then for each λ both α_1 and α_2 are in C_λ. Since C_λ is convex, the segment joining α_1 and α_2 is in C_λ. Thus the segment is in $\cap_{\lambda \in \Lambda} C_\lambda$ and the intersection is convex. $\square$

As a slight generalization of the expression we gave for the line segment joining two points, we define a *convex linear combination* of elements of a subset S to be a vector β expressible in the form

$$\beta = t_1\alpha_1 + t_2\alpha_2 + \cdots + t_r\alpha_r, \tag{1.15}$$

where

$$t_1 + t_2 + \cdots + t_r = 1, \qquad t_i \geq 0, \tag{1.16}$$

and $\{\alpha_1, \alpha_2, \ldots, \alpha_r\}$ is a finite subset of S. If S is a finite subset, a useful and informative picture of the situation is formed in the following way: Imagine the points of S to be contained in a plane or 3-dimensional space. The set of convex linear combinations is then a polygon (in the plane) or polyhedron in which the corners are points of S. Those points of S which are not at the corners are contained in the edges, faces, or interior of the polygon or polyhedron. It is the purpose of Theorem 1.9 to prove that this is a dependable picture.

Theorem 1.8. *A set C is convex if and only if every convex linear combination of vectors in C is in C.*

PROOF. If every convex linear combination of vectors in C is in C, then, in particular, every line segment joining a pair of vectors in C is in C. Thus C is convex.

On the other hand, assume C is convex and let $\beta = \sum_{i=1}^{r} t_i \alpha_i$ be a convex linear combination of vectors in C. For $r = 1$, $\beta = \alpha_1 \in C$; and for $r = 2$, β is on the line segment joining α_1 and α_2 hence $\beta \in C$. Assume that a convex linear combination involving fewer than r elements of C is in C. We can assume that $t_r \neq 1$, for otherwise $\beta = \alpha_r \in C$. Then for each i, $(1 - t_r)\alpha_i + t_r \alpha_r \in C$ and

$$\beta = \sum_{i=1}^{r-1} \frac{t_i}{1 - t_r} \{(1 - t_r)\alpha_i + t_r \alpha_r\} \tag{1.17}$$

is a convex linear combination of $r - 1$ elements of C, and is therefore in C. □

Let S be any subset of V. The *convex hull* H(S) of S is the smallest convex set containing S. Since V is a convex set containing S, and the intersection of all convex sets containing S is a convex set containing S, such a smallest set always exists.

Theorem 1.9. *The convex hull of a set S is the set of all convex linear combinations of vectors in S.*

PROOF. Let T be the set of all convex linear combinations of vectors in S. Clearly, $S \subset T$ and, since H(S) contains all convex linear combinations of vectors in S, $T \subset H(S)$. Thus the theorem will be established if we show that T is convex. Let α, $\beta \in T$. These vectors can be expressed in the form

$$\alpha = \sum_{i=1}^{r} t_i \alpha_i, \qquad \sum_{i=1}^{r} t_i = 1, \qquad t_i \geq 0, \qquad \alpha_i \in S,$$

$$\beta = \sum_{i=1}^{r} s_i \alpha_i, \qquad \sum_{i=1}^{r} s_i = 1, \qquad s_i \geq 0, \qquad \alpha_i \in S,$$

where both expressions involve the same finite set of elements of S. This can be done by adjoining, where necessary, some terms with zero coefficients. Then for $0 \leq t \leq 1$,

$$(1 - t)\alpha + t\beta = \sum_{i=1}^{r} \{(1 - t)t_i + ts_i\}\alpha_i.$$

Since $(1 - t)t_i + ts_i \geq 0$ and $\sum_{i=1}^{r} \{(1 - t)t_i + ts_i\} = (1 - t) + t = 1$, $(1 - t)\alpha + t\beta \in T$ and T is a convex set. Thus $T = H(S)$. □

BIBLIOGRAPHICAL NOTES

The connections between linear algebra and geometry appear to some extent in almost all expository material on matrix theory. For an excellent classical treatment see M. Bocher, *Introduction to Higher Algebra*. For an elegant modern treatment see R. W. Gruenberg and A. J. Weir, *Linear Geometry*. A very readable exposition that starts from first principles and treats some classical problems with modern tools is available in N. H. Kuiper, *Linear Algebra and Geometry*.

EXERCISES

1. For each of the following linear manifolds, write down a parametric representation and a determining set of linear conditions:

(3) $L_1 = (1, 0, 1) + \langle (1, 1, 1), (2, 1, 0) \rangle$.

(2) $L_2 = (1, 2, 2) + \langle (2, 1, -2), (2, -2, 1) \rangle$.

(3) $L_3 = (1, 1, 1, 2) + \langle (0, 1, 0, -1), (2, 1, -2, 3) \rangle$.

2. For L_1 and L_2 given in Exercise 1, find a parametric representation and linear conditions for $L_1 \cap L_2$.

3. In R_3 find the smallest linear manifold L containing $\{(2, 1, 2), (2, 2, 1), (-1, 1, \frac{7}{2})\}$. Show that L is parallel to $L_1 \cap L_2$, where L_1 and L_2 are given in Exercise 1.

4. Determine whether $(0, 0)$ is in the convex hull of $S = \{(1, 1), (-6, 7), (5, -6)\}$. (This can be determined reasonably well by careful plotting on a coordinate system. At least it can be done with sufficient accuracy to make a guess which can be verified. For higher dimensions the use of plotting points is too difficult and inaccurate. An effective method is given in Section 3.)

5. Determine whether $(0, 0, 0)$ is in the convex hull of $T = \{(6, -5, -2), (3, -8, 6), (-4, 8, -5), (-9, 2, 8), (-7, -2, 8), (-5, 5, 1)\}$.

6. Show that the intersection of two linear manifolds is either empty or a linear manifold.

7. If L_1 and L_2 are linear manifolds, the *join* of L_1 and L_2 is the smallest linear manifold containing both L_1 and L_2 which we denote by $L_1 \, J \, L_2$. If $L_1 = \alpha_1 + S_1$ and $L_2 = \alpha_2 + S_2$, show that the join of L_1 and L_2 is $\alpha_1 + \langle \alpha_2 - \alpha_1 \rangle + S_1 + S_2$.

8. Let $L_1 = \alpha_1 + S_1$ and $L_2 = \alpha_2 + S_2$. Show that if $L_1 \cap L_2$ is not empty, then $L_1 \, J \, L_2 = \alpha_1 + S_1 + S_2$.

9. Let $L_1 = \alpha_1 + S_1$ and $L_2 = \alpha_2 + S_2$. Show that if $L_1 \cap L_2$ is empty, then $L_1 \, J \, L_2 \neq \alpha_1 + S_1 + S_2$, that is, $\alpha_2 - \alpha_1 \neq S_1 + S_2$.

10. Show that $\dim L_1 \, J \, L_2 = \dim (S_1 + S_2)$ if $L_1 \cap L_2 \neq \emptyset$ and $\dim L_1 \, J \, L_2 = \dim (S_1 + S_2) + 1$ if $L_1 \cap L_2 = \emptyset$.

2 | Finite Cones and Linear Inequalities

This section requires Section 1 for background and, of course, those sections required for Section 1. Although some material is independently developed here, Section 10 of Chapter II would also be helpful.

In this section and the following section we assume that F is the field of real numbers R, or a subfield of R.

If a set is closed under multiplication by non-negative scalars, it is called a *cone*. This is in analogy with the familiar cones of elementary geometry with vertex at the origin which contain with any point not at the vertex all points on the same half-line from the vertex through the point. If the cone is also closed under addition, it is called a *convex cone*. It is easily seen that a convex cone is a convex set.

If C is a convex cone and there exists a finite set of vectors $\{\alpha_1, \ldots, \alpha_p\}$ in C such that every vector in C can be represented as a linear combination of the α_i with non-negative coefficients, a *non-negative linear combination*, we call $\{\alpha_1, \ldots, \alpha_p\}$ the *generators* of C and call C a *finite cone*. The cone generated by a single non-zero vector is called a *half-line*. A dependable picture of a finite cone is formed by considering the half-lines formed by each of the generators as constituting an edge of a pointed cone as in Fig. 4. By considering a solid circular cone in R^3 it should be clear that there are convex cones that are not finite. A finite cone is the convex hull of a finite number of half-lines.

Let S be the largest subspace contained in C. If $S = \{0\}$, then S contains no line through the origin. In this case we say that C is *pointed*. If S is of dimension 1, then C is wedge shaped with S forming the edge of the wedge.

Given any subset $W \subset V$, let W^+ denote the set of all linear functionals that take on non-negative values for all $\alpha \in W$; that is, $W^+ = \{\phi \mid \phi\alpha \geq 0$ for all $\alpha \in W\}$. W^+ is closed under non-negative linear combinations and is a convex cone in $\hat{V}$. W^+ is called the *dual cone* or *polar cone* of W. Similarly, if $W \subset \hat{V}$, then W^+ is the set of all vectors which have non-negative values for

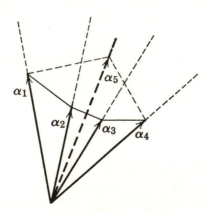

Fig. 4

all linear functionals in W. In this case, too, W^+ is called the dual cone of W. For the dual of the dual $(W^+)^+$ we write W^{++}.

Theorem 2.1. (1) If $W_1 \subset W_2$, then $W_1^+ \supset W_2^+$.

(2) $(W_1 + W_2)^+ = W_1^+ \cap W_2^+$ if $0 \in W_1 \cap W_2$.

(3) $W_1^+ + W_2^+ \subset (W_1 \cap W_2)^+$.

PROOF. (1) is obvious.

(2) If $\phi \in W_1^+ \cap W_2^+$, then for all $\alpha = \alpha_1 + \alpha_2$ where $\alpha_1 \in W_1$ and $\alpha_2 \in W_2$ we have $\phi\alpha = \phi\alpha_1 + \phi\alpha_2 \geq 0$. Hence, $W_1^+ \cap W_2^+ \subset (W_1 + W_2)^+$. On the other hand, $W_1 \subset W_1 + W_2$ so that $W_1^+ \supset (W_1 + W_2)^+$. Similarly, $W_2^+ \supset (W_1 + W_2)^+$. Hence, $W_1^+ \cap W_2^+ \supset (W_1 + W_2)^+$. It follows then that $W_1^+ \cap W_2^+ = (W_1 + W_2)^+$.

(3) $W_1 \supset W_1 \cap W_2$ so that $W_1^+ \subset (W_1 \cap W_2)^+$. Similarly, $W_2^+ \subset (W_1 \cap W_2)^+$. It then follows that $W_1^+ + W_2^+ \subset (W_1 \cap W_2)^+$. $\square$

Theorem 2.2. $W \subset W^{++}$ and $W^+ = W^{+++}$.

PROOF. Let $W \subset V$. If $\alpha \in W$, then $\phi\alpha \geq 0$ for all $\phi \in W^+$. This means that $W \subset W^{++}$. It then follows that $W^+ \subset (W^+)^{++} = W^{+++}$. On the other hand from Proposition 2.1 we have $W^+ \supset (W^{++})^+ = W^{+++}$. Thus $W^+ = W^{+++}$. The situation is the same for $W \subset \hat{V}$. $\square$

A cone C is said to be *reflexive* if $C = C^{++}$.

Theorem 2.3. *A cone is reflexive if and only if it is the dual cone of a set in the dual space.*

PROOF. Suppose C is reflexive. Then $C = C^{++}$ is the dual cone of C^+. On the other hand, if C is the dual cone of $W \subset \hat{V}$, then $C = W^+ = W^{+++} = C^{++}$ and C is reflexive. $\square$

The dual cone of a finite cone is called a *polyhedral cone*. If C is a finite cone in $\hat{V}$ generated by the finite set $G = \{\phi_1, \ldots, \phi_q\}$, then $C^+ = D = \{\alpha \mid \phi_i\alpha \geq 0$ for all $\phi_i \in G\}$. A dependable picture of a polyhedral cone can be formed by considering a finite cone, for we soon show that the two types of cones are equivalent. Each face of the cone is a part of one of the hyperplanes $\{\alpha \mid \phi_i\alpha = 0\}$, and the cone is on the positive side of each of these hyperplanes. In a finite cone the emphasis is on the edges as generating the cone; in a polyhedral cone the emphasis is on the faces as bounding the cone.

Theorem 2.4. *Let σ be a linear transformation of U into V. If C is a finite cone in U, then $\sigma(C)$ is a finite cone. If D is a polyhedral cone in V, then $\sigma^{-1}(D)$ is a polyhedral cone.*

PROOF. If $\{\alpha_1, \ldots, \alpha_p\}$ generates C, then $\{\sigma(\alpha_1), \ldots, \sigma(\alpha_p)\}$ generates $\sigma(C)$. Let D be a polyhedral cone dual to the finite cone E in $\hat{V}$. The following statements are equivalent: $\alpha \in \sigma^{-1}(D)$; $\sigma(\alpha) \in D$; $\psi\sigma(\alpha) \geq 0$ for all $\psi \in E$; $\hat{\sigma}(\psi)\alpha \geq 0$ for all $\psi \in E$; $\alpha \in (\hat{\sigma}(E))^+$. Thus $\sigma^{-1}(D)$ is dual to the finite cone $\hat{\sigma}(E)$ in $\hat{U}$ and is therefore polyhedral. $\square$

Theorem 2.5. *The sum of a finite number of finite cones is a finite cone and the intersection of a finite number of polyhedral cones is a polyhedral cone.*

PROOF. The first assertion of the theorem is obvious. Let $D_1, \ldots, D_r$ be polyhedral cones, and let $C_1, \ldots, C_r$ be the finite cones of which they are the duals. Then $C_1 + \cdots + C_r$ is a finite cone, and by Theorem 2.1 $D_1 \cap \cdots \cap D_r = C_1^+ \cap \cdots \cap C_r^+ = (C_1 + \cdots + C_r)^+$ is polyhedral. $\square$

Theorem 2.6. *Every finite cone is polyhedral.*

PROOF. The theorem is obviously true in a vector space of dimension 1. Let $\dim V = n$ and assume the theorem is true in vector spaces of dimension less than n.

Let $A = \{\alpha_1, \ldots, \alpha_p\}$ be a finite set generating the finite cone C. We can assume that each $\alpha_k \neq 0$. For each α_k let W_k be a complementary subspace of $\langle \alpha_k \rangle$; that is, $V = W_k \oplus \langle \alpha_k \rangle$. Let π_k be the projection of V onto W_k along $\langle \alpha_k \rangle$. $\pi_k(C)$ is a finite cone in W_k. By the induction assumption it is polyhedral since $\dim W_k = n - 1$. Then $\pi_k^{-1}(\pi_k(C)) = C_k$ is polyhedral by Theorem 2.4. Since $C \subset C_k$ for each k, C is contained in the polyhedral cone $C_1 \cap \cdots \cap C_p$.

We must now show that if $\alpha_0 \notin C$ and C is not a half-line, then there is a C_j such that $\alpha_0 \notin C_j$. If not, then suppose $\alpha_0 \in C_j$ for $j = 1, \ldots, p$. Then $\pi_j(\alpha_0) \in \pi_j(C)$ so that there is an $a_j \in F$ such that $\alpha_0 + a_j\alpha_j = \sum_{i=1}^p b_{ij}\alpha_i$ where $b_{ij} \geq 0$. We cannot obtain such an expression with $a_i \leq 0$ for then α_0 would be in C. But we can modify these expressions step by step and remove all the terms on the right sides.

Suppose $b_{ij} = 0$ for $i < k$ and $j = 1, \ldots, p$; that is, $\alpha_0 + a_j\alpha_j = \sum_{i=k}^p b_{ij}\alpha_i$. This is already true for $k = 1$. Then

$$\alpha_0 + (a_k - b_{kk})\alpha_k = \sum_{i=k+1}^p b_{ik}\alpha_i.$$

As before, we cannot have $a_k - b_{kk} \leq 0$. Set $a_k - b_{kk} = a_k' > 0$. Then for $j \neq k$ we have

$$\left(1 + \frac{b_{kj}}{a_k'}\right)\alpha_0 + a_j\alpha_j = \sum_{i=k+1}^p \left(b_{ij} + \frac{b_{kj}}{a_k'} b_{ik}\right)\alpha_i.$$

Upon division by $1 + \dfrac{b_{kj}}{a_k'}$ we get expressions of the form

$$\alpha_0 + a_j'\alpha_j = \sum_{i=k+1}^p b_{ij}'\alpha_j, \qquad j = 1, 2, \ldots, p,$$

with $a'_j > 0$ and $b'_{ij} \geq 0$. Continuing in this way we eventually get $\alpha_0 + c_j\alpha_j = 0$ with $c_j > 0$ for all j. This would imply $\alpha_j = -\dfrac{1}{c_j}\alpha_0$ for $j = 1, \ldots, p$. Thus C is generated by $\{-\alpha_0\}$; that is, C is a half-line, which is polyhedral. This would imply that C is polyhedral. If C is polyhedral there is nothing to prove. If C is not half-line, the assumption that $\alpha_0 \in C_j$ for all j is untenable. But this would mean that $C = C_1 \cap \cdots \cap C_p$, in which case C is polyhedral. $\square$

Theorem 2.7. *A polyhedral cone is finite.*

PROOF. Let $C = D^+$ be a polyhedral cone dual to the finite cone D. We have just proven that a finite cone is polyhedral, so there is a finite cone E such that $D = E^+$. But then E is also polyhedral so that $E = E^{++} = D^+ = C$. Since E is finite, C is also. $\square$

Although polyhedral cones and finite cones are identical, we retain both terms and use whichever is most suitable for the point of view we wish to emphasize. A large number of interesting and important results now follow very easily. The sum and intersection of finite cones are finite cones. The dual cone of a finite cone is finite. A finite cone is reflexive. Also, for finite cones part (3) of Theorem 2.1 can be improved. If C_1 and C_2 are finite cones, then $(C_1 \cap C_2)^+ = (C_1^{++} \cap C_2^{++})^+ = (C_1^+ + C_2^+)^{++} = C_1^+ + C_2^+$.

Our purpose in introducing this discussion of finite cones was to obtain some theorems about linear inequalities, so we now turn our attention to that subject. The following theorem is nothing but a paraphrase of the statement that a finite cone is reflexive.

Theorem 2.8. *Let*

$$a_{11}x_1 + \cdots + a_{1n}x_n \geq 0$$

$$\cdot$$
$$\cdot \qquad\qquad\qquad\qquad (2.1)$$
$$\cdot$$

$$a_{m1}x_1 + \cdots + a_{mn}x_n \geq 0$$

be a system of linear inequalities. If

$$a_1x_1 + \cdots + a_nx_n \geq 0$$

is a linear inequality which is satisfied whenever the system (2.1) *is satisfied, then there exist non-negative scalars* $(y_1, \ldots, y_m)$ *such that* $\sum_{i=1}^{m} y_i a_{ij} = a_j$ *for* $j = 1, \ldots, n$.

PROOF. Let ϕ_i be the linear functional represented by $[a_{i1} \cdots a_{in}]$, and let ϕ be the linear functional represented by $[a_1 \cdots a_n]$. If ξ represented by $(x_1, \ldots, x_n)$ satisfies the system (2.1), then ξ is in the cone C^+ dual to the finite cone C generated by $\{\phi_1, \ldots, \phi_m\}$. Since $\phi\xi \geq 0$ for all $\xi \in C^+$, $\phi \in C^{++} = C$. Thus there exist non-negative y_i such that $\phi = \sum_{i=1}^{m} y_i\phi_i$. The conclusion of the theorem then follows. $\square$

Theorem 2.9. *Let* $A = \{\alpha_1, \ldots, \alpha_n\}$ *be a basis of the vector space* U *and let* P *be the finite cone generated by* A. *Let* σ *be a linear transformation of* U *into* V *and let* β *be a given vector in* V. *Then one and only one of the following two alternatives holds: either*

(1) *there is a* $\xi \in P$ *such that* $\sigma(\xi) = \beta$, *or*

(2) *there is a* $\psi \in \hat{V}$ *such that* $\hat{\sigma}(\psi) \in P^+$ *and* $\psi\beta < 0$.

PROOF. Suppose (1) and (2) are satisfied at the same time. Then $0 > \psi\beta = \psi\sigma(\xi) = \hat{\sigma}(\psi)\xi \geq 0$, which is a contradiction.

On the other hand, suppose (1) is not satisfied. Since P is a finite cone, $\sigma(P)$ is a finite cone. The insolvability of (1) means that $\beta \notin \sigma(P)$. Since $\sigma(P)$ is also a polyhedral cone, there is a $\psi \in \hat{V}$ such that $\psi\beta < 0$ and $\psi\sigma(P) \geq 0$. But then $\hat{\sigma}(\psi)(P) \geq 0$ so that $\hat{\sigma}(\psi) \in P^+$. $\square$

It is apparent that the assumption that A is a basis of U is not used in the proof of Theorem 2.9. We wish, however, to translate this theorem into matrix notation. If ξ is represented by $X = (x_1, \ldots, x_n)$, then $\xi \in P$ if and only if each $x_i \geq 0$. To simplify notation we write "$X \geq 0$" to mean each $x_i \geq 0$, and we refer to P as the *positive orthant*. Since the generators of P form a basis of U, the generators of P^+ are the elements of the dual basis $\hat{A} = \{\phi_1, \ldots, \phi_n\}$. It thus turns out that P^+ is the positive orthant of $\hat{U}$.

Let $\beta = \{\beta_1, \ldots, \beta_m\}$ be a basis of V and $\hat{B} = \{\hat{\beta}_1, \ldots, \hat{\beta}_m\}$ the dual basis in $\hat{V}$. Let $A = [a_{ij}]$ represent σ with respect to A and B, $B = (b_1, \ldots, b_m)$ represent β, and $Y = [y_1 \cdots y_m]$ represent ψ. Then $\hat{\sigma}(\psi)$ is represented by YA and $\hat{\sigma}(\psi) \in P^+$ if and only if $YA \geq 0$. In this notation Theorem 2.9 becomes

Theorem 2.10. *One and only one of the following two alternatives holds: either*

(1) *there is an* $X \geq 0$ *such that* $AX = B$, *or*

(2) *there is a* Y *such that* $YA \geq 0$ *and* $YB < 0$. $\square$

Rather than continue to make these translations we adopt notational conventions which will make such translations more evident. We write $\xi \geq 0$ to mean $\xi \in P$, $\xi \geq \zeta$ to mean $\xi - \zeta \in P$, $\hat{\sigma}(\psi) \geq 0$ to mean $\hat{\sigma}(\psi) \in P^+$, etc.

Theorem 2.11. *With the notation of Theorem 2.9, let* ϕ *be a linear functional in* $\hat{U}$, *let* g *be an arbitrary scalar, and assume* $\beta \in \sigma(P)$. *Then one and only one of the following two alternatives holds; either*

(1) *there is a* $\xi \geq 0$ *such that* $\sigma(\xi) = \beta$ *and* $\phi\xi \geq g$, *or,*

(2) *there is* $\psi \in V$ *such that* $\hat{\sigma}(\psi) \geq \phi$ *and* $\psi\beta < g$.

PROOF. Suppose (1) and (2) are satisfied at the same time. Then $g > \psi\beta = \psi\sigma(\xi) = \hat{\sigma}(\psi)\xi \geq \phi\xi \geq g$, which is a contradiction.

On the other hand, suppose (2) is not satisfied. We wish to find a $\xi \in P$ satisfying the conditions $\sigma(\xi) = \beta$ and $\phi\xi \geq g$ at the same time. We have seen before that vectors and linear transformations can be used to express systems of equations as a single vector equation. A similar technique works here.

Let $U_1 = U \oplus F$ be the set of all pairs (ξ, x) where $\xi \in U$ and $x \in F$. U_1 is made into a vector space over F by defining vector addition and scalar multiplication according to the rules

$$(\xi_1, x_1) + (\xi_2, x_2) = (\xi_1 + \xi_2, x_1 + x_2), a(\xi, x) = (a\xi, ax).$$

Let $\tilde{P}$ be the set of all (ξ, x) where $\xi \in P$ and $x \geq 0$. It is easily seen that $\tilde{P}$ is a finite cone in U_1. In a similar way we construct the vector space $V_1 = V \oplus F$.

We then define Σ to be the mapping of U_1 into V_1 which maps (ξ, x) onto $\Sigma(\xi, x) = (\sigma(\xi), \phi\xi - x)$. It can be checked that Σ is linear. It is now seen that $(\xi, x) \in \tilde{P}$ and $\Sigma(\xi, x) = (\beta, g)$ are equivalent to the conditions $\xi \in P$, $\sigma(\xi) = \beta$, and $\phi\xi = g + x \geq g$.

To use Theorem 2.9 we must describe $\hat{U}_1$ and $\hat{V}_1$ and determine the adjoint transformation $\hat{\Sigma}$. It is not difficult to see that $\widehat{U \oplus F}$ is isomorphic to $\hat{U} \oplus F$ where (ϕ, y) is the linear functional defined by the formula $(\phi, y)(\xi, x) = \phi\xi + yx$. In a similar way $\widehat{V \oplus F}$ is isomorphic to $\hat{V} \oplus F$. Then $\hat{\Sigma}(\psi, y)$ applied to (ξ, x) must have the effect

$$\begin{aligned}
\hat{\Sigma}(\psi, y)(\xi, x) &= (\psi, y)\,\Sigma\,(\xi, x) \\
&= (\psi, y)(\sigma(\xi), \phi\xi - x) \\
&= \psi\sigma(\xi) + y(\phi\xi - x) \\
&= \hat{\sigma}(\psi)\xi + y\phi\xi - yx \\
&= (\hat{\sigma}(\psi) + y\phi)\xi - yx.
\end{aligned}$$

This means that $\hat{\Sigma}(\psi, y) = (\hat{\sigma}(\psi) + y\phi, -y)$.

Now suppose there exist $\psi \in \hat{V}$ and $y \in F$ for which $\hat{\Sigma}(\psi, y) \in \tilde{P}^+$ and $(\psi, y)(\beta, g) = \psi\beta + yg < 0$. This is in the form of condition (2) of Theorem 2.9 and we wish to show that it cannot hold. $\hat{\Sigma}(\psi, y) \in \tilde{P}^+$ means $\hat{\sigma}(\psi) + y\phi \geq 0$ and $-y \geq 0$. If $-y > 0$, then $\hat{\sigma}\left(\dfrac{\psi}{-y}\right) \geq \phi$. Since (2) of this theorem is assumed not to hold this means $\left(\dfrac{\psi}{-y}\right)\beta \geq g$, or $\psi\beta + yg \geq 0$. If $y = 0$, then $\hat{\sigma}(\psi) \geq 0$. Since $\beta \in \sigma(P)$ by assumption, $\psi\beta = \psi\beta + yg < 0$ would contradict Theorem 2.9. Thus (2) of Theorem 2.9 cannot be satisfied. This implies there is a $(\xi, x) \in \tilde{P}$ such that $\Sigma(\xi, x) = (\sigma(\xi). \; \phi\xi - x) = (\beta, g)$, which proves the theorem. $\square$

Theorem 2.12. *Let* P_1 *be the positive orthant in* U *generated by the basis* $A = \{\alpha_1, \ldots, \alpha_n\}$ *and* P_2 *the positive orthant in* V *generated by the basis* $B = \{\beta_1, \ldots, \beta_m\}$. *Let* σ, β, *and* ϕ *be given and assume* $\beta \in \sigma(P_1) + P_2$. *For each scalar* g *one and only one of the following two alternatives holds, either*
(1) *there is a* $\xi \geq 0$ *such that* $\sigma(\xi) \leq \beta$ *and* $\phi\xi \geq g$, *or*
(2) *there is a* $\psi \geq 0$ *such that* $\hat{\sigma}(\psi) \geq \phi$ *and* $\psi\beta < g$.
PROOF.   Construct the vector space $U \oplus V$ and define the mapping Σ of $U \oplus V$ into V by the rule

$$\Sigma(\xi, \eta) = \sigma(\xi) + \eta.$$

Then the condition $\Sigma(\xi, \eta) = \beta$ with $(\xi, \eta) \geq 0$ is equivalent to $\beta - \sigma(\xi) = \eta \geq 0$ with $\xi \geq 0$. Since $\hat{\Sigma}(\psi) = (\hat{\sigma}(\psi), \psi)$, the condition $\hat{\Sigma}(\psi) \geq (\phi, 0)$ is equivalent to $\hat{\sigma}(\psi) \geq \psi$ and $\psi \geq 0$. With this interpretation, the two conditions of this theorem are equivalent to the corresponding conditions of Theorem 2.11. $\square$

Theorem 2.13. *Let* σ, β, *and* ϕ *be given and assume* $\beta \in \sigma(U)$. *For each scalar* g *one and only one of the following two alternatives holds, either*
(1) *there is a* $\xi \in U$ *such that* $\sigma(\xi) = \beta$ *and* $\phi\xi \geq g$, *or*
(2) *there is a* $\psi \in \hat{V}$ *such that* $\hat{\sigma}(\psi) = \phi$ *and* $\psi\beta < g$.
PROOF.   Construct the vector space $U \oplus U$ and define the mapping Σ of $U \oplus U$ into V by the rule

$$\Sigma(\xi_1, \xi_2) = \sigma(\xi_1) - \sigma(\xi_2) = \sigma(\xi_1 - \xi_2).$$

Let $\xi = \xi_1 - \xi_2$. Then the condition $\Sigma(\xi_1, \xi_2) = \beta$ with $(\xi_1, \xi_2) \geq 0$ is equivalent to $\sigma(\xi) = \beta$ with no other restriction on ξ since every $\xi \in U$ can be represented in the form $\xi = \xi_1 - \xi_2$ with $\xi_1 \geq 0$ and $\xi_2 \geq 0$. Since $\hat{\Sigma}(\psi) = (\hat{\sigma}(\psi), -\hat{\sigma}(\psi))$, the condition $\hat{\Sigma}(\psi) \geq (\phi, -\phi)$ is equivalent to $\hat{\sigma}(\psi) = \phi$. With this interpretation, the two conditions of this theorem are equivalent to the corresponding conditions of Theorem 2.11. $\square$

Notice, in Theorems 2.11, 2.12, and 2.13, how an inequality for one variable corresponds to the condition that the other variable be non-negative, while an equation for one of the variables leaves the other variable unrestricted. For example, we have the conditions $\sigma(\xi) = \beta$ and $\psi \in \hat{V}$ in Theorem 2.11 replaced by $\sigma(\xi) \leq \beta$ and $\psi \geq 0$ in Theorem 2.12.

Theorem 2.14. *One and only one of the following two alternatives holds: either*
(1) *there is a* $\xi \geq 0$ *such that* $\sigma(\xi) = 0$ *and* $\phi\xi > 0$, *or*
(2) *there is a* $\psi \in \hat{V}$ *such that* $\hat{\sigma}(\psi) \geq \phi$.

PROOF. This theorem follows from Theorem 2.11 by taking $\beta = 0$ and $g > 0$. The assumption that $\beta \in \sigma(P)$ is then satisfied automatically and the condition $\psi\beta < g$ is not a restriction. $\square$

Theorem 2.15. *One and only one of the following two alternatives holds:* *either*

(1) *there is a* $\xi \geq 0$ *such that* $\sigma(\xi) \leq \beta$, *or*
(2) *there is a* $\psi \geq 0$ *such that* $\hat{\sigma}(\psi) \geq 0$ *and* $\psi\beta < 0$.

PROOF. This theorem follows from Theorem 2.12 by taking $\phi = 0$ and $g < 0$. In this case the assumption that $\beta \in \sigma(P_1) + P_2$ is not satisfied automatically. However, in Theorem 2.12 conditions (1) and (2) are symmetric and this assumption could have been replaced by the dual assumption that $-\phi \in P_1^+ + \hat{\sigma}(-P_2^+)$, and this assumption is satisfied. $\square$

Theorem 2.16. *One and only one of the following two alternatives holds:* *either*

(1) *there is a* $\xi \in U$ *such that* $\sigma(\xi) = \beta$, *or*
(2) *there is a* $\psi \in \hat{V}$ *such that* $\hat{\sigma}(\psi) = 0$ *and* $\psi\beta < 0$.

PROOF. This theorem follows from Theorem 2.13 by taking $\phi = 0$ and $g < 0$. Again, although the condition $\beta \in \sigma(U)$ is not satisfied automatically, the equally sufficient dual condition $\phi \in \hat{\sigma}(V)$ is satisfied. $\square$

It is sometimes convenient to express condition (2) of Theorem 2.16 in a slightly different form. It is equivalent to assert that there is a $\psi \in \hat{V}$ such that $\hat{\sigma}(\psi) = 0$ and $\psi\beta = 1$. If ψ satisfies condition (2), then $\dfrac{\psi}{\psi\beta}$ satisfies this condition. In this form Theorem 2.16 is equivalent to Theorem 7.2 of Chapter II, and identical to Theorem 5.5 of Chapter IV.

An application of these theorems to the problem of linear programming is made in Section 3.

BIBLIOGRAPHICAL NOTES

An excellent expository treatment of this subject with numerous examples is given by D. Gale, *The Theory of Linear Economic Models*. A number of expository and research papers and an extensive bibliography are available in Kuhn and Tucker, *Linear Inequalities and Related Systems, Annals of Mathematics Studies*, Study 38.

EXERCISES

1. In R^3 let C_1 be the finite cone generated by $\{(1, 1, 0), (1, 0, -1), (0, -1, 1)\}$. Wrtie down the inequalities that characterize the polyhedral cone C_1^+.

2. Find the linear functionals which generate C_1^+ as a finite cone.

3. In R^3 let C_2 be the finite cone generated by $\{(0, 1, 1), (1, -1, 0), (1, 1, 1)\}$. Write down a set of generators of $C_1 + C_2$, where C_1 is the finite cone given in Exercise 1.

4. Find a minimum set of generators of $C_1 + C_2$, where C_1 and C_2 are the cones given in Exercises 1 and 3.

5. Find the generators of $C_1^+ + C_2^+$, where C_1 and C_2 are the cones given in Exercises 1 and 3.

6. Determine the generators of $C_1 \cap C_2$.

7. Let $A = \begin{bmatrix} 0 & 1 & 1 \\ 1 & -1 & 1 \\ 1 & 0 & 1 \end{bmatrix}$ and $B = \begin{bmatrix} 1 \\ 1 \\ 0 \end{bmatrix}$.

Determine whether there is an $X \geq 0$ such that $AX = B$. (Since the columns of A are the generators of C_2, this is a question of whether $(1, 1, 0)$ is in C_2. Why?)

8. Use Theorem 2.16 (or the matrix equivalent) to show that the following system of equations has no solution:

$$-x_1 + 2x_2 = 2$$
$$2x_1 - x_2 = 2$$
$$2x_1 + 2x_2 = -1.$$

9. Use Theorem 2.9 (or 2.10) to show that the following system of equations has no non-negative solution:

$$2x_1 + 5x_2 - 7x_3 = 4$$
$$x_1 - 5x_2 - 6x_3 = 3.$$

10. Prove the following theorem: One and only one of the following two alternatives holds: either

(1) there is a $\xi \in U$ such that $\sigma(\xi) \geq \beta$, or
(2) there is a $\psi \geq 0$ such that $\acute{\sigma}(\psi) = 0$ and $\psi\beta > 0$.

11. Use the previous exercise to show that the following system of inequalities has no solution.

$$x_1 - 2x_2 \geq 2$$
$$-2x_1 + x_2 \geq 2$$
$$2x_1 + 2x_2 \geq 1.$$

12. A vector ξ is said to be *positive* if $\xi = \sum_{i=1}^{n} x_i \xi_i$ where each $x_i > 0$ and $\{\xi_1, \ldots, \xi_n\}$ is a basis generating the positive orthant. We use the notation $\xi > 0$ to denote the fact that ξ is positive. A vector ξ is said to be *semi-positive* if $\xi \geq 0$ and $\xi \neq 0$. Use Theorem 2.11 to prove the following theorem. One and only one of the following two alternatives holds: either

(1) there is a semi-positive ξ such that $\sigma(\xi) = 0$, or
(2) there is a $\psi \in \hat{V}$ such that $\acute{\sigma}(\psi)$ is positive.

13. Let W be a subspace of U, and let $W^\perp$ be the annihilator of W. Let $\{\eta_1, \ldots,$ $\eta_r\}$ be a basis of $W^\perp$. Let V be any vector space of dimension r over the same coefficient field, and let $\{\beta_1, \ldots, \beta_r\}$ be a basis of V. Define the linear transformation σ of U into V by the rule, $\sigma(\xi) = \sum_{i=1}^{r} \eta_i(\xi)\beta_i$. Show that W is the kernel of σ. Show that $W^\perp = \hat{\sigma}(\hat{V})$.

14. Show that if W is a subspace of U, then one and only one of the following two alternatives holds: either
(1) there is a semi-positive vector in W, or
(2) there is a positive linear functional in $W^\perp$.

15. Use Theorem 2.12 to prove the following theorem. One and only one of the following two alternatives holds: either
(1) there is a semi-positive ξ such that $\sigma(\xi) \leq 0$, or
(2) there is a $\psi \geq 0$ such that $\hat{\sigma}(\psi) > 0$.

3 | Linear Programming

This section requires Section 2 for background. Specifically, Theorem 2.12 is required. If we were willing to accept that theorem without proof, the required background would be reduced to Chapter I, the first eight sections of Chapter II, and the first four sections of Chapter IV.

Given $A = [a_{ij}]$, $B = (b_1, \ldots, b_m)$, and $C = [c_1 \cdots c_n]$, the *standard maximum linear programming problem* is to find any or all non-negative $X = (x_1, \ldots, x_n)$ which maximize

$$CX \tag{3.1}$$

subject to the condition

$$AX \leq B. \tag{3.2}$$

CX is called the *objective function* and the linear inequalities contained in $AX \leq B$ are called the *linear constraints*.

There are many practical problems which can be formulated in these terms. For example, suppose that a manufacturing plant produces n different kinds of products and that x_j is the amount of the jth product that is produced. Such an interpretation imposes the condition $x_j \geq 0$. If c_j is the income from a unit amount of the jth product, then $\sum_{j=1}^{n} c_j x_j = CX$ is the total income. Assume that the objective is to operate this business in such a manner as to maximize CX.

In this particular problem it is likely that each c_j is positive and that CX can be made large by making each x_j large. However, there are usually practical considerations which limit the quantities that can be produced. For example, supppse that limited quantities of various raw materials to make these products are available. Let b_i be the amount of the ith ingredient available. If a_{ij} is the amount of the ith ingredient consumed in producing one unit of the jth product, then we have the condition $\sum_{j=1}^{n} a_{ij}x_j \le b_i$. These constraints mean that the amount of each product produced must be chosen carefully if CX is to be made as large as possible.

We cannot enter into a discussion of the many interesting and important problems that can be formulated as linear programming problems. We confine our attention to the theory of linear programming and practical methods for finding solutions. Linear programming problems often involve large numbers of variables and constraints and the importance of an efficient method for obtaining a solution cannot be overemphasized. The simplex method presented by G. B. Dantzig in 1949 was the first really practical method given for solving such problems, and it provided the stimulus for the development of an extensive theory of linear inequalities. It is the computational method we describe here.

The simplex method is deceptively simple and it is possible to solve problems of moderate complexity by hand with it. The rationale behind the method is more subtle, however. We establish necessary and sufficient conditions for the linear programming problem to have solutions and determine procedures by which a proposed solution can be tested for optimality. We describe the simplex method and show why it works before giving the details of the computational procedures.

We must first translate the statement of the linear programming problem into the terminology and notation of vector spaces. Let U and V be vector spaces over F of dimensions n and m, respectively. Let $A = \{\alpha_1, \ldots, \alpha_n\}$ be a fixed basis of U and let $B = \{\beta_1, \ldots, \beta_m\}$ be a basis of V. If $A = [a_{ij}]$ is a given $m \times n$ matrix, we let σ be the linear transformation of U into V represented by A with respect to A and B. Let P_1 be the finite cone in U generated by A, and P_2 the finite cone in V generated by B.

If β is the vector in V represented by $B = (b_1, \ldots, b_m)$, the condition $AX \le B$ is equivalent to saying that $\sigma(\xi) \le \beta$. Let $\hat{A}$ be the basis in $\hat{U}$ dual to A and let $\hat{B}$ be the basis in $\hat{V}$ dual to B. Let ϕ be the linear functional in $\hat{U}$ represented by $C = [c_1, \ldots c_n]$. In these terms the standard maximum linear programming problem is to find any or all $\xi \ge 0$ which maximize $\phi\xi$ subject to the constraint $\sigma(\xi) \le \beta$.

Let $\hat{\sigma}$ be the dual of σ. The *standard dual linear programming problem* is to find any or all $\psi \ge 0$ which minimize $\psi\beta$ subject to the constraint $\hat{\sigma}(\psi) \ge \phi$. If we take $\sigma' = -\sigma$, $\beta' = -\beta$, and $\phi' = -\phi$, then the dual problem is to find a $\psi \ge 0$ which maximizes $\psi\beta'$ subject to the constraint $\hat{\sigma}'(\psi) \le \phi'$. Thus, the relation between the original problem, which we call the *primal* problem, and the dual problem is symmetric. We could have taken a minimum problem as the primal problem, in which case the dual problem would have been a maximum problem. In this discussion, however, we consistently take the primal problem to be a maximum problem.

Any $\xi \geq 0$ such that $\sigma(\xi) \leq \beta$ is called a *feasible vector* for the standard primal linear programming problem. If a feasible vector exists, the primal problem is said to be *feasible*. Any $\psi \geq 0$ such that $\hat{\sigma}(\psi) \geq \phi$ is called a *feasible vector* for the dual problem, and if such a vector exists, the dual problem is said to be *feasible*. A feasible vector ξ_0 such that $\phi\xi_0 \geq \phi\xi$ for all feasible ξ is called an *optimal vector* for the primal problem.

Theorem 3.1. *The standard linear programming problem has a solution if and only if both the primal problem and the dual problem are feasible. The dual problem has a solution if and only if the primal problem has a solution, and the maximum value of $\phi\xi$ for the primal problem is equal to the minimum value of $\psi\beta$ for the dual problem.*

PROOF. If the primal linear programming problem is infeasible, then certainly no optimum vector exists. If the primal problem is feasible, then the assumption $\beta \in \sigma(P_1) + P_2$ of Theorem 2.12 is satisfied. If the dual problem is infeasible, then condition (2) of Theorem 2.12 cannot be satisfied. Thus for every g there is a $\xi \geq 0$ such that $\sigma(\xi) \leq \beta$ and $\phi\xi \geq g$. This means the values of $\phi\xi$ are unbounded and the primal problem has no solution.

Now, assume that both the primal problem and dual problem are feasible. If ξ is feasible for the primal problem and ψ is feasible for the dual problem, then $0 \leq \psi\{\beta - \sigma(\xi)\} = \psi\beta - \psi\sigma(\xi) = \psi\beta - \hat{\sigma}(\psi)\xi = \psi\beta - \phi\xi + \{\phi - \hat{\sigma}(\psi)\} \leq \psi\beta - \phi\xi$. Thus $\phi\xi$ is a lower bound for the values of $\psi\beta$. Assume, for now, that F is the field of real numbers and let g be the greatest lower bound of the values of $\psi\beta$ for feasible ψ. With this value of g condition (2) of Theorem 2.12 cannot be satisfied, so that there exists a feasible ξ_0 such that $\phi\xi_0 \geq g$. Since $\phi\xi_0$ is also a lower bound for the values of $\psi\beta$, $\phi\xi_0 > g$ is impossible. Thus $\phi\xi_0 = g$ and $\phi\xi_0 \geq \phi\xi$ for all feasible ξ. Because of the symmetry between the primal and dual problems the dual problem has a solution under exactly the same conditions. Furthermore, since g is the greatest lower bound for the values of $\psi\beta$, g is also the minimum value of $\psi\beta$ for feasible ψ.

If we permit F to be a subfield of the real numbers, but do not require that it be the field of real numbers, then we cannot assert that the value of g chosen as a greatest lower bound must be in F. Actually, it is true that g is in F, and with a little more effort we could prove it at this point. However, if A, B, and C have components in a subfield of the real numbers we can consider them as representing linear transformations and vectors in spaces over the real numbers. Under these conditions the argument given above is valid. Later, when we describe the simplex method, we shall see that the components of ξ_0 will be computed rationally in terms of the components of A and B and will lie in any field containing the components of A, B, and C. We then see that g is in F. $\square$

Theorem 3.2. *If ξ is feasible for the standard primal problem and ψ is feasible for the dual problem, then ξ is optimal for the primal problem and ψ is optimal for the dual problem if and only if $\psi\{\beta - \sigma(\xi)\} = 0$ and $\{\hat{\sigma}(\psi) - \phi\}\xi = 0$, or if and only if $\phi\xi = \psi\beta$.*

PROOF.   Suppose that ξ is feasible for the primal problem and ψ is feasible for the dual problem. Then $0 \le \psi\{\beta - \sigma(\xi)\} = \psi\beta - \psi\sigma(\xi) = \psi\beta - \hat{\sigma}(\psi)\xi = \psi\beta - \phi\xi + \{\phi - \hat{\sigma}(\psi)\}\xi \le \psi\beta - \phi\xi$. It is clear that $\psi\{\beta - \sigma(\xi)\} = 0$ and $\{\hat{\sigma}(\psi) - \phi\}\xi = 0$ if and only if $\psi\beta = \phi\xi$.

If ξ_0 and ψ_0 are feasible and $\phi\xi_0 = \psi_0\beta$ then $\phi\xi \le \psi_0\beta = \phi\xi_0$ for all feasible ξ. Thus ξ_0 is optimal. A similar argument shows that ψ_0 is optimal. On the other hand, suppose ξ_0 and ψ_0 are optimal. Let $\psi_0\beta = g$. Then condition (2) of Theorem 2.12 cannot be satisfied for this choice of g. Thus, there is a feasible ξ such that $\phi\xi \ge g$. Since ξ_0 is optimal, we have $\phi\xi_0 \ge \phi\xi \ge g = \psi_0\beta$. Since $\phi\xi_0 \le \psi_0\beta$, this means $\phi\xi_0 = \psi_0\beta$. $\square$

Theorem 3.2 has an important interpretation in terms of the inequalities of the linear programming problem as originally stated. Let $\zeta = \beta - \sigma(\xi)$ be represented by $Z = (z_1, \ldots, z_m)$ and let $\eta = \hat{\sigma}(\psi) - \phi$ be represented by $W = [w_1 \cdots w_n]$. Then the feasibility of ξ and ψ implies $z_i \ge 0$ and $w_i \ge 0$. The condition $\psi\zeta = \sum_{i=1}^{n} y_i z_i = 0$ means $y_i z_i = 0$ for each i. Thus $z_i > 0$ implies $y_i = 0$, and $y_i > 0$ implies $z_i = 0$. This means that if $\sum_{j=1}^{n} a_{ij}x_j \le b_i$ is satisfied as a strict inequality, then $y_i = 0$, and if $y_i > 0$, this inequality must be satisfied as an equality. A similar relation holds between the x_j and the dual constraints. This gives an effective test for the optimality of a proposed feasible pair $X = (x_1, \ldots, x_n)$ and $Y = [y_1 \cdots y_m]$.

It is more convenient to describe the simplex method in terms of solving the equation $\sigma(\xi) = \beta$ instead of the inequality $\sigma(\xi) \le \beta$. Although these two problems are not equivalent, the two types of problems are equivalent. In other words, to every problem involving an inequality there is an equivalent problem involving an equation, and to every problem involving an equation there is an equivalent problem involving an inequality. To see this, construct the vector space $U \oplus V$ and define the mapping σ_1 of $U \oplus V$ into V by the rule $\sigma_1(\xi, \eta) = \sigma(\xi) + \eta$. Then the equation $\sigma_1(\xi, \eta) = \beta$ with $\xi \ge 0$ and $\eta \ge 0$ is equivalent to the inequality $\sigma(\xi) \le \beta$ with $\xi \ge 0$. This shows that to each problem involving an inequality there is an equivalent problem involving an equation.

To see the converse, construct the vector space $V \oplus V$ and define the mapping σ_2 of U into $V \oplus V$ by the rule $\sigma_2(\xi) = (\sigma(\xi), -\sigma(\xi))$. Then the inequality $\sigma_2(\xi) \le (\beta, -\beta)$ with $\xi \ge 0$ is equivalent to the equation $\sigma(\xi) = \beta$ with $\xi \ge 0$.

Given a linear transformation σ of U into V, $\beta \in V$, and $\phi \in \hat{U}$, the *canonical maximum linear programming problem* is to find any or all $\xi \ge 0$ which maximize $\phi\xi$ subject to the constraint $\sigma(\xi) = \beta$. With this formulation of the linear programming problem it is necessary to see what becomes of the dual problem. Referring to σ_2 above, for $(\psi_1, \psi_2) \in \hat{V} \oplus \hat{V}$ we have $\hat{\sigma}_2(\psi_1, \psi_2)\xi = (\psi_1, \psi_2)\sigma_2(\xi) = (\psi_1, \psi_2)(\sigma(\xi), -\sigma(\xi)) = \psi_1\sigma(\xi) - \psi_2\sigma(\xi) = \hat{\sigma}(\psi_1 - \psi_2)\xi$. Thus, if we let $\psi = \psi_1 - \psi_2$, we see that we must have $\hat{\sigma}(\psi) = \hat{\sigma}_2(\psi_1, \psi_2) \geq \phi$ and $(\psi_1, \psi_2) \in P_2^+ \oplus P_2^+$. But the condition $(\psi_1, \psi_2) \in P_2^+ \oplus P_2^+$ is not a restriction on ψ since any $\psi \in \hat{V}$ can be written in the form $\psi = \psi_1 - \psi_2$ where $\psi_1 \geq 0$ and $\psi_2 \geq 0$. Thus, the *canonical dual linear programming problem* is to find any or all $\psi \in \hat{V}$ which minimize $\psi\beta$ subject to the constraint $\hat{\sigma}(\psi) \geq \phi$.

It is readily apparent that condition (1) and (2) of Theorem 2.11 play the same roles with respect to the canonical primal and dual problems that the corresponding conditions (1) and (2) of Theorem 2.12 play with respect to the standard primal and dual problems. Thus, theorems like Theorem 3.1 and 3.2 can be stated for the canonical problems.

The canonical primal problem is *feasible* if there is a $\xi \geq 0$ such that $\sigma(\xi) = \beta$ and the canonical dual problem is *feasible* if there is a $\psi \in \hat{V}$ such that $\hat{\sigma}(\psi) \geq \phi$.

Theorem 3.3. *The canonical linear programming problem has a solution if and only if both the primal problem and the dual problem are feasible. The dual problem has a solution if and only if the primal problem has a solution, and the maximum value of $\phi\xi$ for the primal problem is equal to the minimum value of $\psi\beta$ for the dual problem.* $\square$

Theorem 3.4. *If ξ is feasible for the canonical primal problem and ψ is feasible for the dual problem, then ξ is optimal for the primal problem and ψ is optimal for the dual problem if and only if $\{\hat{\sigma}(\psi) - \phi\}\xi = 0$, or if and only if $\phi\xi = \psi\beta$.* $\square$

From now on, assume that the canonical primal linear programming problem is feasible; that is, $\beta \in \sigma(P_1)$. There is no loss of generality in assuming that $\sigma(P_1)$ spans V, for in any case β is in the subspace of V spanned by $\sigma(P_1)$ and we could restrict our attention to that subspace. Thus $\{\sigma(\alpha_1), \ldots, \sigma(\alpha_n)\} = \sigma(A)$ also spans V and $\sigma(A)$ contains a basis of V. A feasible vector ξ is called a *basic feasible vector* if ξ can be expressed as a linear combination of m vectors in A which are mapped onto a basis of V. The corresponding subset of A is said to be *feasible*.

Since A is a finite set there are only finitely many feasible subsets of A. Suppose that ξ is a basic feasible vector expressible in terms of the feasible subset $\{\alpha_1, \ldots, \alpha_m\}$, $\xi = \sum_{i=1}^{m} x_i\alpha_i$. Then $\sigma(\xi) = \sum_{i=1}^{m} x_1\sigma(\alpha_i) = \beta$. Since $\{\sigma(\alpha_1), \ldots, \sigma(\alpha_m)\}$ is a basis the representation of β in terms of that basis is unique. Thus, to each feasible subset of A there is one and only one basic feasible vector; that is, there are only finitely many basic feasible vectors.

Theorem 3.5. *If the canonical primal linear programming problem is feasible, then there exists a basic feasible vector. If the canonical primal problem has a solution, then there exists a basic feasible vector which is optimal.*

PROOF. Let $\xi = \sum_{i=1}^{k} x_i \alpha_i$ be feasible. If $\{\sigma(\alpha_1), \ldots, \sigma(\alpha_k)\}$ is linearly independent, then ξ is a basic feasible vector since $\{\alpha_1, \ldots, \alpha_k\}$ can be extended to a feasible subset of A. Suppose this set is linearly dependent; that is, $\sum_{i=1}^{k} t_i \sigma(\alpha_i) = 0$ where at least one $t_i > 0$. Then $\sum_{i=1}^{k} (x_i - at_i) \sigma(\alpha_i) = \beta$ for every $a \in F$. Let a be the minimum of x_i/t_i for those $t_i > 0$. For notational convenience, let $x_k/t_k = a$. Then $x_k - at_k = 0$ and $\sum_{i=1}^{k-1} (x_i - at_i)\sigma(\alpha_i) = \beta$. If $t_i \leq 0$ we have $x_i - at_i \geq 0$ because $a \geq 0$. If $t_i > 0$, then $x_i - at_i \geq x_i - (x_i/t_i)t_i = 0$. Thus $\xi' = \sum_{i=1}^{k-1} (x_i - at_i)\alpha_i$ is also feasible and expressible in terms of fewer elements of A. We can continue in this way until we obtain a basic feasible vector.

Now, suppose the canonical problem has a solution and that ξ as given above is an optimal vector. If $\{\alpha_1, \ldots, \alpha_k\}$ is not a feasible subset of A, we can assume $x_i > 0$ for $i = 1, \ldots, k$ since otherwise ξ could be expressed in terms of a smaller subset of A. Let ψ be an optimum vector for the dual problem and let $\eta = \hat{\sigma}(\psi) - \phi$ be represented by $W = [w_1 \cdots w_n]$. By Theorem 3.4 $w_i = 0$ for $i = 1, \ldots, k$. It then follows that ξ' obtained as above is also optimal. We can continue in this way until we obtain a basic feasible vector which is optimal. $\square$

If a linear programming problem has a solution, there remains the problem of finding it. Since there are only finitely many basic feasible vectors and at least one of them must be optimal, in concept we could try them all and take the one that yields the largest value of $\phi\xi$. It is not easy, however, to find even one basic feasible vector, or even one feasible vector, and for problems with large numbers of variables there would still be an enormous number of vectors to test. It is convenient to divide the process of finding a solution into two parts: to find a basic feasible vector and to find an optimum vector when a basic feasible vector is known. We take up the second of these problems first because it is easier to describe the simplex method for it. It is then easy to modify the simplex method to handle the first problem.

Now, suppose that $B = \{\beta_1, \ldots, \beta_m\}$ is a basis of V where $\beta_i = \sigma(\alpha_i)$ and $\{\alpha_1, \ldots, \alpha_m\}$ is a feasible subset of A. Let $\beta = \sum_{i=1}^{m} b_i \beta_i$. Then $\xi = \sum_{i=1}^{m} b_i \alpha_i$ is the corresponding basic feasible vector and $\phi\xi = \sum_{i=1}^{m} c_i b_i$. Suppose that a new coordinate system in V is chosen in which only one basis element is replaced with a new one. Let $\beta_r = \sigma(\alpha_r)$ be replaced by

$\beta_k = \sigma(\alpha_k)$. Since σ is represented by $A = [a_{ij}]$ with respect to the bases A and B we have $\beta_k = \sum_{i=1}^{m} a_{ik}\beta_i$. Since $a_{rk} \neq 0$ we can solve for β_r and obtain

$$\beta_r = \frac{1}{a_{rk}}\left(\beta_k - \sum_{\substack{i=1 \\ i \neq r}}^{m} a_{ik}\beta_i\right). \tag{3.3}$$

Then

$$\beta = \sum_{i=1}^{m} b_i\beta_i = \frac{b_r}{a_{rk}}\beta_k + \sum_{\substack{i=1 \\ i \neq r}}^{m}\left(b_i - b_r\frac{a_{ik}}{a_{rk}}\right)\beta_i. \tag{3.4}$$

If we let

$$b'_k = \frac{b_r}{a_{rk}} \qquad \text{and} \qquad b'_i = b_i - b_r\frac{a_{ik}}{a_{rk}}, \tag{3.5}$$

then another solution to the equation $\sigma(\xi) = \beta$ is

$$\xi' = b'_k\alpha_k + \sum_{\substack{i=1 \\ i \neq r}}^{m} b'_i\alpha_i. \tag{3.6}$$

Notice that although β remains fixed and only its coordinates change, each particular choice of a basis leads to a different choice for the solution of the equation $\sigma(\xi) = \beta$.

We now wish to impose conditions so that ξ' will be feasible. Since $b_r \geq 0$ we must have $a_{rk} > 0$ and, since $b_i \geq 0$ either $a_{ik} \leq 0$ or $b_i/a_{ik} \geq b_r/a_{rk}$. This means r is an index for which $a_{rk} > 0$ and b_r/a_{rk} is the minimum of all b_i/a_{ik} for which $a_{ik} > 0$. For the moment, suppose this is the case. Then ξ' is also a basic feasible vector.

Now,

$$\phi\xi' = c_k b'_k + \sum_{\substack{i=1 \\ i \neq r}}^{m} c_i b'_i$$

$$= c_k\frac{b_r}{a_{rk}} + \sum_{\substack{i=1 \\ i \neq r}}^{m} c_i\left(b_i - b_r\frac{a_{ik}}{a_{rk}}\right)$$

$$= \sum_{i=1}^{m} c_i b_i - c_r b_r + c_k\frac{b_r}{a_{rk}} - \sum_{\substack{i=1 \\ i \neq r}}^{m} c_i b_r\frac{a_{ik}}{a_{rk}}$$

$$= \phi\xi + \frac{b_r}{a_{rk}}\left(c_k - \sum_{i=1}^{m} c_i a_{ik}\right)$$

$$= \phi\xi + \frac{b_r}{a_{rk}}(c_k - d_k). \tag{3.7}$$

Thus $\phi\xi' \geq \phi\xi$ if $c_k - \sum_{i=1}^{m} c_i a_{ik} > 0$, and $\phi\xi' > \phi\xi$ if also $b_r > 0$.

The simplex method specifies the choice of the basis element $\beta_r = \sigma(\alpha_r)$ to be removed and the vector $\beta_k = \sigma(\alpha_k)$ to be inserted into the new basis in the following manner:

(1) Compute $\sum_{i=1}^{m} c_i a_{ij} = d_j$ for $j = 1, \ldots, n$.

(2) Select an index k for which $c_k - d_k > 0$.

(3) For that k select an index r for which $a_{rk} > 0$ and b_r/a_{rk} is the minimum of all b_i/a_{ik} for which $a_{ik} > 0$.

(4) Replace $\beta_r = \sigma(\alpha_r)$ by $\beta_k = \sigma(\alpha_k)$.

(5) Express β in terms of the new basis and determine ξ'.

(6) Determine the new matrix A' representing σ.

There are two ways these replacement rules may fail to operate. There may be no index k for which $c_k - d_k > 0$, and if there is such a k, there may be no index r for which $a_{rk} > 0$. Let us consider the second possibility first. Suppose there is an index k for which $c_k - d_k > 0$ and $a_{ik} \leq 0$ for $i = 1, \ldots, m$. In other words, if this situation should occur, we choose to ignore any other choice of the index k for which the selection rules would operate. Then $\zeta = \alpha_k - \sum_{i=1}^{m} a_{ik}\alpha_i \geq 0$ and $\sigma(\zeta) = \sigma(\alpha_k) - \sum_{i=1}^{m} a_{ik}\sigma(\alpha_i) = 0$. Also $\phi\zeta = c_k - \sum_{i=1}^{m} c_i a_{ik} > 0$. Thus ζ satisfies condition (1) of Theorem 2.14. Since condition (2) cannot then be satisfied, the dual problem is infeasible and the problem has no solution.

Let us now assume the dual problem is feasible so that the selection prescribed in step (3) is always possible. With the new basic feasible vector ξ' obtained the replacement rules can be applied again, and again. Since there are only finitely many basic feasible vectors, this sequence of replacements must eventually terminate at a point where the selection prescribed in step (2) is not possible, or a finite set of basic feasible vectors may be obtained over and over without termination.

It was pointed out above that $\phi\xi' \geq \phi\xi$, and $\phi\xi' > \phi\xi$ if $b_r > 0$. This means that if $b_r > 0$ at any replacement, then we can never return to that particular feasible vector. There are finitely many subspaces of V spanned by $m - 1$ or fewer vectors in $\sigma(A)$. If β lies in one of these subspaces, the linear programming problem is said to be *degenerate*. If the problem is not degenerate, then no basic feasible vector can be expressed in terms of $m - 1$ or fewer vectors in A. Under these conditions, for each basic feasible vector every $b_i > 0$. Thus, if the problem is not degenerate, infinite repetition is not possible and the replacements must terminate.

Unfortunately, many practical problems are degenerate. A special replacement procedure can be devised which makes infinite repetition impossible. However, a large amount of experience indicates that it is very difficult to devise a problem for which the replacement procedure given above will not terminate.

Now, suppose $c_j - d_j \leq 0$ for $j = 1, \ldots, n$. Let $\hat{B} = \{\psi_1, \ldots, \psi_m\}$ be the basis in $\hat{V}$ dual to B, and consider $\psi = \sum_{i=1}^{m} c_i \psi_i$. Then $\hat{\sigma}(\psi) = \sum_{i=1}^{m} c_i \hat{\sigma}(\psi_i) = \sum_{i=1}^{m} c_i \{\sum_{j=1}^{n} a_{ij} \phi_j\} = \sum_{j=1}^{n} \{\sum_{i=1}^{m} c_i a_{ij}\} \phi_j = \sum_{j=1}^{n} d_j \phi_j \geq \sum_{j=1}^{n} c_j \phi_j = \phi$. Thus ψ is feasible for the canonical dual linear programming problem. But with this ψ we have $\psi\beta = \{\sum_{i=1}^{m} c_i \psi_i\}\{\sum_{j=1}^{m} b_j \beta_j\} = \sum_{i=1}^{m} c_i b_i$ and $\phi\xi = \{\sum_{i=1}^{n} c_i \phi_i\}\{\sum_{j=1}^{m} b_j \xi_j\} = \sum_{i=1}^{m} c_i b_i$. This shows that $\psi\beta = \phi\xi$. Since both ξ and ψ are feasible, this means that both are optimal. Thus optimal solutions to both the primal and dual problems are obtained when the replacement procedure prescribed by the simplex method terminates.

It is easy to formulate the steps of the simplex method into an effective computational procedure. First, we shall establish formulas by which we can compute the components of the new matrix A' representing σ.

$$\sigma(\alpha_j) = \sum_{i=1}^{m} a_{ij} \beta_i$$

$$= \sum_{\substack{i=1 \\ i \neq r}}^{m} a_{ij} \beta_i + \frac{a_{rj}}{a_{rk}} \left(\beta_k - \sum_{\substack{i=1 \\ i \neq r}}^{m} a_{ik} \beta_i \right)$$

$$= \frac{a_{rj}}{a_{rk}} \beta_k + \sum_{\substack{i=1 \\ i \neq r}}^{m} \left(a_{ij} - \frac{a_{rj}}{a_{rk}} a_{ik} \right) \beta_i. \tag{3.8}$$

Thus,

$$a'_{kj} = \frac{a_{rj}}{a_{rk}}, \qquad a'_{ij} = a_{ij} - \frac{a_{rj}}{a_{rk}} a_{ik}. \tag{3.9}$$

It turns out to be more convenient to compute the new $d'_j - c_j$ directly.

$$c_j - d'_j = -\sum_{\substack{i=1 \\ i \neq r}}^{m} c_i a'_{ij} - c_k a'_{kj} + c_j$$

$$= -\sum_{i=1}^{m} c_i \left(a_{ij} - \frac{a_{rj}}{a_{rk}} a_{ik} \right) - c_k \frac{a_{rj}}{a_{rk}} + c_j$$

$$= (c_j - d_j) - \frac{a_{rj}}{a_{rk}} (c_k - d_k). \tag{3.10}$$

For immediate comparison, we rewrite formulas (3.5) and (3.7),

$$b'_k = \frac{b_r}{a_{rk}}, \qquad b'_i = b_i - \frac{b_r}{a_{rk}} a_{ik}, \tag{3.5}$$

$$\phi\xi' = \phi\xi + \frac{b_r}{a_{rk}} (c_k - d_k). \tag{3.7}$$

The similarity between formulas (3.5), (3.7), (3.9), and (3.10) suggests that simple rules can be devised to include them as special cases. It is convenient to write all the relevant numbers in an array of the following form:

$$
\begin{array}{c}
\cdots \qquad c_k \qquad \cdots \qquad c_j \qquad \cdots
\end{array}
$$

$$\phi\xi = \sum_{i=1}^{m} c_i b.$$

(3.11)

The array within the rectangle is the augmented matrix of the system of equations $AX = B$. The first column in front of the rectangle gives the identity of the basis element in the feasible subset of A, and the second column contains the corresponding values of c_i. These are used to compute the values of $d_j (j = 1, \ldots, n)$ and $\phi\xi$ below the rectangle. The row $[\cdots \ c_k \ \cdots \ c_j \ \cdots]$ is placed above the rectangle to facilitate computation of the $(c_j - d_j)(j = 1, \ldots, n)$. Since this top row does not change when the basis is changed, it is usually placed at the top of a page of work for reference and not carried along with the rest of the work. This array has become known as a *tableau*.

The selection rules of the simplex method and the formulas (3.5), (3.7), (3.9), and (3.10) can now be formalized as follows:

(1) Select an index k for which $c_k - d_k > 0$.
(2) For that k select an index r for which $a_{rk} > 0$ and b_r/a_{rk} is the minimum of all b_i/a_{rk} for which $a_{rk} > 0$.
(3) Divide row r within the rectangle by a_{rk}, relabel this row "row k," and replace c_r by c_k.
(4) Multiply the new row k by a_{ik} and subtract the result from row i.

Similarly, multiply row k by $(c_k - d_k)$ and subtract from the bottom row outside the rectangle.

Similar rules do not apply to the row $[\cdots \ d_k \ \cdots \ d_j \ \cdots]$. Once the bottom row has been computed this row can be omitted from subsequent tableaux, or computed independently each time as a check on the accuracy of the computation. The operation described in steps (3) and (4) is known as the *pivot operation*, and the element a_{rk} is called the *pivot element*. In the tableau above the pivot element has been encircled, a practice that has been found helpful in keeping the many steps of the pivot operation in order.

The elements $c_j - d_j (j = 1, \ldots, n)$ appearing in the last row are called *indicators*. The simplex method terminates when all indicators are ≤ 0.

Suppose a solution has been obtained in which $B' = \{\beta_1', \ldots, \beta_m'\}$ is the final basis of V obtained. Let $\hat{B}' = \{\psi_1', \ldots, \psi_m'\}$ be the corresponding dual basis. As we have seen, an optimum vector for the dual problem is obtained by setting $\psi = \sum_{k=1}^m c_{i(k)} \psi_k'$ where $i(k)$ is the index of the element of A mapped onto β_k', that is, $\sigma(\alpha_{i(k)}) = \beta_k'$. By definition of the matrix $A' = [a_{ij}']$ representing σ with respect to the bases A and B', we have $\beta_j = \sigma(\alpha_j) = \sum_{k=1}^m a_{i(k),j}' \beta_k'$. Thus, the elements in the first m columns of A' are the elements of the matrix of transition from the basis B' to the basis B. This means that $\psi_k' = \sum_{j=1}^m a_{i(k),j}' \psi_j$. Hence,

$$\psi = \sum_{k=1}^m c_{i(k)} \psi_k' = \sum_{k=1}^m c_{i(k)} \left(\sum_{j=1}^m a_{i(k),j}' \psi_j \right)$$

$$= \sum_{j=1}^m \left(\sum_{k=1}^m c_{i(k)} a_{i(k),j}' \right) \psi_j$$

$$= \sum_{j=1}^m d_j' \psi_j. \tag{3.12}$$

This means that $D' = [d_1' \cdots d_m']$ is the representation of the optimal vector for the solution to the dual problem in the original coordinate system.

All this discussion has been based on the premise that we can start with a known basic feasible vector. However, it is not a trivial matter to obtain even one feasible vector for most problems. The ith equation in $AX = B$ is

$$\sum_{j=1}^n a_{ij} x_j = b_i. \tag{3.13}$$

Since we can multiply both sides of this equation by -1 if necessary, there is no loss of generality to assume that each $b_i \geq 0$. We then replace equation (3.13) by

$$\sum_{i=1}^n a_{ij} x_j + v_i = b_i. \tag{3.14}$$

It is very easy to obtain a basic feasible solution of the corresponding system of linear equations; take $x_1 = \cdots = x_n = 0$ and $v_i = b_i$. We then construct a new objective function

$$\sum_{j=1}^n c_j x_j - M \sum_{i=1}^m v_i, \tag{3.15}$$

where M is taken as a number very much larger than any number to be considered; that is, so large that this new objective function cannot be maximized unless $v_1 = \cdots = v_m = 0$. The natural working of the simplex method will soon bring this about if the original problem is feasible. At this point the columns of the tableaux associated with the newly introduced variables could be dropped from further consideration, since a basic feasible solution to the original problem will be obtained at this point. However, it is better to retain these columns since they will provide the matrix of transition by which the coordinates of the optimum vector for the dual problem can be computed. This provides the best check on the accuracy of the computation since the optimality of the proposed solution can be tested unequivocally by using Theorem 3.4 or comparing the values of $\phi\xi$ and $\psi\beta$.

BIBLIOGRAPHICAL NOTES

Because of the importance of linear programming in economics and industrial engineering books and articles on the subject are very numerous. Most are filled with bewildering, tedious numerical calculations which add almost nothing to understanding and make the subject look difficult. Linear programming is not difficult, but it is subtle and requires clarity of exposition. D. Gale, *The Theory of Linear Economic Models*, is particularly recommended for its clarity and interesting, though simple, examples.

EXERCISES

1. Formulate the standard primal and dual linear programming problems in matric form.

2. Formulate the canonical primal and dual linear programming problems in matric form.

3. Let $A = [a_{ij}]$ be an $m \times n$ matrix. Let A_{11} be the matrix formed from the first r rows and first s columns of A; let A_{12} be the matrix formed from the first r rows and last $n - s$ columns of A; let A_{21} be the matrix formed from the last $m - r$ rows and first s columns of A; and let A_{22} be the matrix formed from the last $m - r$ rows and last $n - s$ columns of A. We then write

$$A = \begin{bmatrix} A_{11} & A_{12} \\ A_{21} & A_{22} \end{bmatrix}.$$

We say that we have partitioned A into the designated submatrices. Using this notation, show that the matrix equation $AX = B$ is equivalent to the matrix inequality

$$\begin{bmatrix} A \\ -A \end{bmatrix} X \leq \begin{bmatrix} B \\ -B \end{bmatrix}.$$

4. Use Exercise 3 to show the equivalence of the standard primal linear programming problem and the canonical primal linear programming problem.

5. Using the notation of Exercise 3, show that the dual canonical linear programming problem is to minimize

$$(Z_1 - Z_2)B = [Z_1 \quad Z_2] \begin{bmatrix} B \\ -B \end{bmatrix}$$

subject to the constraint

$$[Z_1 \quad Z_2] \begin{bmatrix} A \\ -A \end{bmatrix} \geq C$$

and $[Z_1 \quad Z_2] \geq 0$. Show that if we let $Y = Z_1 - Z_2$, the dual canonical linear programming problem is to minimize YB subject to the constraint $YA \geq C$ without the condition $Y \geq 0$.

6. Let A, B, and C be the matrices given in the standard maximum linear programming problem. Let F be the smallest field containing all the elements appearing in A, B, and C. Show that if the problem has an optimal solution, the simplex method gives an optimal vector all of whose components are in F, and that the maximum value of CX is in F.

7. How should the simplex method be modified to handle the canonical minimum linear programming problem in the form: minimize CX subject to the constraints $AX = B$ and $X \geq 0$?

8. Find $(x_1, x_2) \geq 0$ which maximizes $5x_1 + 2x_2$ subject to the conditions

$$2x_1 + x_2 \leq 6$$
$$4x_1 + x_2 \leq 10$$
$$-x_1 + x_2 \leq 3.$$

9. Find $[y_1 \quad y_2 \quad y_3] \geq 0$ which minimize $6y_1 + 10y_2 + 3y_3$ subject to the conditions

$$y_1 + y_2 + y_3 \geq 2$$
$$2y_1 + 4y_2 - y_3 \geq 5.$$

(In this exercise take advantage of the fact that this is the problem dual to the problem in the previous exercise, which has already been worked.)

10. Sometimes it is easier to apply the simplex method to the dual problem than it is to apply the simplex method to the given primal problem. Solve the problem in Exercise 9 by applying the simplex method to it directly. Use this work to find a solution to the problem in Exercise 8.

11. Find $(x_1, x_2) \geq 0$ which maximizes $x_1 + 2x_2$ subject to the conditions

$$-2x_1 + x_2 \leq 2$$
$$-x_1 + x_2 \leq 3$$
$$x_1 + x_2 \leq 7$$
$$2x_1 + x_2 \leq 11.$$

12. Draw the lines

$$-2x_1 + x_2 = 2$$
$$-x_1 + x_2 = 3$$
$$x_1 + x_2 = 7$$
$$2x_1 + x_2 = 11$$

in the x_1, x_2-plane. Notice that these lines are the extremes of the conditions given in Exercise 11. Locate the set of points which satisfy all the inequalities of Exercise 11 and the condition $(x_1, x_2) \geq 0$. The corresponding canonical problem involves the linear conditions

$$
\begin{array}{rcl}
-2x_1 + x_2 + x_3 & & = 2 \\
-x_1 + x_2 \quad\quad + x_4 & & = 3 \\
x_1 + x_2 \quad\quad\quad\quad + x_5 & & = 7 \\
2x_1 + x_2 \quad\quad\quad\quad\quad\quad + x_6 & = & 11.
\end{array}
$$

The first feasible solution is $(0, 0, 2, 3, 7, 11)$ and this corresponds to the point $(0, 0)$ in the x_1, x_2-plane. In solving the problem of Exercise 11, a sequence of feasible solutions is obtained. Plot the corresponding points in the x_1, x_2-plane.

13. Show that the geometric set of points satisfying all the linear constraints of a standard linear programming is a convex set. Let this set be denoted by C. A vector in C is called an *extreme vector* if it is not a convex linear combination of other vectors in C. Show that if ϕ is a linear functional and β is a convex linear combination of the vectors $\{\alpha_1, \ldots, \alpha_r\}$, then $\phi(\beta) \leq \max \{\phi(\alpha_j)\}$. Show that if β is not an extreme vector in C, then either $\phi(\xi)$ does not take on its maximum value at β in C, or there are other vectors in C at which $\phi(\xi)$ takes on its maximum value. Show that if C is closed and has only a finite number of extreme vectors, then the maximum value of $\phi(\frac{1}{3})$ occurs at an extreme vector of C.

14. Theorem 3.2 provides an easily applied test for optimality. Let $X = (x_1, \ldots, x_n)$ be feasible for the standard primal problem and let $Y = [y_1 \cdots y_m]$ be feasible for the standard dual problem. Show that X and Y are both optimal if and only if $x_j > 0$ implies $\sum_{i=1}^{m} y_i a_{ij} = c_j$ and $y_i > 0$ implies $\sum_{j=1}^{n} a_{ij} x_j = b_i$.

15. Consider the problem of maximizing $2x_1 + x_2$ subject to the conditions

$$
\begin{array}{rcl}
x_1 - 3x_2 & \leq & 4 \\
x_1 - 2x_2 & \leq & 5 \\
x_1 & \leq & 9 \\
2x_1 + x_2 & \leq & 20 \\
x_1 + 3x_2 & \leq & 30 \\
-x_1 + x_2 & \leq & 5.
\end{array}
$$

Consider $X = (8, 4)$, $Y = [0 \; 0 \; 0 \; 1 \; 0 \; 0]$. Test both for feasibility for the primal and dual problems, and test for optimality.

16. Show how to use the simplex method to find a non-negative solution of $AX = B$. This is also equivalent to the problem of finding a feasible solution for a canonical linear programming problem. (*Hint:* Take $Z = (z_1, \ldots, z_m)$, $F = [1 \; 1 \cdots 1]$, and consider the problem of minimizing FZ subject to $AX + Z = B$. What is the resulting necessary and sufficient condition for the existence of a solution to the original problem?)

17. Apply the simplex method to the problem of finding a non-negative solution of

$$6x_1 + 3x_2 - 4x_3 - 9x_4 - 7x_5 - 5x_6 = 0$$
$$-5x_1 - 8x_2 + 8x_3 + 2x_4 - 2x_5 + 5x_6 = 0$$
$$-2x_1 + 6x_2 - 5x_3 + 8x_4 + 8x_5 + x_6 = 0$$
$$x_1 + x_2 + x_3 + x_4 + x_5 + x_6 = 1.$$

This is equivalent to Exercise 5 of Section 1.

4 | Applications to Communication Theory

This section requires no more linear algebra than the concepts of a basis and the change of basis. The material in the first four sections of Chapter I and the first four sections of Chapter II is sufficient. It is also necessary that the reader be familiar with the formal elementary properties of Fourier series.

Communication theory is concerned largely with signals which are uncertain, uncertain to be transmitted and uncertain to be received. Therefore, a large part of the theory is based on probability theory. However, there are some important concepts in the theory which are purely of a vector space nature. One is the sampling theorem, which says that in a certain class of signals a particular signal is completely determined by its values (samples) at an equally spaced set of times extending forever.

Although it is usually not stated explicitly, the set of functions considered as signals form a vector space over the real numbers; that is, if $f(t)$ and $g(t)$ are signals, then $(f + g)(t) = f(t) + g(t)$ is a signal and $(af)(t) = af(t)$, where a is a real number, is also a signal. Usually the vector space of signals is infinite dimensional so that while many of the concepts and theorems developed in this book apply, there are also many that do not. In many cases the appropriate tool is the theory of Fourier integrals. In order to bring the topic within the context of this book, we assume that the signals persist for only a finite interval of time and that there is a bound for the highest frequency that will be encountered. If the time interval is of length 1, this assumption has the implication that each signal $f(t)$ can be represented as a finite series of the form

$$f(t) = \tfrac{1}{2}a_0 + \sum_{k=1}^{N} a_k \cos 2\pi kt + \sum_{k=1}^{N} b_k \sin 2\pi kt. \tag{4.1}$$

Formula (4.1) is in fact just a precise formulation of the vague statement that the highest frequency to be encountered is bounded. Since the coefficients can be taken to be arbitrary real numbers, the set of signals under consideration forms a real vector space V of dimension $2N + 1$. We show that $f(t)$ is determined by its values at $2N + 1$ points equally spaced in time. This statement is known as the *finite sampling theorem*.

The classical infinite sampling theorem from communication theory requires an assumption analogous to the assumption that the highest frequencies are bounded. Only the assumption that the signal persists for a finite interval of time is relaxed. In any practical problem some bound can be placed on the duration of the family of signals under consideration. Thus, the restriction on the length of the time interval does not alter the significance or spirit of the theorem in any way.

Consider the function

$$\psi(t) = \frac{1}{2N+1}\left(1 + 2\sum_{k=1}^{N}\cos 2\pi kt\right) \in V. \tag{4.2}$$

$$\psi(t) = \frac{\sin \pi t + \sum_{k=1}^{N} 2\cos 2\pi kt \sin \pi t}{(2N+1)\sin \pi t}$$

$$= \frac{\sin \pi t + \sum_{k=1}^{N}\sin(2\pi kt + \pi t) - \sin(2\pi kt - \pi t)}{(2N+1)\sin \pi t}$$

$$= \frac{\sin(2N+1)\pi t}{(2N+1)\sin \pi t} \tag{4.3}$$

From (4.2) we see that $\psi(0) = 1$, and from (4.3) we see that $\psi(j/2N + 1) = 0$ for $0 < |j| \leq N$.

Consider the functions

$$\psi_k(t) = \psi\left(t - \frac{k}{2N+1}\right), \qquad \text{for } k = -N, -N+1, \ldots, N. \tag{4.4}$$

These $2N + 1$ functions are all members of V. Furthermore, for $t_j = j/(2N + 1)$ we see that $\psi_j(t_j) = 1$ while $\psi_k(t_j) = 0$ for $k \neq j$. Thus, the $2N + 1$ functions obtained are linearly independent. Since V is of dimension $2N + 1$, it follows that the set $\{\psi_k(t) \mid k = -N, \ldots, N\}$ is a basis of V. These functions are called the *sampling functions*.

If $f(t)$ is any element of V it can be written in the form

$$f(t) = \sum_{k=-N}^{N} d_k\psi_k(t). \tag{4.5}$$

However,

$$f(t_j) = \sum_{k=-N}^{N} d_k\psi_k(t_j) = d_j, \tag{4.6}$$

or

$$f(t) = \sum_{k=-N}^{N} f(t_k)\psi_k(t). \tag{4.7}$$

Thus, the coordinates of $f(t)$ with respect to the basis $\{\psi_k(t)\}$ are $(f(t_{-N}), \ldots, f(t_N))$, and we see that these samples are sufficient to determine $f(t)$.

It is of some interest to express the elements of the basis $\{\frac{1}{2}, \cos 2\pi t, \ldots, \sin 2\pi Nt\}$ in terms of the basis $\{\psi_k(t)\}$.

$$\frac{1}{2} = \sum_{k=-N}^{N} \frac{1}{2}\psi_k(t)$$

$$\cos 2\pi jt = \sum_{k=-N}^{N} \cos 2\pi jt_k \psi_k(t) \qquad (4.8)$$

$$\sin 2\pi jt = \sum_{k=-N}^{N} \sin 2\pi jt_k \psi_k(t).$$

Expressing the elements of the basis $\{\psi_k(t)\}$ in terms of the basis $\{\frac{1}{2}, \cos 2\pi t, \ldots, \sin 2\pi Nt\}$ is but a matter of the definition of the $\psi_k(t)$:

$$\psi_k(t) = \psi\left(t - \frac{k}{2N+1}\right)$$

$$= \frac{1}{2N+1}\left[1 + 2\sum_{j=1}^{N}\cos 2\pi j\left(t - \frac{k}{2N+1}\right)\right]$$

$$= \frac{1}{2N+1}\left(1 + 2\sum_{j=1}^{N}\cos\frac{2\pi jk}{2N+1}\cos 2\pi jt + 2\sum_{j=1}^{N}\sin\frac{2\pi jk}{2N+1}\sin 2\pi jt\right).$$

$$(4.9)$$

With this interpretation, formula (4.1) is a representation of $f(t)$ in one coordinate system and (4.7) is a representation of $f(t)$ in another. To express the coefficients in (4.1) in terms of the coefficients in (4.7) is but a change of coordinates. Thus, we have

$$a_j = \frac{2}{2N+1}\sum_{k=-N}^{N} f(t_k)\cos\frac{2\pi jk}{2N+1} = \frac{2}{2N+1}\sum_{k=-N}^{N} f(t_k)\cos 2\pi jt_k$$

$$(4.10)$$

$$b_j = \frac{2}{2N+1}\sum_{k=-N}^{N} f(t_k)\sin\frac{2\pi jk}{2N+1} = \frac{2}{2N+1}\sum_{k=-N}^{N} f(t_k)\sin 2\pi jt_k.$$

There are several ways to look at formulas (4.10). Those familiar with the theory of Fourier series will see the a_j and b_j as Fourier coefficients with formulas (4.10) using finite sums instead of integrals. Those familiar with probability theory will see the a_j as covariance coefficients between the samples of $f(t)$ and the samples of $\cos 2\pi jt$ at times t_k. And we have just viewed them as formulas for a change of coordinates.

If the time interval had been of length T instead of 1, the series corresponding to (4.1) would be of the form

$$f(t) = \frac{1}{2}a_0 + \sum_{k=1}^{N}a_k\cos\frac{2\pi k}{T}t + \sum_{k=1}^{N}b_k\sin\frac{2\pi k}{T}t. \qquad (4.11)$$

The vector space would still be of dimension $2N + 1$ and we would need $2N + 1$ samples spread equally over an interval of length T, or $(2N + 1)/T$ samples per unit time. Since $N/T = W$ is the highest frequency present in the series (4.11), we see that for large intervals of time approximately $2W$ samples per unit time are required to determine the signal. The infinite sampling theorem, referred to at the beginning of this section, says that if W is the highest frequency present, then $2W$ samples per second suffice to determine the signal. The spirit of the finite sampling theorem is in keeping with the spirit of the infinite sampling theorem, but the finite sampling theorem has the practical advantage of providing effective formulas for determining the function $f(t)$ and the Fourier coefficients from the samples.

BIBLIOGRAPHY NOTES

For a statement and proof of the infinite sampling theorem see P. M. Woodward, *Probability and Information Theory, with Applications to Radar.*

EXERCISES

1. Show that $\psi(t_r - t_s) = \psi\left(\dfrac{r - s}{2N + 1}\right) = \delta_{rs}$ for $-N \leq r, s \leq N$.

2. Show that if $f(t)$ can be represented in the form of (4.1), then

$$a_k = 2 \int_{-\frac{1}{2}}^{\frac{1}{2}} f(t) \cos 2\pi kt \, dt, \qquad k = 0, 1, \ldots, N,$$

$$b_k = 2 \int_{-\frac{1}{2}}^{\frac{1}{2}} f(t) \sin 2\pi kt \, dt, \qquad k = 1, \ldots, N.$$

3. Show that $\displaystyle\int_{-\frac{1}{2}}^{\frac{1}{2}} \psi(t) \, dt = \dfrac{1}{2N + 1}$.

4. Show that $\displaystyle\int_{-\frac{1}{2}}^{\frac{1}{2}} \psi_k(t) \, dt = \dfrac{1}{2N + 1}$.

5. Show that if $f(t)$ can be represented in the form (4.7), then

$$\int_{-\frac{1}{2}}^{\frac{1}{2}} f(t) \, dt = \sum_{k=-N}^{N} \frac{1}{2N + 1} f(t_k).$$

This is a formula for expressing an integral as a finite sum. Such a formula is called a *mechanical quadrature*. Such formulas are characteristic of the theory of orthogonal functions.

6. Show that

$$\sum_{k=-N}^{N} \frac{1}{2N + 1} \cos 2\pi rt_k = \delta_{rk},$$

and

$$\sum_{k=-N}^{N} \frac{1}{2N+1} \sin 2\pi r t_k = 0$$

7. Show that

$$\sum_{k=-N}^{N} \frac{1}{2N+1} \cos 2\pi r(t - t_k) = \delta_{rk}$$

and

$$\sum_{k=-N}^{N} \frac{1}{2N+1} \sin 2\pi r(t - t_k) = 0.$$

8. Show that

$$\sum_{k=-N}^{N} \psi_k(t) = 1.$$

9. Show that

$$f(t) = \sum_{k=-N}^{N} f(t)\psi_k(t).$$

10. If $f(t)$ and $g(t)$ are integrable over the interval $[-\frac{1}{2}, \frac{1}{2}]$, let

$$(f, g) = 2 \int_{-1/2}^{1/2} f(t)g(t) \, dt.$$

Show that if $f(t)$ and $g(t)$ are elements of V, then (f, g) defines an inner product in V. Show that $\left\{ \frac{1}{\sqrt{2}}, \cos 2\pi t, \ldots, \sin 2\pi N t \right\}$ is an orthonormal set.

11. Show that if $f(t)$ can be represented in the form (4.1), then

$$2 \int_{-1/2}^{1/2} f(t)^2 \, dt = \frac{a_0^2}{2} + \sum_{k=1}^{N} a_k^2 + \sum_{k=1}^{N} b_k^2.$$

Show that this is Parseval's equation of Chapter V.

12. Show that

$$\int_{-1/2}^{1/2} \cos 2\pi r t \, \psi(t) \, dt = \frac{1}{2N+1}.$$

13. Show that

$$\int_{-1/2}^{1/2} \cos 2\pi r t \, \psi_k(t) \, dt = \frac{1}{2N+1} \cos 2\pi r t_k.$$

14. Show that

$$\int_{-1/2}^{1/2} \sin 2\pi r t \, \psi_k(t) \, dt = \frac{1}{2N+1} \sin 2\pi r t_k.$$

15. Show that

$$\int_{-1/2}^{1/2} \psi_r(t)\psi_s(t) \, dt = \frac{1}{2N+1} \delta_{rs}.$$

16. Using the inner product defined in Exercise 10 show that $\{\psi_k(t) \,|\, k = -N, \ldots, N\}$ is an orthonormal set.

17. Show that if $f(t)$ can be represented in the form (4.7), then

$$2 \int_{-\frac{1}{2}}^{\frac{1}{2}} f(t)^2 \, dt = \sum_{k=-N}^{N} \frac{2}{2N+1} f(t_r)^2.$$

Show that this is Parseval's equation of Chapter V.

18. Show that if $f(t)$ can be represented in the form (4.7), then

$$\int_{-\frac{1}{2}}^{\frac{1}{2}} f(t)\psi_k(t) \, dt = \frac{1}{2N+1} f(t_k).$$

Show how the formulas in Exercises 12, 13, 14, and 15 can be treated as special cases of this formula.

19. Let $f(t)$ be any function integrable on $[-\frac{1}{2}, \frac{1}{2}]$. Define

$$d_k = (2N+1) \int_{-\frac{1}{2}}^{\frac{1}{2}} f(t)\psi_k(t) \, dt.$$

Then $f_N(t) = \sum_{k=-N}^{N} d_k \psi_k(t) \in V$. Show that

$$\int_{-\frac{1}{2}}^{\frac{1}{2}} f(t)\psi_r(t) \, dt = \int_{-\frac{1}{2}}^{\frac{1}{2}} f_N(t)\psi_r(t) \, dt, \qquad r = -N, \ldots, N.$$

Show that if $g(t) \in V$, then

$$\int_{-\frac{1}{2}}^{\frac{1}{2}} f(t)g(t) \, dt = \int_{-\frac{1}{2}}^{\frac{1}{2}} f_N(t)g(t) \, dt.$$

20. Show that if $f_N(t)$ is defined as in Exercise 19, then

$$\int_{-\frac{1}{2}}^{\frac{1}{2}} [f(t) - f_N(t)]^2 \, dt = \int_{-\frac{1}{2}}^{\frac{1}{2}} f(t)^2 \, dt - \int_{-\frac{1}{2}}^{\frac{1}{2}} f_N(t)^2 \, dt.$$

Show that

$$\int_{-\frac{1}{2}}^{\frac{1}{2}} f_N(t)^2 \, dt \le \int_{-\frac{1}{2}}^{\frac{1}{2}} f(t)^2 \, dt.$$

21. Let $g(t)$ be any function in V. Show that

$$\int_{-\frac{1}{2}}^{\frac{1}{2}} [f(t) - g(t)]^2 \, dt - \int_{-\frac{1}{2}}^{\frac{1}{2}} [f(t) - f_N(t)]^2 \, dt = \int_{-\frac{1}{2}}^{\frac{1}{2}} [f_N(t) - g(t)]^2 \, dt.$$

Show that

$$\int_{-\frac{1}{2}}^{\frac{1}{2}} [f(t) - f_N(t)]^2 \, dt \le \int_{-\frac{1}{2}}^{\frac{1}{2}} [f(t) - g(t)]^2 \, dt.$$

Because of this inequality we say that, of all functions in V, $f_N(t)$ is the best approximation of $f(t)$ *in the mean*; that is, it is the approximation which minimizes the integrated square error. $f_N(t)$ is the function closest to $f(t)$ in the metric defined by the inner product of Exercise 10.

22. Again, let $f(t)$ be a function integrable on $[-\frac{1}{2}, \frac{1}{2}]$. Define

$$F_N(t) = \sum_{k=-N}^{N} f(t_k)\psi_k(t).$$

Let $\epsilon > 0$ be given. Show that there is an $N(\epsilon)$ such that if $N \geq N(\epsilon)$, then

$$\int_{-\frac{1}{2}}^{\frac{1}{2}} f(t)\,dt - \epsilon \leq \int_{-\frac{1}{2}}^{\frac{1}{2}} F_N(t)\,dt \leq \int_{-\frac{1}{2}}^{\frac{1}{2}} f(t)\,dt + \epsilon.$$

5 | Vector Calculus

We assume that the reader is familiar with elementary calculus, which considers scalar-valued functions of a scalar variable and scalar-valued functions of several scalar variables. Functions of several variables can also be taken up from a vector point of view in which a set of scalar variables is replaced by a vector. The background in linear algebra required is covered in this text in the first three sections of Chapter IV and Sections 1 and 3 of Chapter V.

In this section we consider vector-valued functions of a vector variable. We assume that the reader is acquainted in a different setting with most of the topics mentioned in this section. We emphasize the algebraic aspects of these topics and state or prove only those assertions of an analytic nature that are intimately linked with the algebraic structure.

In this section we assume that V is a vector space of finite dimension n over the real numbers or the complex numbers, and that a positive definite inner product is defined in V. We shall write (α, β) for the inner product of α, $\beta \in V$. In Section 3 of Chapter V we showed that for any linear functional $\phi \in \hat{V}$, there exists a unique $\eta \in V$ such that $\phi(\beta) = (\eta, \beta)$ for all $\beta \in V$. We showed there that the mapping of ϕ onto $\eta(\phi) = \eta$ defined in this way is one-to-one and onto.

We can use this mapping to define an inner product in $\hat{V}$. Thus, we define

$$(\phi, \psi) = \overline{(\eta(\phi), \eta(\psi))} = (\eta(\psi), \eta(\phi)). \tag{5.1}$$

The conjugate appears in this definition because the mapping η is conjugate linear and we require that the inner product be linear in the second variable. It is not difficult to show that this does, in fact, define an inner product in $\hat{V}$. For the norm in $\hat{V}$ we have

$$\|\phi\|^2 = (\phi, \phi) = (\eta(\phi), \eta(\phi)) = \|\eta(\phi)\|^2. \tag{5.2}$$

From Schwarz's inequality we obtain

$$|\phi(\beta)| = |(\eta(\phi), \beta)| \leq \|\eta(\phi)\| \cdot \|\beta\| = \|\phi\| \cdot \|\beta\|. \tag{5.3}$$

Theorem 5.1. $\|\phi\|$ *is the smallest value of* M *for which* $|\phi(\beta)| \leq M \|\beta\|$ *for all* $\beta \in V$.

PROOF. (5.3) shows that $|\phi(\beta)| \leq M \|\beta\|$ holds for all β if $M = \|\phi\|$. Let $\beta = \eta(\phi)$. Then $|\phi(\beta)| = |(\eta(\phi), \beta)| = (\beta, \beta) = \|\beta\|^2 = \|\phi\| \cdot \|\beta\|$. Thus, the inequality $|\phi(\beta)| \leq M \|\beta\|$ cannot hold for all values of β if $M < \|\phi\|$. □

Note. Although it was not pointed out explicitly, we have also shown that for each ϕ such a smallest value of M exists. When any value of M exists such that $|\phi(\beta)| \leq M \|\beta\|$ for all β, we say that ϕ is *bounded*. Therefore, we have shown that every linear functional is bounded. In infinite dimensional vector spaces there may be linear functionals that are not bounded.

If f is any function mapping U into V, we define

$$\lim_{\xi \to \xi_0} f(\xi) = \alpha. \tag{5.4}$$

to be equivalent to the following statement: "For any $\epsilon > 0$, there is a $\delta > 0$ such that $\|\xi - \xi_0\| < \delta$ implies $\|f(\xi) - \alpha\| < \epsilon$." The function f is said to be *continuous* at ξ_0 if

$$\lim_{\xi \to \xi_0} f(\xi) = f(\xi_0). \tag{5.5}$$

These definitions are the usual definitions from elementary calculus with the interpretations of the words extended to the terminology of vector spaces. These definitions could be given in other equivalent forms, but those given will suffice for our purposes.

Theorem 5.2. *Every* (bounded) *linear functional in* V *is continuous on all of* V.

PROOF. Let M be any positive real number such that $|\phi(\beta)| \leq M \|\beta\|$ holds for all $\beta \in V$. Then, for the given $\epsilon > 0$, it suffices (uniformly) to take $\delta = \epsilon/M$. For any β_0 we have

$$|\phi(\beta) - \phi(\beta_0)| = |\phi(\beta - \beta_0)| \leq M \|\beta - \beta_0\| < \epsilon \tag{5.6}$$

whenever $\|\beta - \beta_0\| < \delta$. □

Theorem 5.3. *Let* A $= \{\alpha_1, \ldots, \alpha_n\}$ *be any basis in* V. *There exist positive real numbers* C *and* D, *depending only on the inner product and the chosen basis, such that for any* $\xi = \sum_{i=1}^n x_i \alpha_i \in V$ *we have*

$$C \sum_{i=1}^n |x_i| \leq \|\xi\| \leq D \sum_{i=1}^n |x_i|. \tag{5.7}$$

PROOF. By the triangle inequality

$$\|\xi\| = \left\| \sum_{i=1}^n x_i \alpha_i \right\| \leq \sum_{i=1}^n \|x_i \alpha_i\| = \sum_{i=1}^n |x_i| \cdot \|\alpha_i\|.$$

Let $D = \max \{\|\alpha_i\|\}$. Then

$$\|\xi\| \leq \sum_{i=1}^{n} |x_i| \cdot D = D \sum_{i=1}^{n} |x_i|.$$

On the other hand, let $\hat{A} = \{\phi_1, \ldots, \phi_n\}$ be the dual basis of A. Then $|x_i| = |\phi_i(\xi)| \leq \|\phi_i\| \cdot \|\xi\|$. Taking $C^{-1} = \sum_{i=1}^{n} \|\phi_i\| > 0$, we have

$$\sum_{i=1}^{n} |x_i| \leq \sum_{i=1}^{n} \|\phi_i\| \cdot \|\xi\| = C^{-1} \|\xi\|. \quad \square$$

The interesting and significant thing about Theorem 5.3 is it implies that even though the limit was defined in terms of a given norm the resulting limit is independent of the particular norm used, provided it is derived from a positive definite inner product. The inequalities in (5.7) say that a vector is small if and only if its coordinates are small, and a bound on the size of the vector is expressed in terms of the ordinary absolute values of the coordinates.

Let ξ be a vector-valued function of a scalar variable t. We write $\xi = \xi(t)$ to indicate the dependence of ξ on t. A useful picture of this concept is to think of ξ as a position vector, a vector with its tail at the origin and its head locating a point or position in space. As t varies, the head and the point it determines moves. The picture we wish to have in mind is that of ξ tracing out a curve as t varies over an interval (in the real case).

If the limit

$$\lim_{h \to 0} \frac{\xi(t + h) - \xi(t)}{h} = \frac{d\xi}{dt} \tag{5.8}$$

exists, the vector valued function ξ is said to be *differentiable* at t. This limit is usually called the derivative, but we wish to give this name to a different concept. At this moment the reasons for making the proposed distinction would seem artificial and hard to explain. Therefore, we shall re-examine this idea after we consider functions of vector variables.

Since $\xi(t + h) - \xi(t)$ is a vector and h is a scalar, $\lim_{h \to 0} \dfrac{\xi(t + h) - \xi(t)}{h}$ is a vector. It is interpreted as a vector tangent to the curve at the point $\xi(t)$.

Now, let f be a scalar valued function defined on V. Often, f is not defined on all of V, but only on some subdomain. We do not wish to become involved in such questions. We assume that whenever we refer to the behavior of f at some $\xi_0 \in V$, $f(\xi)$ is also defined for all points in a sphere around ξ_0 of radius sufficiently generous to include all other vectors under discussion.

Let ξ_0 be an arbitrary vector in V, which we take to be fixed for the moment. Let η be any other vector in V. For the given ξ_0 and η, we assume the expression

$$\frac{f(\xi_0 + h\eta) - f(\xi_0)}{h} \tag{5.9}$$

is defined for all $h \neq 0$ in an interval around 0. If

$$\lim_{h \to 0} \frac{f(\xi_0 + h\eta) - f(\xi_0)}{h} = f'(\xi_0, \eta) \tag{5.10}$$

exists for each η, the function f is said to be *differentiable* at ξ_0. It is *continuously differentiable* at ξ_0 if f is differentiable in a neighborhood around ξ_0 and $f'(\xi, \eta)$ is continuous at ξ_0 for each η; that is, $\lim_{\xi \to \xi_0} f'(\xi, \eta) = f'(\xi_0, \eta)$.

We wish to show that $f'(\xi, \eta)$ is a linear function of η. However, in order to do this it is necessary that $f'(\xi, \eta)$ satisfy some analytic conditions. The following theorems lead to establishing conditions sufficient to make this conclusion.

Theorem 5.4 (*Mean value theorem*). *Assume that $f'(\xi + h\eta, \eta)$ exists for all $h, 0 < h < 1$, and that $f(\xi + h\eta)$ is continuous for all $h, 0 \leq h \leq 1$. Then there exists a real number $\theta, 0 < \theta < 1$, such that*

$$f(\xi + \eta) - f(\xi) = f'(\xi + \theta\eta, \eta). \tag{5.11}$$

PROOF. Let $g(h) = f(\xi + h\eta)$ for ξ and η fixed. Then $g(h)$ is a real valued function of a real variable and

$$g'(h) = \lim_{\Delta h \to 0} \frac{g(h + \Delta h) - g(h)}{\Delta h}$$

$$= \lim_{\Delta h \to 0} \frac{f(\xi + h\eta + \Delta h\eta) - f(\xi + h\eta)}{\Delta h}$$

$$= f'(\xi + h\eta, \eta) \tag{5.12}$$

exists by assumption for $0 < h < 1$. By the mean value theorem for $g(h)$ we have

$$g(1) - g(0) = g'(\theta), \quad 0 < \theta < 1, \tag{5.13}$$

or

$$f(\xi + \eta) - f(\xi) = f'(\xi + \theta\eta, \eta). \ \square$$

Theorem 5.5. *If $f'(\xi, \eta)$ exists, then for $a \in F$. $f'(\xi, a\eta)$ exists and*

$$f'(\xi, a\eta) = af'(\xi, \eta). \tag{5.14}$$

PROOF. $f'(\xi\ a\eta) = \lim_{h \to 0} \frac{f(\xi + ha\eta) - f(\xi)}{h}$

$$= a \lim_{ah \to 0} \frac{f(\xi + ah\eta) - f(\xi)}{ah}$$

$$= af'(\xi, \eta)$$

for $a \neq 0$, and $f'(\xi, a\eta) = 0$ for $a = 0$. $\square$

Lemma 5.6. *Assume* $f'(\xi_0, \eta_1)$ *exists and that* $f'(\xi, \eta_2)$ *exists in a neighborhood of* ξ_0. *If* $f'(\xi, \eta_2)$ *is continuous at* ξ_0, *then* $f'(\xi_0, \eta_1 + \eta_2)$ *exists and*

$$f'(\xi_0, \eta_1 + \eta_2) = f'(\xi_0, \eta_1) + f'(\xi_0, \eta_2). \qquad (5.15)$$

PROOF.

$$f'(\xi_0, \eta_1 + \eta_2) = \lim_{h \to 0} \frac{f(\xi_0 + h(\eta_1 + \eta_2)) - f(\xi_0)}{h}$$

$$= \lim_{h \to 0} \frac{f(\xi_0 + h\eta_1 + h\eta_2) - f(\xi_0 + h\eta_1) + f(\xi_0 + h\eta_1) - f(\xi_0)}{h}$$

$$= \lim_{h \to 0} \frac{f'(\xi_0 + h\eta_1 + \theta h\eta_2, h\eta_2)}{h} + f'(\xi_0, \eta_1)$$

by Theorem 5.4,

$$= \lim_{h \to 0} f'(\xi_0 + h\eta_1 + \theta h\eta_2, \eta_2) + f'(\xi_0 \ \eta_1)$$

by Theorem 5.5,

$$= f'(\xi_0, \eta_2) + f'(\xi_0, \eta_1)$$

by continuity at ξ_0. $\square$

Theorem 5.7. *Let* $A = \{\alpha_1, \ldots, \alpha_n\}$ *be a basis of* V *over* F. *Assume* $f'(\xi_0, \alpha_i)$ *exists for all i, and that* $f'(\xi, \alpha_i)$ *exists in a neighborhood of* ξ_0 *and is continuous at* ξ_0 *for* $n - 1$ *of the elements of* A. *Then* $f'(\xi_0, \eta)$ *exists for all* η *and is linear in* η.

PROOF. Suppose $f'(\xi, \alpha_i)$ exists in a neighborhood of ξ_0 and is continuous for $i = 2, 3, \ldots, n$. Let $S_k = \langle \alpha_1, \ldots, \alpha_k \rangle$. Theorem 5.5 says that $f'(\xi_0, \eta)$ is linear in η for $\eta \in S_1$.

By induction, assume $f'(\xi_0, \eta)$ is linear in η for $\eta \in S_k$. Then by Theorem 5.5, $f'(\xi_0, a_{k+1}\alpha_{k+1})$ exists in a neighborhood of ξ_0 and is continuous at ξ_0 for all $a_{k+1} \in F$. By Lemma 5.6, $f'(\xi_0, \eta + a_{k+1}\alpha_{k+1}) = f'(\xi_0, \eta) + f'(\xi_0, a_{k+1}\alpha_{k+1}) = f'(\xi_0, \eta) + a_{k+1}f'(\xi_0, \alpha_{k+1})$. Since all vectors in S_{k+1} are of the form $\eta + a_{k+1}\alpha_{k+1}$, $f'(\xi_0, \eta)$ is linear for all $\eta \in S_{k+1}$. Finally, $f'(\xi_0, \eta)$ is linear for all $\eta \in S_n = V$. $\square$

Theorem 5.7 is usually applied under the assumption that $f'(\xi, \eta)$ is continuously differentiable in a neighborhood of ξ_0 for all $\eta \in V$. This is certainly true if $f'(\xi, \alpha_i)$ is continuously differentiable in a neighborhood of ξ_0 for all α_i in a basis of V. Under these conditions, $f'(\xi, \eta)$ is a scalar valued linear function of η defined on V. The linear functional thus determined depends on ξ and we denote it by $df(\xi)$. Thus $df(\xi)$ is a linear functional such that $df(\xi)(\eta) = f'(\xi, \eta)$ for all $\eta \in V$. $df(\xi)$ is called the *differential* of f at ξ.

If $A = \{\alpha_1, \ldots, \alpha_n\}$ is any basis of V, any vector $\xi \in V$ can be represented in the form $\xi = \sum_i x_i \alpha_i$. Thus, any function of ξ also depends on the coordinates $(x_1, \ldots, x_n)$. To avoid introducing a new symbol for this function, we write

$$f(\xi) = f(x_1, \ldots, x_n). \tag{5.16}$$

Since $df(\xi)$ is a linear functional it can be expressed in terms of the dual basis $\hat{A} = \{\phi_1, \ldots, \phi_n\}$. The coordinates of $df(\xi)$ are easily computed by evaluating $df(\xi)$ for the basis elements $\alpha_i \in A$. We see that

$$df(\xi)(\alpha_i) = f'(\xi, \alpha_i) = \lim_{h \to 0} \frac{f(\xi + h\alpha_i) - f(\xi)}{h}$$

$$= \lim_{h \to 0} \frac{f(x_1, \ldots, x_i + h, \ldots, x_n) - f(x_1, \ldots, x_n)}{h}$$

$$= \frac{\partial f}{\partial x_i}. \tag{5.17}$$

Thus,

$$df(\xi) = \sum_i \frac{\partial f}{\partial x_i} \phi_i. \tag{5.18}$$

For any

$$\eta = \sum_i y_i \alpha_i,$$

$$df(\xi)(\eta) = f'(\xi, \eta) = \sum_i \frac{\partial f}{\partial x_i} y_i. \tag{5.19}$$

From (5.17), the assumption that $f'(\xi, \eta)$ is a continuous function of ξ implies that the partial derivatives are continuous. Conversely, the continuity of the partial derivatives implies that the conditions of Theorem 5.7 are satisfied and, therefore, that $f'(\xi, \eta)$ is a continuous function of ξ. In either case, $f'(\xi, \eta)$ is a linear function of η and formula (5.19) holds.

Theorem 5.8. *If $f'(\xi, \eta)$ is a continuous function of ξ for all η, then $df(\xi)$ is a continuous function of ξ.*

PROOF. By formula (5.17), if $f'(\xi, \eta)$ is a continuous function of ξ, then $\dfrac{\partial f}{\partial x_i} = f'(\xi, \alpha_i)$ is a continuous function of ξ. Because of formula (5.18) it then follows that $df(\xi)$ is a continuous function of ξ. □

If η is a vector of unit length, $f'(\xi, \eta)$ is called the *directional derivative* of $f(\xi)$ in the direction of η.

Consider the function that has the value x_i for each $\xi = \sum_i x_i \alpha_i$. Denote this function by X_i. Then

$$dX_i(\xi)(\alpha_j) = \lim_{h \to 0} \frac{X_i(\xi + h\alpha_j) - X_i(\xi)}{h}$$

$$= \delta_{ij}.$$

Since

$$\phi_i \alpha_j = \delta_{ij},$$

we see that

$$dX_i(\xi) = \phi_i.$$

It is more suggestive of traditional calculus to let x_i denote the function X_i, and to denote ϕ_i by dx_i. Then formula (5.18) takes the form

$$df = \sum_i \frac{\partial f}{\partial x_i} dx_i. \tag{5.18}$$

Let us turn our attention for a moment to vector-valued functions of vector variables. Let U and V be vector spaces over the same field (either real or complex). Let U be of dimension n and V of dimension m. We assume that positive definite inner products are defined in both U and V. Let F be a function defined on U with values in V. For ξ and $\eta \in U$, $F'(\xi, \eta)$ is defined by the limit,

$$F'(\xi, \eta) = \lim_{h \to 0} \frac{F(\xi + h\eta) - F(\xi)}{h}, \tag{5.20}$$

if this limit exists. If $F'(\xi, \eta)$ exists, F is said to be *differentiable* at ξ. F is *continuously differentiable* at ξ_0 if F is differentiable in a neighborhood around ξ_0 and $F'(\xi, \eta)$ is continuous at ξ_0 for each η; that is, $\lim_{\xi \to \xi_0} F'(\xi, \eta) = F'(\xi_0, \eta)$.

In analogy to the derivative of a scalar valued function of a vector variable, we wish to show that under appropriate conditions $F'(\xi, \eta)$ is linear in η.

Theorem 5.9. *If for each $\eta \in U$, $F'(\xi, \eta)$ is defined in a neighborhood of ξ_0 and continuous at ξ_0, then $F'(\xi_0, \eta)$ is linear in η.*

PROOF.   Let ψ be a linear functional in $\hat{V}$. Let f be defined by the equation

$$f(\xi) = \psi\{F(\xi)\}. \tag{5.21}$$

Since ψ is linear and continuous,

$$f'(\xi, \eta) = \lim_{h \to 0} \frac{f(\xi + h\eta) - f(\xi)}{h}$$

$$= \lim_{h \to 0} \psi\left\{\frac{F(\xi + h\eta) - F(\xi)}{h}\right\}$$

$$= \psi\left\{\lim_{h \to 0} \frac{F(\xi + h\eta) - F(\xi)}{h}\right\}$$

$$= \psi\{F'(\xi, \eta)\}. \tag{5.22}$$

Since ψ is continuous and defined for all of V, $f'(\xi, \eta)$ is defined in a neighborhood of ξ_0 and continuous at ξ_0. By Theorem 5.7, $f'(\xi_0, \eta)$ is linear in η. Thus,

$$\psi\{F'(\xi_0, a_1\eta_1 + a_2\eta_2)\} = a_1\psi\{F'(\xi_0, \eta_1)\} + a_2\psi\{F'(\xi_0, \eta_2)\}$$
$$= \psi\{a_1F'(\xi_0, \eta_1) + a_2F'(\xi_0, \eta_2)\}.$$

Since $F'(\xi_0, a_1\eta_1 + a_2\eta_2) - a_1F'(\xi_0, \eta_1) - a_2F'(\xi_0, \eta_2)$ is annihilated by all $\psi \in \hat{V}$, it must be 0; that is,

$$F'(\xi_0, a_1\eta_1 + a_2\eta_2) = a_1F'(\xi_0, \eta_1) + a_2F'(\xi_0, \eta_2). \quad \Box \qquad (5.23)$$

For each ξ, the mapping of $\eta \in U$ onto $F'(\xi, \eta) \in V$ is a linear transformation which we denote by $F'(\xi)$. $F'(\xi)$ is called the *derivative* of F at ξ.

It is of some interest to introduce bases in U and V and find the corresponding matrix representation of $F'(\xi)$. Let $A = \{\alpha_1, \ldots, \alpha_n\}$ be a basis in U and $B = \{\beta_1, \ldots, \beta_m\}$ be a basis in V. Let $\xi = \sum_j x_j\alpha_j$ and $F(\xi) = \sum_j y_j(\xi)\beta_j$. Let $\hat{B} = \{\psi_1, \ldots, \psi_m\}$ be the dual bais of B. Then $\psi_kF(\xi) = y_k(\xi)$. If $F'(\xi)$ is represented by the matrix $J = [a_{ij}]$, we have

$$F'(\xi)(\alpha_j) = \sum_i a_{ij}\beta_i. \qquad (5.24)$$

Then

$$a_{kj} = \psi_kF'(\xi)(\alpha_j) = \psi_kF'(\xi, \alpha_j)$$
$$= \psi_k \lim_{h \to 0} \frac{F(\xi + h\alpha_j) - F(\xi)}{h}$$
$$= \lim_{h \to 0} \frac{\psi_kF(\xi + h\alpha_j) - \psi_kF(\xi)}{h}$$
$$= \lim_{h \to 0} \frac{y_k(\xi + h\alpha_j) - y_k(\xi)}{h}$$
$$= \frac{\partial y_k}{\partial x_j}, \qquad (5.25)$$

according to formula (5.17). Thus $F'(\xi)$ is represented by the matrix

$$J(\xi) = \left[\frac{\partial y_k}{\partial x_j}\right]. \qquad (5.26)$$

$J(\xi)$ is known as the *Jacobian matrix* of $F'(\xi)$ with respect to the basis A and B.

The case where $U = V$—that is, where F is a mapping of V into itself—is of special interest. Then $F'(\xi)$ is a linear transformation of V into itself and the corresponding Jacobian matrix $J(\xi)$ is a square matrix. Since the trace of $J(\xi)$ is invariant under similarity transformation, it depends on $F'(\xi)$ alone and not on the matrix representation of $F'(\xi)$. This trace is called the *divergence* of F at ξ.

$$\text{Tr}(F'(\xi)) = \text{div } F(\xi) = \sum_{i=1}^{n} \frac{\partial y_i}{\partial x_i}. \tag{5.27}$$

Let us re-examine the differentiation of functions of scalar variables, which at this point seem to be treated in a way essentially different from functions of vector variables. It would also be desirable if the treatment of such functions could be made to appear as a special case of functions of vector variables without distorting elementary calculus to fit the generalization.

Let W be a 1-dimensional vector space and let $\{\gamma_1\}$ be a basis of W. We can identify the scalar t with the vector $t\gamma_1$, and consider $\xi(t)$ as just a short-handed way of writing $\xi(t\gamma_1)$. In keeping with formula (5.10), we consider

$$\lim_{h \to 0} \frac{\xi(t\gamma_1 + h\eta) - \xi(t\gamma_1)}{h} = \xi'(t\gamma_1, \eta). \tag{5.28}$$

Since W is 1-dimensional it will be sufficient to take the case $\eta = \gamma_1$. Then

$$\xi'(t\gamma_1)(\gamma_1) = \xi'(t\gamma_1, \gamma_1) = \lim_{h \to 0} \frac{\xi(t\gamma_1 + h\gamma_1) - \xi(t\gamma_1)}{h}$$

$$= \lim_{h \to 0} \frac{\xi(t + h) - \xi(t)}{h}$$

$$= \frac{d\xi}{dt}. \tag{5.29}$$

Thus $d\xi/dt$ is the value of the linear transformation $\xi'(t\gamma_1)$ applied to the basis vector γ_1.

Theorem 5.10. *Let F be a mapping of U into V. If F is linear, $F'(\xi) = F$ for all $\xi \in U$.*

PROOF. $F'(\xi)(\eta) = F'(\xi, \eta) = \lim_{h \to 0} \dfrac{F(\xi + h\eta) - F(\xi)}{h}$

$$= \lim_{h \to 0} F(\eta)$$

$$= F(\eta). \;\square \tag{5.30}$$

Finally, let us consider the differentiation of composite functions. Let F be a mapping of U into V, and G a mapping of V into W. Then $GF = H$ is a mapping of U into W.

Theorem 5.11. *If F is linear and G is differentiable, then $(GF)'(\xi) = G'(F(\xi))F$. If G is linear and F is differentiable, then $(GF)'(\xi) = GF'(\xi)$.*

PROOF.   Assume F is linear and G is differentiable. Then

$$(GF)'(\xi)(\eta) = (GF)'(\xi, \eta) = \lim_{h \to 0} \frac{GF(\xi + h\eta) - GF(\xi)}{h}$$

$$= \lim_{h \to 0} \frac{G[F(\xi) + hF(\eta)] - G[F(\xi)]}{h}$$

$$= G'[F(\xi), F(\eta)]$$

$$= G'(F(\xi))F(\eta). \tag{5.31}$$

Assume G is linear and F is differentiable. Then

$$(GF)'(\xi)(\eta) = (GF)'(\xi, \eta) = \lim_{h \to 0} \frac{GF(\xi + h\eta) - GF(\xi)}{h}$$

$$= \lim_{h \to 0} G\left[\frac{F(\xi + h\eta) - F(\xi)}{h}\right]$$

$$= G\left[\lim_{h \to 0} \frac{F(\xi + h\eta) - F(\xi)}{h}\right]$$

$$= G[F'(\xi, \eta)]$$

$$= G[F'(\xi)(\eta)]$$

$$= [GF'(\xi)](\eta). \ \square \tag{5.32}$$

Theorem 5.12.   *Let F be a mapping of U into V, and G a mapping of V into W. Assume that F is continuously differentiable at ξ_0 and that G is continuously differentiable in a neighborhood of $\tau_0 = F(\xi_0)$. Then GF is continuously differentiable at ξ_0 and $(GF)'(\xi_0) = G'(F(\xi_0))F'(\xi_0)$.*

PROOF.   For notational convenience let $F(\xi_0 + h\eta) - F(\xi_0) = \omega$. Let ψ be any linear functional in $\hat{\mathsf{W}}$. Then ψG is a scalar-valued function on U continuously differentiable in a neighborhood of τ_0. Hence

$$\psi GF(\xi_0 + h\eta) - \psi GF(\xi_0) = \psi G(\tau_0 + \omega) - \psi G(\tau_0)$$

$$= (\psi G)'(\tau_0 + \theta\omega, \omega)$$

for some real $\theta, 0 < \theta < 1$. Now

$$\lim_{h \to 0} \frac{\omega}{h} = \lim_{h \to 0} \frac{F(\xi_0 + h\eta) - F(\xi_0)}{h}$$

$$= F'(\xi_0, \eta)$$

and

$$\lim_{h \to 0} (\tau_0 + \theta\omega) = \tau_0.$$

Since $G'(\tau, \omega)$ is continuous in τ in a neighborhood of τ_0 and bounded linear in ω,

$$\psi(GF)'(\xi_0)(\eta) = (\psi GF)'(\xi_0, \eta)$$

$$= \lim_{h \to 0} \frac{1}{h} (\psi G)'(\tau_0 + \theta\omega, \omega)$$

$$= \lim_{h \to 0} \psi G'\left(\tau_0 + \theta\omega, \frac{\omega}{h}\right)$$

$$= \psi G'(\tau_0, F'(\xi_0, \eta))$$

$$= \psi G'(F(\xi_0), F'(\xi_0)(\eta))$$

$$= \psi[G'(F(\xi_0))F'(\xi_0)](\eta). \tag{5.33}$$

Since (5.33) holds for all $\psi \in \hat{W}$, we have

$$(GF)'(\xi_0)(\eta) = G'(F(\xi_0))F'(\xi_0)(\eta). \;\square \tag{5.34}$$

This gives a very reasonable interpretation of the chain rule for differentiation. The derivative of the composite function GF is the linear function obtained by taking the composite function of the derivatives of F and G.

Notice, also, that by combining Theorem 5.10 with Theorem 5.12 we can see that Theorem 5.11 is a special case of Theorem 5.12. If F is linear $(GF)'(\xi) = G'(F(\xi))F'(\xi) = G'(F(\xi))F$. If G is linear, $(GF)'(\xi) = G'(F(\xi))F'(\xi) = GF'(\xi)$.

Example 1. For $\xi = x_1\xi_1 + x_2\xi_2$, let $f(\xi) = x_1^2 + 3x_2^2 - 2x_1x_2$. Then for $\eta = y_1\xi_1 + y_2\xi_2$ we have

$$f'(\xi, \eta) = \lim_{h \to 0} \frac{f(\xi + h\eta) - f(\xi)}{h}$$

$$= 2x_1y_1 + 6x_2y_2 - 2x_1y_2 - 2x_2y_1$$

$$= (2x_1 - 2x_2)y_1 + (6x_2 - 2x_1)y_2.$$

If $\{\phi_1, \phi_2\}$ is the basis of $\hat{R}^2$ dual to $\{\xi_1, \xi_2\}$, then

$$f'(\xi)(\eta) = f'(\xi, \eta)$$

$$= [(2x_1 - 2x_2)\phi_1 + (6x_2 - 2x_1)\phi_1](\eta),$$

and hence

$$f'(\xi) = (2x_1 - 2x_2)\phi_1 + (6x_2 - 2x_1)\phi_1$$

$$= \frac{\partial L}{\partial x_1} \phi_1 + \frac{\partial L}{\partial x_2} \phi_2.$$

Example 2. For $\xi = x_1\xi_1 + x_2\xi_2 + x_3\xi_3$, let $F(\xi) = \sin x_2\xi_1 + x_1x_2x_3\xi_2 + (x_1^2 + x_3^2)\xi_3$. Then for $\eta = y_1\xi_1 + y_2\xi_2 + y_2\xi_3$ we have

$$F'(\xi, \eta) = y_2 \cos x_2\xi_1 + (x_2x_3y_1 + x_1x_3y_2 + x_1x_2y_3)\xi_2 + (2x_1y_1 + 2x_3y_3)\xi_3$$
$$= F'(\xi)(\eta).$$

We see that $F'(\xi)$ is a linear transformation of R^3 into itself represented with respect for the basis $\{\xi_1, \xi_2, \xi_3\}$ by the matrix

$$\begin{bmatrix} 0 & \cos x_2 & 0 \\ x_2 x_3 & x_1 x_3 & x_1 x_2 \\ 2x_1 & 0 & 2x_3 \end{bmatrix}.$$

In both of these examples rather conventional notation has been used. However, the interpretation of the context of these computations is quite different from the conventional interpretation. In Example 1, $f(\xi)$ should be regarded as a function mapping R^2 into R, and $f'(\xi)$ should be regarded as a linear approximation of f at the point ξ. Thus $f'(\xi) \in \text{Hom}(R^2, R) = \hat{R}^2$. In Example 2, $F(\xi)$ is a function of R^3 into R^3 and $F'(\xi)$ is a linear approximation of F at the point ξ. Thus $F'(\xi) \in \text{Hom}(R^3, R^3)$.

6 | Spectral Decomposition of Linear Transformations

Most of this section requires no more than the material through Section 7 of Chapter III. However, familiarity with the Jordan normal form as developed in Section 8 of Chapter III is required for the last part of this section and very helpful for the first part.

Let σ be a linear transformation of an n-dimensional vector space V into itself. The set of eigenvalues of σ is called the *spectrum* of σ. Assume that V has a basis $\{\alpha_1, \ldots, \alpha_n\}$ of eigenvectors of σ. Let λ_i be the eigenvalue corresponding to α_i. Let S_i be the subspace spanned by α_i, and let π_i be the projection of V onto S_i along $S_1 \oplus \cdots \oplus S_{i-1} \oplus S_{i+1} \oplus \cdots \oplus S_n$. These projections have the properties

$$\pi_i^2 = \pi_i, \tag{6.1}$$

and

$$\pi_i \pi_j = 0 \quad \text{for} \quad i \neq j. \tag{6.2}$$

Any linear transformation σ for which $\sigma^2 = \sigma$ is said to be *idempotent*. If σ and τ are two linear transformations such that $\sigma\tau = \tau\sigma = 0$, we say they are *orthogonal*. Similar terminology is applied to matrices representing linear transformations with these properties.

Every $\xi \in V$ can be written in the form

$$\xi = \xi_1 + \cdots + \xi_n, \tag{6.3}$$

where $\xi_i \in S_i$. Then $\pi_i(\xi) = \xi_i$ so that

$$\xi = \pi_1(\xi) + \cdots + \pi_n(\xi)$$
$$= (\pi_1 + \cdots + \pi_n)(\xi). \tag{6.4}$$

Since (6.4) holds for every $\xi \in V$ we have

$$1 = \pi_1 + \pi_2 + \cdots + \pi_n. \tag{6.5}$$

A formula like (6.5) in which the identity transformation is expressed as a sum of mutually orthogonal projections is called a *resolution of the identity*. From (6.1) and (6.2) it follows that $(\pi_1 + \pi_2)^2 = (\pi_1 + \pi_2)$ so that a sum of projections orthogonal to each other is a projection. Conversely, it is sometimes possible to express a projection as a sum of projections. If a projection cannot be expressed as a sum of non-zero projections, it is said to be *irreducible*. Since the projections given are onto 1-dimensional subspaces, they are irreducible. If the projections appearing in a resolution of the identity are irreducible, the resolution is called *irreducible* or *maximal*.

Now, for $\xi = \sum_{i=1}^{n} \xi_i$ as in (6.3) we have

$$\sigma(\xi) = \sigma\left(\sum_{i=1}^{n} \xi_i \right) = \sum_{i=1}^{n} \sigma(\xi_i)$$

$$= \sum_{i=1}^{n} \lambda_i \xi_i$$

$$= \sum_{i=1}^{n} \lambda_i \pi_i(\xi)$$

$$= \left(\sum_{i=1}^{n} \lambda_i \pi_i \right)(\xi). \tag{6.6}$$

Since (6.6) holds for every $\xi \in V$ we have

$$\sigma = \sum_{i=1}^{n} \lambda_i \pi_i. \tag{6.7}$$

A representation of σ in the form of (6.7), where each λ_i is an eigenvalue of σ and each π_i is a projection, is called a *spectral decomposition*. If the eigenvalues are each of multiplicity 1, the decomposition is unique. If some of the eigenvalues are of higher multiplicity, the choice of projections is not unique, but the number of times each eigenvalue occurs in the decomposition is equal to its multiplicity and therefore unique.

An advantage of a spectral decomposition like (6.7) is that because of (6.1) and (6.2) we have

$$\sigma^2 = \sum_{i=1}^{n} \lambda_i^{2} \pi_i, \tag{6.8}$$

$$\sigma^k = \sum_{1=i}^{n} \lambda_i^{k} \pi_i, \tag{6.9}$$

and

$$f(\sigma) = \sum_{i=1}^{n} f(\lambda_i)\pi_i \qquad (6.10)$$

for any polynomial $f(x)$ with coefficients in F.

Given a matrix representing a linear transformation σ, there are several effective methods for finding the matrices representing the projections π_i, and, from them, the spectral decomposition. Any computational procedure which yields the eigenvectors of σ must necessarily give the projections since with the eigenvectors as basis the projections have very simple representations. However, what is usually wanted is the representations of the projections in the original coordinate system. Let $S_j = (s_{1j}, s_{2j}, \ldots, s_{nj})$ be the representation of α_j in the original coordinate system. Since we have assumed there is a basis of eigenvectors, the matrix $S = [s_{ij}]$ is non-singular. Let $T = [t_{ij}]$ be the inverse of S. Then

$$\cdot\ P_k = [s_{ik}t_{kj}] \qquad (6.11)$$

represents π_k, as can easily be checked.

We give another method for finding the projections which does not require finding the eigenvectors first, although they will appear in the end result. We introduce this method because it is useful in situations where a basis of eigenvectors does not exist. We, however, assume that the characteristic polynomial factors into linear factors. Let $\{\lambda_1, \ldots, \lambda_p\}$ be the distinct eigenvalues of σ, and let

$$m(x) = (x - \lambda_1)^{s_1} \cdots (x - \lambda_p)^{s_p} \qquad (6.12)$$

be the minimum polynomial for σ. Set

$$h_i(x) = \frac{m(x)}{(x - \lambda_i)^{s_i}}. \qquad (6.13)$$

We wish to show now that there exist polynomials $g_1(x), \ldots, g_p(x)$ such that

$$1 = g_1(x)h_1(x) + \cdots + g_p(x)h_p(x). \qquad (6.14)$$

Consider the set of all possible non-zero polynomials that can be written in the form $p_1(x)h_1(x) + \cdots + p_p(x)h_p(x)$ where the $p_i(x)$ are polynomials (not all zero since the resulting expression must be non-zero). At least one polynomial, for example, $h_1(x)$, can be written in this form. Hence, there is a non-zero polynomial of lowest degree that can be written in this form. Let $d(x)$ be such a polynomial and let $g_1(x), \ldots, g_p(x)$ be the corresponding coefficient polynomials,

$$d(x) = g_1(x)h_1(x) + \cdots + g_p(x)h_p(x). \qquad (6.15)$$

We assert that $d(x)$ divides all $h_i(x)$. For example, let us try to divide $h_1(x)$ by $d(x)$. Either $d(x)$ divides $h_1(x)$ exactly or there is a remainder $r_1(x)$ of degree less than the degree of $d(x)$. Thus,

$$h_1(x) = d(x)q_1(x) + r_1(x) \tag{6.16}$$

where $q_1(x)$ is the quotient. Suppose, for the moment, that the remainder $r_1(x)$ is not zero. Then

$$r_1(x) = h_1(x) - d(x)q_1(x)$$
$$= h_1(x)\{1 - g_1(x)q_1(x)\} - g_2(x)q_1(x)h_2(x) - \cdots - g_p(x)q_1(x)h_p(x).$$
$$\tag{6.17}$$

But this contradicts the selection of $d(x)$ as a non-zero polynomial of smallest degree which can be written in this form. Thus $d(x)$ must divide $h_1(x)$. Similarly, $d(x)$ divides each $h_i(x)$.

Since the factorization of $m(x)$ is unique, the $h_i(x)$ have no common non-constant factor and $d(x)$ must be a constant. Since we can divide any of these expressions by a non-zero constant without altering its form, we can take $d(x)$ to be 1. Thus, we have an expression in the form of (6.14).

If we divide (6.14) by $m(x)$ we obtain

$$\frac{1}{m(x)} = \frac{g_1(x)}{(x - \lambda_1)^{s_1}} + \cdots + \frac{g_p(x)}{(x - \lambda_p)^{s_p}}. \tag{6.18}$$

This is the familiar partial fractions decomposition and the polynomials $g_i(x)$ can be found by any of several effective techniques.

Now setting $e_i(x) = g_i(x)h_i(x)$, we see that we have obtained a set of polynomials $\{e_1(x), \ldots, e_p(x)\}$ such that

(1) $1 = e_1(x) + \cdots + e_p(x)$,
(2) $e_i(x)e_j(x)$ is divisible by $m(x)$ for $i \neq j$.
(3) $(x - \lambda_i)^{s_i} e_i(x)$ is divisible by $m(x)$.

Now, we use these polynomials to form polynomial expressions of the linear transformation σ. Then $\{e_1(\sigma), \ldots, e_p(\sigma)\}$ is a set of linear transformations with the properties that

(1) $1 = e_1(\sigma) + \cdots + e_p(\sigma)$,
(2) $e_i(\sigma)e_j(\sigma) = 0$ for $i \neq j$,
(3) $(\sigma - \lambda_i)^{s_i}e_i(\sigma) = 0$. $\tag{6.19}$

From (1) and (2) it follows, also, that

$$\begin{aligned}(4)\ e_i(\sigma) = 1 \cdot e_i(\sigma) &= (e_1(\sigma) + \cdots + e_p(\sigma))e_i(\sigma) \\ &= e_1(\sigma)e_i(\sigma) + \cdots + e_p(\sigma)e_i(\sigma) \\ &= [e_i(\sigma)]^2.\end{aligned}$$

These four properties suffice to show that the $e_i(\sigma)$ are mutually orthogonal projections. From (3) we see that $e_i(\sigma)(V)$ is in the kernel of $(\sigma - \lambda_i)^{s_i}$. As in Chapter III-8, we denote the kernel of $(\sigma - \lambda_i)^{s_i}$ by M_i. Since $e_j(x)$ is divisible by $(x - \lambda_i)^{s_i}$ for $j \neq i$, $e_j(\sigma)M_i = 0$. Hence, if $\beta_i \in M_i$, we have

$$\beta_i = (e_1(\sigma) + \cdots + e_p(\sigma))(\beta_i)$$
$$= e_i(\sigma)(\beta_i). \tag{6.20}$$

This shows that $e_i(\sigma)$ acts like the identity on M_i. Then $M_i = e_i(\sigma)(M_i) \subset e_i(\sigma)(V) \subset M_i$ so that $e_i(\sigma)(V) = M_i$. By (6.19), for any $\beta \in V$ we have

$$\beta = (e_i(\sigma) + \cdots + e_p(\sigma))(\beta)$$
$$= \beta_1 + \cdots + \beta_p, \tag{6.21}$$

where $\beta_i = e_i(\sigma)(\beta) \in M_i$. If $\beta = 0$, then $\beta_i = e_i(\sigma)(\beta) = e_i(\sigma)(0) = 0$ so that the representation of β in the form of a sum like (6.21) with $\beta_i \in M_i$ is unique. Thus,

$$V = M_i \oplus \cdots \oplus M_p. \tag{6.22}$$

This provides an independent proof of Theorem 8.3 of Chapter III.

Let $\sigma_i = \sigma \cdot e_i(\sigma)$. Then

(1) $\sigma = \sigma_1 + \cdots + \sigma_p$,
(2) $\sigma_i \sigma_j = 0$ for $i \neq j$, and
(3) $f(\sigma) = f(\sigma_1) + \cdots + f(\sigma_p)$, \hfill (6.23)

for any polynomial $f(x)$ with coefficients in F. If $s_i = 1$, then $(\sigma - \lambda_i)e_i(\sigma) = 0$ so that $\sigma_i = \lambda_i e_i(\sigma)$. In this case M_i is the eigenspace corresponding to λ_i. If the multiplicity of λ_i is 1, then $\dim M_i = 1$ and $e_i(\sigma) = \pi_i$ is the projection onto M_i. If A represents σ, then $P_i = e_i(A)$ represents π_j. If the multiplicity of λ_i is greater than 1, then $e_i(\sigma)$ is a reducible projection. $e_i(\sigma)$ can be reduced to a sum of irreducible projections in many ways and there is no reasonable way to select a unique reduction.

If $s_i > 1$, the situation is somewhat more complicated. Let $\nu_i = (\sigma - \lambda_i)e_i(\sigma)$. Then $\nu_i^{s_i} = (\sigma - \lambda_i)^{s_i}e_i(\sigma) = 0$. A linear transformation for which some power vanishes is said to be *nilpotent*. Thus,

$$\sigma_i = \lambda_i e_i(\sigma) + \nu_i \tag{6.24}$$

is the sum of a scalar times an indempotent transformation and a nilpotent transformation.

Since σ commutes with $e_i(\sigma)$, $\sigma(M_i) = \sigma e_i(\sigma)(V) = e_i(\sigma)\sigma(V) \subset M_i$. Thus M_i is invariant under σ. It is also true that $\sigma_i(M_i) \subset M_i$ and $\sigma_i(M_j) = \{0\}$ for $j \neq i$. Each M_i is associated with one of the eigenvalues. It is often possible to reduce each M_i to a direct sum of subspaces such that σ is invariant on each summand. In a manner analogous to what has been done so far, we can find for each summand a linear transformation which has the same effect as σ on that summand and vanishes on all other summands. Then this linear transformation can also be expressed as the sum of a scalar times an idempotent transformation and a nilpotent transformation. The determination of the Jordan normal form involves this kind of reduction in some form or other. The matrix B_i of Theorem 8.5 in Chapter III can be written in the form $\lambda_i E_i + N_i$ where E_i is an idempotent matrix and N_i is a nilpotent matrix. However, in the following discussion (6.23) and (6.24) are sufficient for our purposes and we shall not concern ourselves with the details of a further reduction at this point.

There is a particularly interesting and important application of the spectral decomposition and the decomposition (6.23) to systems of linear differential equations. In order to prepare for this application we have to discuss the meaning of infinite series and sequences of matrices and apply these decompositions to the simplification of these series.

If $\{A_k\}$, $A_k = [a_{ij}(k)]$, is a sequence of real matrices, we say

$$\lim_{k \to \infty} A_k = A = [a_{ij}]$$

if and only if $\lim_{k \to \infty} a_{ij}(k) = a_{ij}$ for each i and j. Similarly, the series $\sum_{k=0}^{\infty} c_k A^k$ is said to converge if and only if the sequence of partial sums converges. It is not difficult to show that the series

$$\sum_{k=0}^{\infty} \frac{1}{k!} A^k \tag{6.25}$$

converges for each A. In analogy with the series representation of e^x, the series (6.25) is taken to be the definition of e^A.

If A and B commute, then $(A + B)^k$ can be expanded by the binomial theorem. It can then be shown that

$$e^{A+B} = e^A e^B.$$

This "law of exponents" is not generally satisfied if A and B do not commute.

If E is an idempotent matrix—that is, $E^2 = E$—then

$$e^{\lambda E} = I + \lambda E + \cdots + \frac{\lambda^k E}{k!} + \cdots$$

$$= I - E + E\left(1 + \lambda + \cdots + \frac{\lambda^k}{k!} + \cdots\right)$$

$$= I + (e^\lambda - 1)E. \tag{6.26}$$

If N is a nilpotent matrix, then the series representation of e^N terminates after a finite number of terms, that is, e^N is a polynomial in N. These observations, together with the spectral decomposition and the decomposition

(6.23) in terms of commuting transformations (and hence matrices) of these types, will enable us to express e^A in terms of finite sums and products of matrices.

Let $\sigma = \sigma_1 + \cdots + \sigma_p$ as in (6.23). Let A represent σ and A_i represent σ_i. Let E_i represent $e_i(\sigma)$ and N_i represent ν_i. Since each of these linear transformations is a polynomial in σ, they (and the matrices that represent them) commute. Assume at first that σ has a spectral decomposition, that is, each $\nu_i = 0$. Then $A_i = \lambda_i E_i$ and

$$e^A = e^{A_1 + \cdots + A_p} = e^{A_1} e^{A_2} \cdots e^{A_p}$$

$$= [I + (e^{\lambda_1} - 1)E_1][I + (e^{\lambda_2} - 1)E_2] \cdots [I + (e^{\lambda_p} - 1)E_p]$$

$$= I + \sum_{i=1}^{p} (e^{\lambda_i} - 1)E_i \tag{6.27}$$

because of the orthogonality of the E_i. Then

$$e^A = \sum_{i=1}^{p} e^{\lambda_i} E_i \tag{6.28}$$

because $\sum_{i=1}^{p} E_i = I$. This is a generalization of formula (6.10) to a function of A which is not a polynomial.

The situation when σ does not have a spectral decomposition is slightly more complicated. Then $A_i = \lambda_i E_i + N_i$ and

$$e^A = \prod_{i=1}^{p} e^{\lambda_i E_i} \prod_{i=1}^{p} e^{N_i}$$

$$= \left(\sum_{i=4}^{p} e^{\lambda_i} E_i \right) \prod_{i=1}^{p} e^{N_i}. \tag{6.29}$$

As an example of formula (6.28) consider $A = \begin{bmatrix} 2 & 2 \\ 2 & -1 \end{bmatrix}$. The characteristic polynomial is $x^2 - x - 6 = (x - 3)(x + 2)$. To fix our notation we take $\lambda_1 = -2$ and $\lambda_2 = 3$. Since

$$\frac{1}{(x - 3)(x + 2)} = \frac{-\frac{1}{5}}{x + 2} + \frac{\frac{1}{5}}{x - 3}$$

we see that $e_1(x) = \dfrac{-(x - 3)}{5}$ and $e_2(x) = \dfrac{(x + 2)}{5}$. Thus,

$$E_1 = -\frac{1}{5} \begin{bmatrix} -1 & 2 \\ 2 & -4 \end{bmatrix}, \qquad E_2 = \frac{1}{5} \begin{bmatrix} 4 & 2 \\ 2 & 1 \end{bmatrix},$$

$$I = E_1 + E_2,$$

$$A = -2E_1 + 3E_2,$$

$$e^A = e^{-2}E_1 + e^3 E_2.$$

BIBLIOGRAPHICAL NOTES

In finite dimensional vector spaces many things that the spectrum of a transformation can be used for can be handled in other ways. In infinite dimensional vector spaces matrices are not available and the spectrum plays a more central role. The treatment in P. Halmos, *Introduction to Hilbert Space*, is recommended because of the clarity with which the spectrum is developed.

EXERCISES

1. Use the method of spectral decomposition to resolve each matrix A into the form $A = AE_1 + AE_2 = \lambda_1 E_1 + \lambda_2 E_2$, where λ_1 and λ_2 are the eigenvalues of A and E_1 and E_2 represent the projections of V onto the invariant subspaces of A.

(a) $\begin{bmatrix} 1 & 2 \\ 2 & 1 \end{bmatrix}$,   (b) $\begin{bmatrix} 1 & 2 \\ 2 & -2 \end{bmatrix}$,   (c) $\begin{bmatrix} 1 & 2 \\ 2 & 4 \end{bmatrix}$,

(d) $\begin{bmatrix} 1 & 3 \\ 3 & -7 \end{bmatrix}$,   (e) $\begin{bmatrix} 1 & 4 \\ 4 & -5 \end{bmatrix}$.

2. Use the method of spectral decomposition to resolve the matrix

$$A = \begin{bmatrix} 3 & -7 & -20 \\ 0 & -5 & -14 \\ 0 & 3 & 8 \end{bmatrix}$$

into the form $A = AE_1 + AE_2 + AE_3 = \lambda_1 E_1 + \lambda_2 E_2 + \lambda_3 E_3$, where E_1, E_2, and E_3 are orthogonal idempotent matrices.

3. For each matrix A in the above exercises compute the matrix form of e^A.

4. Use the method of spectral decomposition to resolve the matrix

$$A = \begin{bmatrix} 1 & -3 & 3 \\ 0 & 3 & -2 \\ -1 & -1 & 1 \end{bmatrix}$$

into the form $A = AE_1 + AE_2$ where E_1 and E_2 are orthogonal idempotent matrices. Furthermore, show that AE_1 (or AE_2) is of the form $AE_1 = \lambda_1 E_1 + N_1$ where λ_1 is an eigenvalue of A and N_1 is nilpotent, and that AE_2 (or AE_1) is of the form $AE_2 = \lambda_2 E_2$ where λ_2 is an eigenvalue of A.

5. Let $A = \lambda E + N$, where $E^2 = E$, $EN = N$, and N is nilpotent of order r; that is, $N^{r-1} \neq 0$ but $N^r = 0$. Show that

$$e^A = I + E(e^\lambda - 1) + e^\lambda \sum_{s=1}^{r-1} \frac{N^s}{s!}$$
$$= (I - E)(1 - e^\lambda) + e^\lambda e^N = I - E + e^\lambda e^N E.$$

6. Compute the matrix form of e^A for the matrix A given in Exercise 4.

7. Let $A = AE_1 + AE_2 + \cdots + AE_p$, where the E_i are orthogonal idempotents and each $AE_i = \lambda_i E_i + N_i$ where $E_i N_i = N_i$. Show that

$$e^A = \sum_{i=1}^{p} e^{\lambda_i} e^{N_i} E_i.$$

7 | Systems of Linear Differential Equations

Most of this section can be read with background material from Chapter I, the first four sections of Chapter II, and the first five sections of Chapter III. However, the examples and exercises require the preceding section, Section 5. Theorem 7.3 requires Section 11 of Chapter V. Some knowledge of the theory of linear differential equations is also needed.

We consider the system of linear differential equations

$$\dot{x}_i = \frac{dx_i}{dt} = \sum_{j=1}^{n} a_{ij}(t) x_j, \qquad i = 1, 2, \ldots, n \tag{7.1}$$

where the $a_{ij}(t)$ are continuous real valued functions for values of t, $t_1 \leq t \leq t_2$. A solution consists of an n-tuple of functions $X(t) = (x_1(t), \ldots, x_n(t))$ satisfying (7.1). It is known from the theory of differential equations that corresponding to any n-tuple $X_0 = (x_{10}, x_{20}, \ldots, x_{n0})$ of scalars there is a unique solution $X(t)$ such that $X(t_0) = X_0$. The system (7.1) can be written in the more compact form

$$\dot{X} = AX, \tag{7.2}$$

where $\dot{X}(t) = (\dot{x}_1(t), \ldots, \dot{x}_n(t))$ and $A = [a_{ij}(t)]$.

The equations in (7.1) are of first order, but equations of higher order can be included in this framework. For example, consider the single nth order equation

$$\frac{d^n x}{dt^n} + a_{n-1} \frac{d^{n-1} x}{dt^{n-1}} + \cdots + a_1 \frac{dx}{dt} + a_0 x = 0. \tag{7.3}$$

This equation is equivalent to the system

$$\dot{x}_1 = x_2$$
$$\dot{x}_2 = x_3$$
$$\cdot$$
$$\cdot$$
$$\cdot$$
$$\dot{x}_{n-1} = x_n$$
$$\dot{x}_n = -a_{n-1} x_n - \cdots - a_1 x_2 - a_0 x_1. \tag{7.4}$$

The formula $x_k = \dfrac{d^{k-1} x}{dt^{k-1}}$ expresses the equivalence between (7.3) and (7.4).

The set of all real-valued differentiable functions defined on the interval $[t_1, t_2]$ forms a vector space over R, the field of real numbers. The set V of n-tuples $X(t)$ where the $x_k(t)$ are differentiable also forms a vector space over R. But V is not of dimension n over R; it is infinite dimensional. It is not profitable for us to consider V as a vector space over the set of differentiable functions. For one thing, the set of differentiable functions does not form a field because the reciprocal of a differentiable function is not always differentiable (where that function vanishes). This is a minor objection because appropriate adjustments in the theory could be made. However, we want the operation of differentiation to be a linear operator. The condition $\dfrac{dax}{dt} = a\dfrac{dx}{dt}$ requires that a must be a scalar. Thus we consider V to be a vector space over R.

Statements about the linear independence of a set of n-tuples must be formulated quite carefully. For example.

$$X_1(t) = (1, 0), \qquad X_2(t) = (t, 0)$$

are linearly independent over R even though the determinant

$$\begin{vmatrix} 1 & t \\ 0 & 0 \end{vmatrix} = 0$$

for each value of t. This is because $X_1(t_0)$ and $X_2(t_0)$ are linearly dependent for each value of t_0, but the relation between $X_1(t_0)$ and $X_2(t_0)$ depends on the value of t_0. In particular, a matrix of functions may have n independent columns and not be invertible and the determinant of the matrix might be 0. However, when the set of n-tuples is a set of solutions of a system of differential equations, the connection between their independence for particular values of t_0 and for all values of t is closer.

Theorem 7.1. *Let t_0 be any real number, $t_1 \leq t_0 \leq t_2$, and let $\{X_1(t), \ldots, X_m(t)\}$ be a finite set of solutions of* (7.1). *A necessary and sufficient condition for the linear independence of $\{X_1(t), \ldots, X_m(t)\}$ is the linear independence of $\{X_1(t_0), \ldots, X_m(t_0)\}$.*

PROOF. If the set $\{X_1(t), \ldots, X_m(t)\}$ is linearly dependent, then certainly $\{X_1(t_0), \ldots, X_m(t_0)\}$ is linearly dependent. On the other hand, suppose there exists a non-trivial relation of the form

$$\sum_{k=1}^{m} c_k X_k(t_0) = 0. \tag{7.5}$$

Consider the function $X(t) = \sum_{k=1}^{m} c_k X_k(t)$. $X(t)$ is a solution of (7.1). But the function $Y(t) = 0$ is also a solution and satisfies the condition $Y(t_0) = 0 = X(t_0)$. Since the solution is unique, we must have $X(t) = 0$.

It follows immediately that the system (7.1) can have no more than n linearly independent solutions. On the other hand, for each k there exists a solution satisfying the condition $X_k(t_0) = (\delta_{1k}, \ldots, \delta_{nk})$. It then follows that the set $\{X_1(t), \ldots, X_n(t)\}$ is linearly independent. $\square$

It is still more convenient for the theory we wish to develop to consider the differential matrix equation

$$\dot{F} = AF \tag{7.6}$$

where $F(t) = [f_{ij}(t)]$ is an $n \times n$ matrix of differentiable functions and $\dot{F} = [f'_{ij}(t)]$. If an F is obtained satisfying (7.6), then each column of F is a solution of (7.2). Conversely, if solutions for (7.2) are known, a solution for (7.6) can be obtained by using the solutions of (7.2) to form the columns of F. Since there are n linearly independent solutions of (7.2) there is a solution of (7.6) in which the columns are linearly independent. Such a solution to (7.6) is called a *fundamental solution*. We shall try to find the fundamental solutions of (7.6).

The differential and integral calculus of matrices is analogous to the calculus of real functions, but there are a few important differences. We have already defined the derivative in the statement of formula (7.6). It is easily seen that

$$\frac{d}{dt}(F + G) = \frac{dF}{dt} + \frac{dG}{dt}, \tag{7.7}$$

and

$$\frac{d}{dt}(FG) = \frac{dF}{dt}G + F\frac{dG}{dt}. \tag{7.8}$$

Since the multiplication of matrices is not generally commutative, the order of the factors in (7.8) is important. For example, if F has an inverse F^{-1}, then

$$0 = \frac{d}{dt}(F^{-1}F) = \frac{d(F^{-1})}{dt}F + F^{-1}\frac{dF}{dt}. \tag{7.9}$$

Hence,

$$\frac{d(F^{-1})}{dt} = -F^{-1}\frac{dF}{dt}F^{-1}. \tag{7.10}$$

This formula is analogous to the formula $(dx^{-1}/dt) = -x^{-2}(dx/dt)$ in elementary calculus, but it cannot be written in this simpler form unless F and dF/dt commute.

Similar restrictions hold for other differentiation formulas. An important example is the derivative of a power of a matrix.

$$\frac{dF^2}{dt} = \frac{dF}{dt}F + F\frac{dF}{dt}. \tag{7.11}$$

Again, this formula cannot be further simplified unless F and dF/dt commute. However, if they do commute we can also show that

$$\frac{dF^k}{dt} = kF^{k-1}\frac{dF}{dt}. \tag{7.12}$$

Theorem 7.2. *If F is a fundamental solution of* (7.6), *then every solution G of* (7.6) *is of the form $G = FC$ where C is a matrix of scalars.*

PROOF. It is easy to see that if F is a solution of (7.6) and C is an $n \times n$ matrix of scalars, then FC is also a solution. Notice that CF is not necessarily a solution. There is nothing particularly deep about this observation, but it does show, again, that due care must be used.

Conversely, let F be a fundamental solution of (7.6) and let G be any other solution. By Theorem 7.1, for any scalar t_0 $F(t_0)$ is a non-singular matrix. Let $C = F(t_0)^{-1}G(t_0)$. Then $H = FC$ is a solution of (7.6). But $H(t_0) = F(t_0)C = G(t_0)$. Since the solution satisfying this condition is unique we have $G = H = FC$. □

Let

$$B(t) = \int_{t_0}^{t} A(s)\,ds \tag{7.13}$$

where $B(t) = [b_{ij}(t)]$ and $b_{ij}(t) = \int_{t_0}^{t} a_{ij}(s)\,ds$. We assume that A and B commute so that $\dfrac{dB^k}{dt} = kAB^{k-1}$. Consider

$$e^B = \sum_{k=0}^{\infty} \frac{B^k}{k!}. \tag{7.14}$$

Since this series converges uniformly in any finite interval in which all the elements of B are continuous, the series can be differentiated term by term. Then

$$\frac{de^B}{dt} = \sum_{k=0}^{\infty} \frac{kAB^{k-1}}{k!} = A\sum_{k=0}^{\infty} \frac{B^k}{k!} = Ae^B; \tag{7.15}$$

that is, e^B is a solution of (6.6). Since $e^{B(t_0)} = e^0 = I$, e^B is a fundamental solution. The general solution is

$$F(t) = e^{B(t)}C. \tag{7.16}$$

In this case, under the assumption that A and B commute, e^B is also a solution of $\dot{F} = FA$.

As an example of the application of these ideas consider the matrix

$$A = \begin{bmatrix} 1 & 2t \\ 2t & 1 \end{bmatrix}$$

and take $t_0 = 0$. Then

$$B = \begin{bmatrix} t & t^2 \\ t^2 & t \end{bmatrix}.$$

It is easily verified that A and B commute. The characteristic polynomial of B is $(x - t - t^2)(x - t + t^2)$. Taking $\lambda_1 = t + t^2$ and $\lambda_2 = t - t^2$, we have $e_1(x) = \dfrac{1}{2t}(x - t + t^2)$ and $e_2(x) = -\dfrac{1}{2t}(x - t - t^2)$. Thus,

$$I = \frac{1}{2}\begin{bmatrix} 1 & 1 \\ 1 & 1 \end{bmatrix} + \frac{1}{2}\begin{bmatrix} 1 & -1 \\ -1 & 1 \end{bmatrix}$$

is the corresponding resolution of the identity. By (6.28) we have

$$e^B = \frac{e^{t+t^2}}{2}\begin{bmatrix} 1 & 1 \\ 1 & 1 \end{bmatrix} + \frac{e^{t-t^2}}{2}\begin{bmatrix} 1 & -1 \\ -1 & 1 \end{bmatrix}$$

$$= \frac{1}{2}\begin{bmatrix} e^{t+t^2} + e^{t-t^2} & e^{t+t^2} - e^{t-t^2} \\ e^{t+t^2} - e^{t-t^2} & e^{t+t^2} + e^{t-t^2} \end{bmatrix}.$$

Since e^B is a fundamental solution, the two columns of e^B are linearly independent solutions of (7.2), for this A.

If A is a matrix of scalars, $B = (t - t_0)A$ certainly commutes with A and the above discussion applies. However, in this case the solution can be obtained more directly without determining e^B. If $C = (c_1, \ldots, c_n)$, $c_i \in R$, represents an eigenvector of A corresponding to λ, that is, $AC = \lambda C$, then $BC = (t - t_0)\lambda C$ so that $(t - t_0)\lambda$ is an eigenvalue of B and C represents an eigenvector of B. Then

$$X(t) = (c_1 e^{\lambda(t-t_0)}, c_2 e^{\lambda(t-t_0)}, \ldots, c_n e^{\lambda(t-t_0)}) \tag{7.17}$$

is a solution of (7.2), as can easily be verified. Notice that $X(t_0) = C$.

Conversely, suppose that $X(t)$ is a solution of (7.2) such that $X(t_0) = C$ is an eigenvector of A. Since $\dot{X}(t)$ is also a solution of (7.2),

$$Y(t) = \dot{X} - \lambda X \tag{7.18}$$

is also a solution of (7.2). But $Y = \dot{X} - \lambda X = AX - \lambda X = (A - \lambda I)X$ for all t, and $Y(t_0) = (A - \lambda I)C = 0$. Since the solution satisfying this condition is unique, $Y(t) = 0$, that is, $\dot{X} = \lambda X$. This means each $x_j(t)$ must satisfy

$$\dot{x}_j(t) = \lambda x_j(t), \qquad x_j(t_0) = c_j. \tag{7.19}$$

Thus $X(t)$ is given by (7.17). This method has the advantage of providing solutions when it is difficult or not necessary to find all eigenvalues of A.

Return now to the case where A is not necessarily a matrix of constants. Let G be a fundamental solution of

$$\dot{G} = -A^T G. \tag{7.20}$$

This system is called the *adjoint* system to (7.6). It follows that

$$\frac{d(G^T F)}{dt} = \dot{G}^T F + G^T \dot{F} = (-A^T G)^T F + G^T A F \tag{7.21}$$

$$= G^T(-A)F + G^T AF = 0.$$

Hence, $G^T F = C$, a matrix of scalars. Since F and G are fundamental solutions, C is non-singular.

An important special case occurs when

$$A^T = -A. \tag{7.22}$$

In this case (7.6) and (7.20) are the same equation and the equation is said to be *self-adjoint*. Then F is also a solution of the adjoint system and we see that $F^T F = C$. In this case C is positive definite and we can find a non-singular matrix D such that $D^T C D = I$. Then

$$D^T F^T F D = I = (FD)^T(FD). \tag{7.23}$$

Let $FD = N$. Here we have $N^T N = I$ so that the columns of N are orthonormal solutions of (7.2).

Conversely, let the columns of F be an orthonormal set of solutions of (7.2). Then

$$0 = \frac{d(F^T F)}{dt} = \dot{F}^T F + F^T \dot{F} = (AF)^T F + F^T AF = F^T(A^T + A)F. \tag{7.24}$$

Since F and F^T have inverses, this means $A^T + A = 0$, so that (7.6) is self-adjoint. Thus,

Theorem 7.3. *The system* (7.2) *has an orthonormal set of n solutions if and only if* $A^T = -A$. $\square$

BIBLIOGRAPHICAL NOTES

E. A. Coddington and N. Levinson, *Theory of Ordinary Differential Equations;* F. R. Gantmacher, *Applications of the Theory of Matrices;* W. Hurewicz, *Lectures on Ordinary Differential Equations;* and S. Lefschetz, *Differential Equations, Geometric Theory,* contain extensive material on the use of matrices in the theory of differential equations. They differ in their emphasis, and all should be consulted.

EXERCISES

1. Consider the matric differential equation

$$\dot{X} = \begin{bmatrix} 1 & 2 \\ 2 & 1 \end{bmatrix} X = AX.$$

Show that $(-1, 1)$ and $(1, 1)$ are eigenvectors of A. Show that $X_1 = (-e^{-(t-t_0)}, e^{-(t-t_0)})$ and $X_2 = (e^{3(t-t_0)}, e^{3(t-t_0)})$ are solutions of the differential equation. Show that

$$F = \begin{bmatrix} -e^{-(t-t_0)} & e^{3(t-t_0)} \\ e^{-(t-t_0)} & e^{3(t-t_0)} \end{bmatrix}$$

is a fundamental solution of $\dot{F} = AF$.

Go through a similar analysis for the equation $\dot{X} = AX$ for each matrix A given in Exercises 1 and 2 of Section 6.

2. Let $A = \lambda E + N$ where N is idempotent and $EN = NE = N$. Show that

$$e^{At} = I - E + e^{\lambda t}e^{Nt}E.$$

3. Let $Y = e^{Nt}$ where N is nilpotent. Show that

$$\dot{Y} = Ne^{Nt}.$$

(This is actually true for any scalar matrix N. However, the fact that N is nilpotent avoids dealing with the question of convergence of the series representing e^N and the question of differentiating a uniformly convergent series term by term.)

4. Let $A = AE_1 + \cdots + AE_p$ where the E_i are orthogonal idempotents and each AE_i is of the form $AE_i = \lambda_i E_i + N_i$ where $E_i N_i = N_i E_i = N_i$ and N_i is nilpotent. Show that

$$B = e^{At} = \sum_{i=1}^{p} e^{\lambda_i t}e^{N_i t}E_i.$$

Show that

$$\dot{B} = AB.$$

8 | Small Oscillations of Mechanical Systems

This section requires a knowledge of real quadratic forms as developed in Section 10 of Chapter IV, and orthogonal diagonalization of symmetric matrices as achieved in Section 11 of Chapter V. If the reader is willing to accept the assertions given without proof, the background in mechanics required is minimal.

Consider a mechanical system consisting of a finite number of particles, $\{P_1, \ldots, P_r\}$. We have in mind describing the motions of such a system for small displacements from an equilibrium position. The position of each particle is specified by its three coordinates. Let $(x_{3i-2}, x_{3i-1}, x_{3i})$ be the coordinates of the particle P_i. Then $(x_1, x_2, \ldots, x_{3r})$ can be considered to be the coordinates of the entire system since these $3r$ coordinates specify the location of each particle. Thus we can represent the configuration of a system of particles by a point in a space of dimension $n = 3r$. A space of this kind is called a *phase space*.

The phase space has not been given enough structure for us to consider it to be a vector space with any profit. It is not the coordinates but the small displacements we have to concentrate our attention upon. This is a typical situation in the applications of calculus to applied problems. For example, even though $y = f(x_1, \ldots, x_n)$ may be a transcendental function of its variables, the relation between the differentials is linear,

$$dy = \frac{\partial y}{\partial x_1} dx_1 + \frac{\partial y}{\partial x_2} dx_2 + \cdots + \frac{\partial y}{\partial x_n} dx_n. \tag{8.1}$$

In order to avoid a cumbersome change of notation we consider the coordinates $(x_1, \ldots, x_n)$ to be the coordinates of a displacement from an equilibrium position rather than the coordinates of a position. Thus the equilibrium point is represented by the origin of the coordinate system. We identify these displacements with the points (or vectors) in an n-dimensional real coordinate space.

The potential energy V of the system is a function of the displacements, $V = V(x_1, \ldots, x_n)$. The Taylor series expansion of V is of the form

$$V = V_0 + \left(\frac{\partial V}{\partial x_1} x_1 + \cdots + \frac{\partial V}{\partial x_n} x_n \right)$$

$$+ \frac{1}{2} \left(\frac{\partial^2 V}{\partial x_1^2} x_1^2 + \frac{\partial^2 V}{\partial x_2^2} x_2^2 + \cdots + \frac{\partial^2 V}{\partial x_n^2} x_n^2 + 2 \frac{\partial^2 V}{\partial x_1 \, \partial x_2} x_1 x_2 \right.$$

$$\left. + \cdots + 2 \frac{\partial^2 V}{\partial x_{n-1} \, \partial x_n} x_{n-1} x_n \right) + \cdots . \tag{8.2}$$

We can choose the level of potential energy so that it is zero in the equilibrium position; that is, $V_0 = 0$. The condition that the origin be an equilibrium position means that $\dfrac{\partial V}{\partial x_1} = \cdots = \dfrac{\partial V}{\partial x_n} = 0$. If we let

$$\frac{\partial^2 V}{\partial x_i \, \partial x_j} = a_{ij} = \frac{\partial^2 V}{\partial x_j \, \partial x_i} = a_{ji}, \tag{8.3}$$

then

$$V = \frac{1}{2} \sum_{i,j=1}^{n} x_i a_{ij} x_j + \cdots . \tag{8.4}$$

If the displacements are small, the terms of degree three or more are small compared with the quadratic terms. Thus,

$$V = \frac{1}{2} \sum_{i,j=1}^{n} x_i a_{ij} x_j \tag{8.5}$$

is a good approximation of the potential energy.

We limit our discussion to conservative systems for which the equilibrium is stable or indifferent. If the equilibrium is stable, any displacement must result in an increase in the potential energy; that is, the quadratic form in (8.5) must be positive definite. If the equilibrium is indifferent, a small displacement will not decrease the potential energy; that is, the quadratic form must be non-negative semi-definite.

The kinetic energy T is also a quadratic form in the velocities,

$$T = \frac{1}{2} \sum_{i,j=1}^{n} \dot{x}_i b_{ij} \dot{x}_j. \tag{8.6}$$

In this case the quadratic form is positive definite since the kinetic energy cannot be zero unless all velocities are zero.

In matrix form we have

$$V = \tfrac{1}{2} X^T A X, \tag{8.7}$$

and

$$T = \tfrac{1}{2} \dot{X}^T B \dot{X}, \tag{8.8}$$

where $A = [a_{ij}]$, $B = [b_{ij}]$, $X = (x_1, \ldots, x_n)$, and $\dot{X} = (\dot{x}_1, \ldots, \dot{x}_n)$.

Since B is positive definite, there is a non-singular real matrix Q such that $Q^T B Q = I$. Since $Q^T A Q = A'$ is symmetric, there is an orthogonal matrix Q' such that $Q'^T A' Q' = A''$ is a diagonal matrix. Let $P = QQ'$. Then

$$P^T A P = Q'^T Q^T A Q Q' = Q'^T A' Q' = A'' \tag{8.9}$$

and

$$P^T B P = Q'^T Q^T B Q Q' = Q'^T I Q' = Q'^T Q' = I. \tag{8.10}$$

Thus P diagonalizes A and B simultaneously. (This is an answer to Exercise 3, Chapter V-11.)

If we set $Y = P^{-1} X$, then (8.7) and (8.8) become

$$V = \tfrac{1}{2} Y^T A'' Y = \frac{1}{2} \sum_{i=1}^{n} a_i y_i^2, \tag{8.11}$$

and

$$T = \tfrac{1}{2} \dot{Y}^T \dot{Y} = \frac{1}{2} \sum_{i=1}^{n} \dot{y}_i^2, \tag{8.12}$$

where a_i is the element in the ith place of the diagonal of A''.

In mechanics it is shown that the *Lagrangian* $L = T - V$ satisfies the differential equation

$$\frac{d}{dt}\left(\frac{\partial L}{\partial \dot{y}_i}\right) - \frac{\partial L}{\partial y_i} = 0, \qquad i = 1, \ldots, n. \tag{8.13}$$

For a reference, see Goldstein, *Classical Mechanics*, p. 18. Applied to (8.11) and (8.12), this becomes

$$\ddot{y}_i + a_i y_i = 0, \qquad i = 1, \ldots, n. \tag{8.14}$$

If A is positive definite, we have $a_i > 0$. If A is non-negative semi-definite, we have $a_i \geq 0$. For those $a_i > 0$, let $a_i = \omega_i^2$ where $\omega_i > 0$. The solutions of (8.14) are then of the form

$$y_j(t) = c_j \cos(\omega_j t + e_j), \qquad j = 1, \ldots, n. \tag{8.15}$$

$P = [p_{ij}]$ is the matrix of transition from the original coordinate system with basis $A = \{\alpha_1, \ldots, \alpha_n\}$ to a new coordinate system with basis $B = \{\beta_1, \ldots, \beta_n\}$; that is, $\beta_j = \sum_{i=1}^{n} p_{ij}\alpha_j$. Thus,

$$x_i(t) = \sum_{j=1}^{n} p_{ij} y_j(t) = \sum_{j=1}^{n} p_{ij} c_j \cos(\omega_j t + \theta_j) \tag{8.16}$$

in the original coordinate system. If the system is displaced to an initial position in which

$$y_k(0) = 1, \ y_j(0) = 0 \quad \text{for} \ \ j \neq k, \tag{8.17}$$
$$\dot{y}_j(0) = 0, \qquad j = 1, \ldots, n,$$

then

$$y_k(t) = \cos \omega_k t, \quad y_j(t) = 0 \quad \text{for} \ \ j \neq k, \tag{8.18}$$

and

$$x_i(t) = p_{ik} \cos \omega_k t, \tag{8.19}$$

or

$$X(t) = (p_{1k}, p_{2k}, \ldots, p_{nk}) \cos \omega_k t. \tag{8.20}$$

β_k is represented by $(p_{1k}, p_{2k}, \ldots, p_{nk})$ in the original coordinate system. This n-tuple represents a configuration from which the system will vibrate in simple harmonic motion with angular frequency ω_k if released from rest. The vectors $\{\beta_1, \ldots, \beta_n\}$ are called the *principal axes* of the system. They represent abstract "directions" in which the system will vibrate in simple harmonic motion. In general, the motion in other directions will not be simple harmonic, or even harmonic since it is not necessary for the ω_j to be commensurable. The coordinates $(y_1, \ldots, y_n)$ in this coordinate system are called the *normal coordinates*.

We have described how the simultaneous diagonalization of A and B

can be carried out in two steps. It is often more convenient to achieve the diagonalization in one step. Consider the matric equation

$$(A - \lambda B)X = 0. \tag{8.21}$$

This is an eigenvalue problem in which we are asked to find a scalar λ for which the equation has a non-zero solution. This means we must find a λ for which

$$\det(A - \lambda B) = 0. \tag{8.22}$$

Using the matrix of transition P given above we have

$$0 = \det P^T \cdot \det(A - \lambda B) \cdot \det P = \det(P^T(A - \lambda B)P)$$
$$= \det(P^T A P - \lambda P^T B P) = \det(A'' - \lambda I). \tag{8.23}$$

Since A'' is positive definite or non-negative semi-definite the eigenvalues are ≥ 0. In fact, these eigenvalues are the a_i of formula (8.11).

Let λ_1 and λ_2 be eigenvalues of equation (8.21) and let X_1 and X_2 be corresponding solutions. If $\lambda_1 \neq \lambda_2$, then

$$X_1{}^T A X_2 = X_1{}^T(\lambda_2 B X_2) = \lambda_2 X_1{}^T B X_2$$
$$= (A^T X_1)^T X_2 = (A X_1)^T X_2$$
$$= (\lambda_1 B X_1)^T X_2 = \lambda_1 X_1{}^T B X_2. \tag{8.24}$$

Thus,

$$X_1{}^T B X_2 = 0. \tag{8.25}$$

This situation is described by saying that X_1 and X_2 are orthogonal with respect to B. This is the same meaning given to this term in Chapter V-1 where an arbitrary positive definite quadratic form was selected to determine the inner product.

This argument shows that if the eigenvalues of (8.21) are distinct, then there exists an orthonormal basis with respect to the inner product defined by B. We must show that such a basis exists even if there are repeated eigenvalues. Let σ be the linear transformation represented by A'' with respect to the basis B, that is, $\sigma(\beta_j) = a_j\beta_j$. Let X_j be the representation of β_j with respect to the basis A, $X_j = (p_{1j}, \ldots, p_{nj})$. The matrix representing σ with respect to A is $PA''P^{-1}$. Thus,

$$PA''P^{-1}X_j = a_j X_j. \tag{8.26}$$

Then

$$AX_j = (P^T)^{-1}P^T A P P^{-1}X_j$$
$$= (P^T)^{-1}A''P^{-1}X_j$$
$$= (P^T)^{-1}P^{-1}PA''P^{-1}X_j$$
$$= (P^T)^{-1}P^{-1}a_j X_j$$
$$= a_j B X_j. \tag{8.27}$$

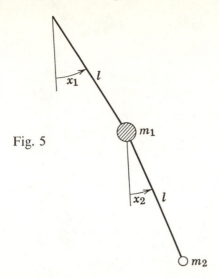

Fig. 5

Since $(A - a_jB)X_j = 0$, we see that a_j is an eigenvalue of (8.21) and X_j is a corresponding eigenvector. Since the columns of P are the X_j, the condition (8.10) is equivalent to the statement that the X_j are orthonormal with respect to the inner product defined by B.

We now have two related methods for finding the principal axes of the given mechanical system: diagonalize B and A in two steps, or solve the eigenvalue problem (8.21). Sometimes it is not necessary to find all the principal axes, in which case the second method is to be preferred. Both methods involve solving an eigenvalue problem. This can be a very difficult task if the system involves more than a few particles. If the system is highly symmetric, there are other methods for finding the principal axes. These methods are discussed in the next sections.

We shall illustrate this discussion with a simple example. Consider the double pendulum in Fig. 5. Although there are two particles in the system, the fact that the pendulum rods are rigid and the system is confined to a plane means that the phase space is only 2-dimensional. This example also illustrates a more general situation in which the phase space coordinates are not rectangular.

The potential energy of the system is

$$V = gm_1(l - l \cos x_1) + gm_2(2l - l \cos x_1 - l \cos x_2)$$
$$= gl[(m_1 + 2m_2) - (m_1 + m_2) \cos x_1 - m_2 \cos x_2].$$

The quadratic term is

$$V = \frac{gl}{2} [(m_1 + m_2)x_1{}^2 + m_2x_2{}^2].$$

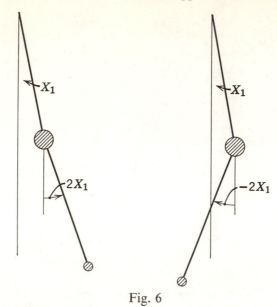

Fig. 6

The kinetic energy is

$$T = \tfrac{1}{2}m_1(l\dot{x}_1)^2 + \tfrac{1}{2}m_2l^2[\dot{x}_1{}^2 + \dot{x}_2{}^2 + 2\cos(x_1 + x_2)\dot{x}_1\dot{x}_2].$$

The quadratic term is

$$T = \tfrac{1}{2}l^2[(m_1 + m_2)\dot{x}_1{}^2 + 2m_2\dot{x}_1\dot{x}_2 + m_2\dot{x}_2{}^2].$$

To simplify the following computation we take $m_1 = 3$ and $m_2 = 1$. Then we must simultaneously diagonalize

$$A = gl\begin{bmatrix} 4 & 0 \\ 0 & 1 \end{bmatrix}, \qquad B = l^2\begin{bmatrix} 4 & 1 \\ 1 & 1 \end{bmatrix}.$$

Solving the equation (8.22), we find $\lambda_1 = (2g/3l)$, $\lambda_2 = 2g/l$. This gives $\omega_1 = \sqrt{2g/3l}$, $\omega_2 = \sqrt{2g/l}$. The coordinates of the normalized eigenvectors are $X_1 = \dfrac{1}{2l\sqrt{3}}(1, 2)$, $X_2 = \dfrac{1}{2l}(1, -2)$. The geometrical configuration for the principal axes are shown in Fig. 6.

The idea behind the concept of principal axes is that if the system is started from rest in one of these two positions, it will oscillate with the angles x_1 and x_2 remaining proportional. Both particles will pass through the vertical line through the pivot point at the same time. The frequencies of these two modes of oscillation are incommensurable, their ratio being $\sqrt{3}$. If the system is started from some other initial configuration, both modes of oscillation will be superimposed and no position will ever be repeated.

BIBLIOGRAPHICAL NOTES

A simple and elegant treatment of the physical principles underlying the theory of small oscillations is given in T. von Karman and M. A. Biot, *Mathematical Methods in Engineering*.

EXERCISES

Consider a conservative mechanical system in the form of an equilateral triangle with mass M at each vertex. Assume the sides of the triangle impose elastic forces according to Hooke's law; that is, if s is the amount that a side is stretched from an equilibrium length, then ks is the restoring force, and $ks^2/2 = \int_0^s ku\,du$ is the potential energy stored in the system by stretching the side through the distance s. Assume the constant k is the same for all three sides of the triangle. Let the triangle be placed on the x, y coordinate plane so that one side is parallel to the x-axis. Introduce a local coordinate system at each vertex so that the that the coordinates of the displacements are as in Fig. 7. We assume that displacements perpendicular to the plane of the triangle change the lengths of the sides by negligible amounts so that it is not necessary to introduce a z-axis. All the following exercises refer to this system.

1. Compute the potential energy as a function of the displacements. Write down the matrix representing the quadratic form of the potential energy in the given coordinate system.

2. Write down the matrix representing the quadratic form of the kinetic energy of the system in the given coordinate system.

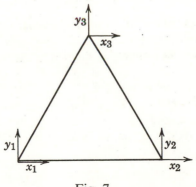

Fig. 7

3. Using the coordinate system,

$$x_1' = x_2 - x_3 \qquad y_1' = y_2 - y_3$$
$$x_2' = x_3 - x_1 \qquad y_2' = y_3 - y_1$$
$$x_3' = x_3 \qquad y_3' = y_3.$$

The quadratic form for the potential energy is somewhat simpler. Determine the matrix representing the quadratic form for the potential energy in this coordinate system.

4. Let the coordinates for the displacements in the original coordinate system be written in the order, $(x_1, y_1, x_2, y_2, x_3, y_3)$. Show that $(1, 0, 1, 0, 1, 0)$ and $(0, 1, 0, 1, 0, 1)$ are eigenvectors of the matrix V representing the potential energy. Give a physical interpretation of this observation.

9 | Representations of Finite Groups by Matrices

For background the material in Chapters I, II, III, and V is required, except for Section 8 of Chapter III. A knowledge of elementary group theory is also needed. Appreciation of some of the results will be enhanced by a familiarity with Fourier transforms.

The theory of representations of finite groups is an elegant theory with its own intrinsic interest. It is also a finite dimensional model of a number of theories which outwardly appear to be quite different; for example, Fourier series, Fourier transforms, topological groups, and abstract harmonic analysis. We introduce the subject here mainly because of its utility in finding the principal axes of symmetric mechanical systems.

We have to assume a modest knowledge of group theory. In order to be specific about what is required we state the required definitions and theorems. We do not give proofs for the theorems. They are not difficult and can be considered to be exercises, or their proofs can be found in the first few pages of any standard treatment of group theory.

Definition. A *group G* is a set of elements in which a law of combination is defined having the following properties:

(1) If **a**, **b** $\in G$, then **ab** is uniquely defined by the law of combination and is an element of G.

(2) **(ab)c** = **a(bc)**, for all **a**, **b**, **c** $\in G$.

(3) There exists an element **e** $\in G$, called the *unit element*, such that **ea** = **ac** = **a** for each **a** $\in G$.

(4) For each **a** $\in G$ there exists an element **a**$^{-1}$, called the *inverse* of **a**, such that **a**$^{-1}$**a** = **aa**$^{-1}$ = **e**.

Although the law of combination is written multiplicatively, this does not mean that it has anything to do with multiplication. For example, the field of rational numbers is a group under addition, and the set of positive rational numbers is a group under multiplication. A vector space is a group under vector addition. If the condition

(5) $\mathbf{ab} = \mathbf{ba}$

is also satisfied, the group is said to be *commutative* or *abelian*.

The number of elements in a group is called the *order* of the group. We restrict our attention to groups of finite order.

A subset of a group which satisfies the group axioms with the same law of combination is called a *subgroup*. Let G be a group and S a subgroup of G. For each $\mathbf{a} \in G$, the set of all $\mathbf{b} \in G$ such that $\mathbf{a}^{-1}\mathbf{b} \in S$ is called a left coset of S defined by $\mathbf{a}$. By $\mathbf{a}S$ we mean the set of all products of the form $\mathbf{ac}$ where $\mathbf{c} \in S$. Then $\mathbf{a}^{-1}\mathbf{b} \in S$ is equivalent to the condition $\mathbf{b} \in \mathbf{a}S$; that is, $\mathbf{a}S$ is the left coset of S determined by $\mathbf{a}$. Two left cosets are equal, $\mathbf{a}S = \mathbf{b}S$, if and only if $S = \mathbf{a}^{-1}\mathbf{b}S$ or $\mathbf{a}^{-1}\mathbf{b} \in S$; that is, if and only if $\mathbf{b} \in \mathbf{a}S$. The number of elements in each coset of S is equal to the order of S. A right coset of S is of the form $S\mathbf{a}$.

Theorem 9.1. *If G is of finite order and S is a subgroup of G, the order of S divides the order of G.* □

If S is a subgroup such that its right and left cosets are equal—that is, $\mathbf{a}S = S\mathbf{a}$ for each $\mathbf{a} \in G$—then S is said to be an *invariant* or *normal* subgroup of G. If S is normal, then $(\mathbf{a}S)(\mathbf{b}S) = \mathbf{a}(S\mathbf{b})S = \mathbf{a}(\mathbf{b}S)S = (\mathbf{ab})SS = \mathbf{ab}S$ so that the product of two cosets is a coset.

Theorem 9.2. *If S is a normal subgroup of G, the cosets of S form a group under the law of combination $(\mathbf{a}S)(\mathbf{b}S) = \mathbf{ab}S$.* □

If S is a normal subgroup of G, the group of cosets is called the *factor group* of G by S and is denoted by G/S.

If G_1 and G_2 are groups, a mapping f of G_1 into G_2 is called a *homomorphism* if $f(\mathbf{ab}) = f(\mathbf{a})f(\mathbf{b})$ for all $\mathbf{a}$, $\mathbf{b} \in G_1$. Notice that the law of combination on the left is that in G_1, while the law of combination on the right is that in G_2. If the homomorphism is one-to-one, it is called an *isomorphism*. If $\mathbf{e}$ is the unit element in G_2, the set of all $\mathbf{a} \in G_1$ such that $f(\mathbf{a}) = \mathbf{e}$ is called the *kernel* of the homomorphism.

Theorem 9.3. *(The homomorphism theorem). If G_2 is the image of G_1 under the homomorphism f and K is the kernel of f, then G_2 is isomorphic to G_1/K, where $f(a) \in G_2$ corresponds to $aK \in G_1/K$.* □

An isomorphism of a group onto itself is called an *automorphism*. For each fixed $\mathbf{a} \in G$, the mapping $f_{\mathbf{a}}$ of G onto itself defined by $f_{\mathbf{a}}(\mathbf{x}) = \mathbf{a}^{-1}\mathbf{x}\mathbf{a}$ is an automorphism. It is called an *inner automorphism*.

Two group elements $\mathbf{b}_1$ and $\mathbf{b}_2$ are said to be *conjugate* if there is an $\mathbf{a} \in G$ such that $\mathbf{a}^{-1}\mathbf{b}_1\mathbf{a} = \mathbf{b}_2$. For each $\mathbf{b} \in G$, the set of conjugates of $\mathbf{b}$ is called the *conjugate class* determined by $\mathbf{b}$. The conjugate class determined by $\mathbf{b}$ is the set of all images of $\mathbf{b}$ under all possible inner automorphisms. The *normalizer* N_b of $\mathbf{b}$ is the set of all $\mathbf{a}$ such that the inner automorphism $f_\mathbf{a}$ leaves $\mathbf{b}$ fixed.

Theorem 9.4. *The normalizer N_b is a subgroup of G. The number of conjugates of $\mathbf{b}$ is equal to the number of left cosets of N_b. The number of elements in conjugate class is a divisor of the finite order of the group.* $\square$

We have already met several important groups. The set of all $n \times n$ matrices with non-zero determinant forms a group under matrix multiplication. This group is infinite. There are several important subgroups; the matrices with determinant ± 1, the orthogonal matrices, etc. These are also infinite. There are also a large number of finite subgroups. Given a finite group, representation theory seeks to find groups of matrices which are models, in some sense, of the given group.

Definition. Let G be a given finite group. A *representation* of G is a group $D(G)$ of square matrices (under the operation of matrix multiplication) with a homomorphism mapping G onto $D(G)$. Notice that the homomorphism must be specified and is part of the representation. If the mapping is an isomorphism, the representation is said to be *faithful*. The matrices in $D(G)$ represent linear transformations on a vector space V over C. The corresponding linear transformation also form a group homomorphic to G, and we also say that this group of transformations represents G. The dimension of V is called the *dimension* of the representation.

If $\mathbf{a} \in G$, we denote by $D(\mathbf{a})$ the matrix in $D(G)$ corresponding to $\mathbf{a}$ under the homomorphism of G onto $D(G)$. $\sigma_\mathbf{a}$ will denote the corresponding linear linear transformation on V.

Since $\sigma_\mathbf{a} = \sigma_\mathbf{a}\sigma_{\mathbf{b}^{-1}}\sigma_\mathbf{b}$ and the rank of a product is less than or equal to the rank of any factor, we have $\rho(\sigma_\mathbf{a}) \leq \rho(\sigma_\mathbf{b})$. Similarly, we must have $\rho(\sigma_\mathbf{b}) \leq \rho(\sigma_\mathbf{a})$, and hence the ranks of all matrices in $D(G)$ must be equal. Let their common rank be r. Then $\sigma_\mathbf{a}(V) = S_\mathbf{a}$ is of dimension r, and $\sigma_{\mathbf{a}^2}(V) = \sigma_\mathbf{a}(\sigma_\mathbf{a}(V)) = \sigma_\mathbf{a}(S_\mathbf{a})$ is also of dimension r. Since $\sigma_\mathbf{a}(S_\mathbf{a}) \subset S_\mathbf{a}$, we have $\sigma_\mathbf{a}(S_\mathbf{a}) = S_\mathbf{a}$. Also $S_\mathbf{a} = \sigma_\mathbf{a}(V) = \sigma_\mathbf{b}(\sigma_{\mathbf{b}^{-1}}\sigma_\mathbf{a}(V)) \subset \sigma_\mathbf{b}(V) = S_\mathbf{b}$. Similarly, $S_\mathbf{b} \subset S_\mathbf{a}$ so that $S_\mathbf{a} = S_\mathbf{b}$. This means that all linear transformations representing G are automorphisms of some subspaces S of dimension r in V, and $\sigma_\mathbf{a}(V) = S$ for all $\mathbf{a} \in G$. Thus, we may as well restrict our attention to S, and in the future we assume, without loss of generality, that the matrices and transformations representing a group are non-singular, that is, $S = V$.

A subspace $U \subset V$ such that $\sigma_\mathbf{a}(U) \subset U$ for all $\mathbf{a} \in G$ is called an *invariant subspace* of V under $D(G)$. Since σ_a is non-singular, actually $\sigma_a(U) = U$. This means that U is also a representation space of G over C. If V has a proper invariant subspace we say that V and the representation are *reducible*. Otherwise the representation and the space are said to be *irreducible*. If there is another proper invariant subspace W such that $V = U \oplus W$, we say that V and the representation are *completely reducible*. Notice that if U_1 and U_2 are invariant under $D(G)$, then $U_1 \cap U_2$ is also invariant under $D(G)$.

Theorem 9.5. *For a finite group G a reducible representation is completely reducible.*

PROOF. Let U be an invariant subspace. Let T be any subspace complementary to U, and let π be the projection of V onto U along T. Since $\sigma_a(U) = U$, $\pi\sigma_a\pi = \sigma_a\pi$. Thus, in a complicated product of σ's and π's, all π's to the left of the last one can be omitted; for example, $\pi\sigma_a\pi\sigma_b\pi\sigma_c = \sigma_a\sigma_a\pi\sigma_c$. Consider

$$\tau = \frac{1}{g}\sum_a \sigma_{a^{-1}}\pi\sigma_a, \tag{9.1}$$

where g is the order of G, and the sum is taken over all $a \in G$.

An expression like (9.1) is called a *symmetrization* of π. τ has the important property that it commutes with each σ_a whereas π might not. Specifically,

$$\tau\sigma_b = \frac{1}{g}\sum_a \sigma_{a^{-1}}\pi\sigma_a\sigma_b = \frac{1}{g}\sum_a \sigma_b\sigma_{b^{-1}a^{-1}}\pi\sigma_{ab}$$

$$= \sigma_b\frac{1}{g}\sum_a \sigma_{(ab)^{-1}}\pi\sigma_{ab} = \sigma_b\tau. \tag{9.2}$$

The reasoning behind this conclusion is worth examining in detail. If $G = \{a_1, a_2, \ldots, a_g\}$ is the list of the elements in G written out explicitly and b is any element in G, then $\{a_1b, a_2b, \ldots, a_gb\}$ must also be a list (in a different order) of the elements in G because of axioms (1) and (4). Thus, as a runs through the group, ab also runs through the group and $\sum_a \sigma_{(ab)^{-1}}\pi\sigma_{ab} = \sum_a \sigma_{a^{-1}}\pi\sigma_a$. Also notice that this conclusion does not depend on the condition that π is a projection. The symmetrization of any linear transformation would commute with each σ_a.

Now,

$$\tau\pi = \frac{1}{g}\sum_a \sigma_{a^{-1}}\pi\sigma_a\pi = \frac{1}{g}\sum_a \sigma_{a^{-1}}\sigma_a\pi = \frac{1}{g}\sum_a \pi = \pi, \tag{9.3}$$

and

$$\pi\tau = \frac{1}{g}\sum_a \pi\sigma_{a^{-1}}\pi\sigma_a = \frac{1}{g}\sum_a \sigma_{a^{-1}}\pi\sigma_a = \tau. \tag{9.4}$$

Among other things, this shows that τ and π have the same rank. Since $\tau(V) = \pi\tau(V) \subset U$, this means that $\tau(V) = U$. Then $\tau^2 = \tau(\pi\tau) = (\tau\pi)\tau = \pi\tau = \tau$, so that τ is a projection of V onto U.

Let W be the kernel of τ. Then, for each $\mathbf{a} \in G$, $\tau \sigma_{\mathbf{a}}(W) = \sigma_{\mathbf{a}} \tau(W) = \sigma_{\mathbf{a}}(0) = 0$. Thus $\sigma_{\mathbf{a}}(W) \subset W$ and W is an invariant subspace. Finally, it is easily seen that $V = U \oplus W$. Thus, the representation is completely reducible. □

The importance of complete reducibility is that the representation on V induces two representations, one on U and one on W. If a basis $\{\alpha_1, \ldots, \alpha_r, \beta_1, \ldots, \beta_{n-r}\}$ of V is chosen so that $\{\alpha_1, \ldots, \alpha_r\}$ is a basis of U and $\{\beta_1, \ldots, \beta_{n-r}\}$ is a basis of W, then $D(\mathbf{a})$ is of the form

$$D(\mathbf{a}) = \begin{bmatrix} D^1(\mathbf{a}) & 0 \\ 0 & D^2(\mathbf{a}) \end{bmatrix}, \tag{9.5}$$

where $D^1(\mathbf{a})$ is an $r \times r$ matrix and $D^2(\mathbf{a})$ is an $n - r \times n - r$ matrix. $D^1(\mathbf{a})$ represents $\sigma_{\mathbf{a}} \pi$ on U, and $D^2(\mathbf{a})$ represents $\sigma_{\mathbf{a}}(1 - \pi)$ on W. The set $D^1(G) = \{D^1(\mathbf{a}) \mid \mathbf{a} \in G\}$ is a representation of G on U, and $D^2(G) = \{D^2(\mathbf{a}) \mid \mathbf{a} \in G\}$ is a representation of G on W. We say that $D(G)$ is the *direct sum* of $D^1(G)$ and $D^2(G)$ and we write $D(G) = D^1(G) + D^2(G)$.

If either the representation on U or the representation on W is reducible, we can decompose it into a direct sum of two others. We can proceed in this fashion step by step, and the process must ultimately terminate since at each step we obtain subspaces of smaller dimensions. When that point is reached we will have decomposed $D(G)$ into a direct sum of irreducible representations. If U is an invariant subspace obtained in one decomposition of V and U' is an invariant subspace obtained in another, then $U \cap U'$ is also invariant. If U and U' are both irreducible, then either $U \cap U' = \{0\}$ or $U = U'$. Thus, the irreducible subspaces obtained are unique and independent of the particular order of steps in decomposing V. We see, then, that the irreducible representations will be the ultimate building blocks of all representations of finite groups.

Although the irreducible invariant subspaces of V are unique, the matrices corresponding to the group elements are not unique. They depend on the choices of the bases in these subspaces. We say that two groups of matrices, $D(G) = \{D(\mathbf{a}) \mid \mathbf{a} \in G\}$ and $D'(G) = \{D'(\mathbf{a}) \mid \mathbf{a} \in G\}$. are *equivalent* representations of G if there is a non-singular matrix P such that $D'(\mathbf{a}) = P^{-1}D(\mathbf{a})P$ for every $\mathbf{a} \in G$. In particular, if the two groups of matrices represent the same group of linear transformations on V, then they are equivalent and P is the matrix of transition. But the definition of equivalence allows another interpretation.

Let V and V' be vector spaces over C both of dimension n. Let f be a one-to-one linear transformation mapping V' onto V, and let P be the matrix representing f. We can define a linear transformation $\tau_{\mathbf{a}}$ on V' by the rule

$$\tau_{\mathbf{a}} = f^{-1}\sigma_{\mathbf{a}}f. \tag{9.6}$$

It is easy to show that the set $\{\tau_a\}$ defined in this way is a group and that it is isomorphic to $\{\sigma_a\}$. The groups of matrices representing these linear transformations are also equivalent.

If a representation $D(G)$ is given arbitrarily, it will not necessarily look like the representation in formula (9.5). However, if $D(G)$ is equivalent to a representation that looks like (9.5), we also call that representation reducible. The importance of this notion of equivalence is that there are only finitely many inequivalent irreducible representations for each finite group. Our task is to prove this fact, to describe at least one representation for each class of equivalent irreducible representations, and to find an effective procedure for decomposing any representation into its irreducible components.

Theorem 9.6 (*Schur's lemma*). *Let $D^1(G)$ and $D^2(G)$ be two irreducible representations. If T is any matrix such that $TD^1(\mathbf{a}) = D^2(\mathbf{a})T$ for all $\mathbf{a} \in G$, then either $T = 0$ or T is non-singular and $D^1(G)$ and $D^2(G)$ are equivalent.*

PROOF. Let V_1 of dimension m be the representation space of $D^1(G)$, and let V_2 of dimension n be the representation space of $D^2(G)$. Then T must have n rows and m columns, and so may be thought of as representing a linear transformation f of V_1 into V_2. Let $\sigma_{1,a}$ be the linear transformation on V_1 represented by $D^1(\mathbf{a})$ for $\mathbf{a} \in G$, and let $\sigma_{2,a}$ be the linear transformation on V_2 represented by $D^2(\mathbf{a})$. Since $f\sigma_{1,a} = \sigma_{2,a}f$, $\sigma_{2,a}[f(V_1)] = f[\sigma_{1,a}(V_1)] = f(V_1)$ so that $f(V_1)$ is an invariant subspace of V_2. Since V_2 is irreducible, either $f(V_1) = 0$ and hence $T = 0$, or else $f(V_1) = V_2$.

If $\alpha \in f^{-1}(0)$, then $f[\sigma_{1,a}(\alpha)] = \sigma_{2,a}[f(\alpha)] = \sigma_{2,a}(0) = 0$, so that $\sigma_{1,a}(\alpha) \in f^{-1}(0)$. Thus $f^{-1}(0)$ is an invariant subspace of V_1. Since V_1 is irreducible, either $f^{-1}(0) = 0$, or else $f^{-1}(0) = V_1$ in which case $T = 0$.

Thus, either $f(V_1) = 0$, $f^{-1}(0) = V_1$, and $T = 0$; or else $f(V_1) = V_2$, $f^{-1}(0) = 0$, and T is non-singular. In the latter case the representations are equivalent. □

Theorem 9.7. *Let $D(G)$ be an irreducible representation over the complex numbers. If T is any matrix such that $TD(\mathbf{a}) = D(\mathbf{a})T$ for all $\mathbf{a} \in G$, then $T = \lambda I$ where λ is a complex number.*

PROOF. Let λ be an eigenvalue of T. Then $(T - \lambda I)D(\mathbf{a}) = D(\mathbf{a})(T - \lambda I)$ and, by Theorem 9.6, $T - \lambda I$ is either 0 or non-singular. Since $T - \lambda I$ is singular, $T = \lambda I$. □

Theorem 9.8. *If $D(G)$ is an irreducible representation such that any two matrices in $D(G)$ commute, then all the matrices in $D(G)$ are of first order.*

PROOF. By Theorem 9.7, all matrices in $D(G)$ must be of the form λI. But a set of matrices of this form can be irreducible only if all the matrices are of first order. □

With the proof of Schur's lemma and the theorems that follow immediately from it, the representation spaces have served their purpose. We have no further need for the linear transformations and we have to be specific about the elements in the representing matrices. Let $D(\mathbf{a}) = [a_{ij}(\mathbf{a})]$.

Theorem 9.9.   *Every representation of a finite group over the field of complex numbers is equivalent to a representation in which all matrices are unitary.*

PROOF.   Let $D(G)$ be the representation of G. Consider

$$H = \sum_a D(\mathbf{a})^*D(\mathbf{a}), \tag{9.7}$$

where $D(\mathbf{a})^*$ is the conjugate transpose of $D(\mathbf{a})$. Each $D(\mathbf{a})^*D(\mathbf{a})$ is a positive definite Hermitian form and H, as the sum of positive definite Hermitian forms, is a positive definite Hermitian form. Thus, there is a non-singular matrix P such that $P^*HP = I$. But then

$$(P^{-1}D(\mathbf{a})P)^*(P^{-1}D(\mathbf{a})P) = P^*D(\mathbf{a})^*P^{*-1}P^{-1}D(\mathbf{a})P$$
$$= P^*D(\mathbf{a})^*HD(\mathbf{a})P$$
$$= P^*\left(\sum_b D(\mathbf{a})^*D(\mathbf{b})^*D(\mathbf{b})D(\mathbf{a})\right)P$$
$$= P^*\left(\sum_b D(\mathbf{ba})^*D(\mathbf{ba})\right)P$$
$$= P^*HP = I.$$

Thus, each $P^{-1}D(\mathbf{a})P$ is unitary, as we wished to show. $\square$

For any matrix $A = [a_{ij}]$, $S(A) = \sum_{i=1}^n a_{ii}$ is called the *trace* of A. Since $S(AB) = \sum_{i=1}^n (\sum_{j=1}^n a_{ij}b_{ji}) = \sum_{j=1}^n (\sum_{i=1}^n b_{ji}a_{ij}) = S(BA)$, the trace of a product does not depend on the order of the factors. Thus $S(P^{-1}AP) = S(APP^{-1}) = S(A)$ and we see that the trace is left invariant under similarity. If $D^1(G)$ and $D^2(G)$ are equivalent representations, then $S(D^1(\mathbf{a})) = S(D^2(\mathbf{a}))$ for all $\mathbf{a} \in G$. Thus, the trace, as a function of the group element, is the same for any of a class of equivalent representations. For simplicity we write $S(D(\mathbf{a})) = S_D(\mathbf{a})$ or, if the representation has an index, $S(D^r(\mathbf{a})) = S^r(\mathbf{a})$. If $\mathbf{a}$ and $\mathbf{a}'$ are conjugate elements, $\mathbf{a}' = \mathbf{b}^{-1}\mathbf{ab}$, then $S_D(\mathbf{a}') = S(D(\mathbf{b}^{-1})D(\mathbf{a})D(\mathbf{b})) = S_D(\mathbf{a})$ so that all elements in the same conjugate class of G have the same trace. For a given representation, the trace is a function of the conjugate class. If the representation is irreducible, the trace is called a *character* of G and is written $S^r(\mathbf{a}) = \chi^r(\mathbf{a})$.

Let $D^r(G)$ and $D^s(G)$ be two irreducible representations. Then consider the matrix

$$T = \sum_a D^r(\mathbf{a}^{-1})XD^s(\mathbf{a}) \tag{9.8}$$

where X is any matrix for which the products in question are defined. Then

$$TD^s(\mathbf{b}) = \sum_\mathbf{a} D^r(\mathbf{a}^{-1})XD^s(\mathbf{a})D^s(\mathbf{b})$$

$$= D^r(\mathbf{b}) \sum_\mathbf{a} D^r(\mathbf{b}^{-1}\mathbf{a}^{-1})XD^s(\mathbf{ab})$$

$$= D^r(\mathbf{b})T. \qquad (9.9)$$

Thus, either $D^r(G)$ and $D^s(G)$ are inequivalent and $T = 0$ for all X, or $D^r(G)$ and $D^s(G)$ are equivalent. We adopt the convention that irreducible representations with different indices are inequivalent. If $r = s$ we have $T = \omega I$, where ω will depend on X. Let $D^r(\mathbf{a}) = [a^r_{ij}(\mathbf{a})]$, $T = [t_{ij}]$, and let X be zero everywhere but for a 1 in the jth row, kth column. Then

$$t_{il} = \sum_\mathbf{a} a^r_{ij}(\mathbf{a}^{-1})a^s_{kl}(\mathbf{a}) = 0 \qquad \text{for} \qquad r \neq s. \qquad (9.10)$$

If $r = s$, then ω is a function of r, j, and k and we have

$$t_{il} = \sum_\mathbf{a} a^r_{ij}(\mathbf{a}^{-1})a^r_{kl}(\mathbf{a}) = \omega^r_{jk}\,\delta_{il}. \qquad (9.11)$$

Notice that ω^d_{jk} is independent of i and l. But we also have

$$t_{il} = \sum_\mathbf{a} a^r_{ij}(\mathbf{a}^{-1})a^r_{kl}(\mathbf{a})$$

$$= \sum_\mathbf{a} a^r_{kl}(\mathbf{a})a^r_{ij}(\mathbf{a}^{-1})$$

$$= \sum_\mathbf{a} a^r_{kl}(\mathbf{a}^{-1})a^r_{ij}(\mathbf{a}) = t_{kj} = \omega^r_{li}\,\delta_{jk} \qquad (9.12)$$

where ω^r_{li} is independent of j and k. Thus, we see that $\omega^r_{jk}\delta^r_{lj} = \omega^r_{li}\delta_{jk} = 0$ unless $k = j$ and $l = i$, in which case $\omega^r_{jj} = \omega^r_{ii}$. But ω^r_{jj} is independent of i and ω^r_{ij} is independent of j. Thus, we may write

$$\sum_\mathbf{a} a^r_{ij}(\mathbf{a}^{-1})a^r_{kl}(\mathbf{a}) = \omega^r\,\delta_{il}\,\delta_{jk}. \qquad (9.13)$$

In order to evaluate ω^r set $k = j$, $l = i$, and sum over j;

$$\sum_{j=1}^{n_r}\sum_\mathbf{a} a^r_{ij}(\mathbf{a}^{-1})p^r_{ji}(\mathbf{a}) = n_r\omega^r \qquad (9.14)$$

where n_r is the dimension of the representation $D^r(G)$. But since $\sum_{j=1}^{n_r}a^r_{ij}(\mathbf{a}^{-1})a^r_{ji}(\mathbf{a})$ is a diagonal element of the product $D^r(\mathbf{a}^{-1})D^r(\mathbf{a}) = D^r(\mathbf{e}) = I^r$ we have

$$n_r\omega^r = \sum_\mathbf{a}\sum_{j=1}^{n} a^r_{ij}(\mathbf{a}^{-1})p^r_{ji}(\mathbf{a}) = \sum_\mathbf{a} 1 = g. \qquad (9.15)$$

Thus,

$$\omega^r = \frac{g}{n_r}. \tag{9.16}$$

All the information so far obtained can be expressed in the single formula

$$\sum_a a_{ij}^r(\mathbf{a}^{-1})a_{kl}^s(\mathbf{a}) = \frac{g}{n_r} \delta_{jk} \delta_{il} \delta_{rs}. \tag{9.17}$$

Multiply (9.17) by $a_{lt}^s(\mathbf{b})$ and sum over l. Then

$$\sum_{l=1}^{n_s} \sum_a a_{ij}^r(\mathbf{a}^{-1})a_{kl}^s(\mathbf{a})a_{lt}^s(\mathbf{b}) = \sum_{l=1}^{n_s} \frac{g}{n_r} a_{lt}^s(\mathbf{b}) \delta_{jk} \delta_{il} \delta_{rs},$$

or   (9.18)

$$\sum_a a_{ij}^r(\mathbf{a}^{-1})a_{kt}^s(\mathbf{ab}) = \frac{g}{n_r} a_{it}^s(\mathbf{b}) \delta_{jk} \delta_{rs}.$$

In (9.18) set $i = j$, $t = k$, and sum over j and k:

$$\sum_{j=1}^{n_r} \sum_{k=1}^{n_s} \sum_a a_{jj}^r(\mathbf{a}^{-1})a_{kk}^s(\mathbf{ab}) = \sum_{j=1}^{n_r} \sum_{k=1}^{n_s} \frac{g}{n_r} a_{jk}^s(\mathbf{b}) \delta_{jk} \delta_{rs},$$

or   (9.19)

$$\sum_a \chi^r(\mathbf{a}^{-1})\chi^s(\mathbf{ab}) = \frac{g}{n_r} \chi^s(\mathbf{b}) \delta_{rs}.$$

In particular, it we take $\mathbf{b} = \mathbf{e}$, we have

$$\sum_a \chi^r(\mathbf{a}^{-1})\chi^s(\mathbf{a}) = \frac{g}{n_r} n_s \delta_{rs} = g \delta_{rs}. \tag{9.20}$$

Actually, formula (9.20) could have been obtained directly from formula (9.17) by setting $i = j$, $l = k$, and summing over j and k.

Let $D(G)$ be a direct sum of a finite number of irreducible representations, the representation $D^r(G)$ occurring c_r times, where c_r is a non-negative integer. Then

$$S_D(\mathbf{a}) = \sum_r c_r \chi^r(\mathbf{a}),$$

from which we obtain

$$\sum_a \chi^r(\mathbf{a}^{-1})S_D(\mathbf{a}) = g c_r. \tag{9.21}$$

Furthermore,

$$\sum_a S_D(\mathbf{a}^{-1})S_D(\mathbf{a}) = g \sum_r c_r^2, \tag{9.22}$$

so that a representation is irreducible if and only if

$$\sum_a S_D(\mathbf{a}^{-1})S_D(\mathbf{a}) = g. \tag{9.23}$$

In case the representation is unitary, $a_{ij}^r(\mathbf{a}^{-1}) = \overline{a_{ji}^r(\mathbf{a})}$, so that the relations (9.17) through (9.23) take on the following forms:

$$\sum_{\mathbf{a}} \overline{a_{ji}^r(\mathbf{a})} a_{kl}^s(\mathbf{a}) = \frac{g}{n_r} \delta_{jk}\, \delta_{il}\, \delta_{rs}, \qquad (9.17)'$$

$$\sum_{\mathbf{a}} \overline{a_{ji}^r(\mathbf{a})} a_{kt}^s(\mathbf{ab}) = \frac{g}{n_r} a_{it}^s(\mathbf{b})\, \delta_{jk}\, \delta_{rs}, \qquad (9.18)'$$

$$\sum_{\mathbf{a}} \overline{\chi^r(\mathbf{a})} \chi^s(\mathbf{ab}) = \frac{g}{n_r} \chi^s(\mathbf{b})\, \delta_{rs}, \qquad (9.19)'$$

$$\sum_{\mathbf{a}} \overline{\chi^r(\mathbf{a})} \chi^s(\mathbf{a}) = g\, \delta_{rs}, \qquad (9.20)'$$

$$\sum_{\mathbf{a}} \overline{\chi^r(\mathbf{a})} S_D(\mathbf{a}) = g c_r, \qquad (9.21)'$$

$$\sum_{\mathbf{a}} \overline{S_D(\mathbf{a})} S_D(\mathbf{a}) = g \sum_r c_r^{\,2}, \qquad (9.22)'$$

$$\sum_{\mathbf{a}} \overline{S_D(\mathbf{a})} S_D(\mathbf{a}) = g \text{ if and only if } D(G) \text{ is irreducible.} \qquad (9.23)'$$

Formulas (9.19)′ through (9.23)′ hold for any form of the representations, since each is equivalent to a unitary representation and the trace is the same for equivalent representations.

We have not yet shown that a single representation exists. We now do this, and even more. We construct a representation whose irreducible components are equivalent to all possible irreducible representations. Let $G = \{\mathbf{a}_1, \mathbf{a}_2, \ldots, \mathbf{a}_g\}$. Consider the set V of formal sums of the form

$$\alpha = x_1 \mathbf{a}_1 + x_2 \mathbf{a}_2 + \cdots + x_g \mathbf{a}_g, \; x_i \in C. \qquad (9.24)$$

Addition is defined by adding corresponding coefficients, and scalar multiplication is defined by multiplying each coefficient by the scalar factor. With these definitions, V forms a vector space over C. For each $\mathbf{a}_i \in G$ we can define a linear transformation on V by the rule

$$\mathbf{a}_i(\alpha) = x_1(\mathbf{a}_i\mathbf{a}_1) + x_2(\mathbf{a}_i\mathbf{a}_2) + \cdots + x_g(\mathbf{a}_i\mathbf{a}_g). \qquad (9.25)$$

$\mathbf{a}_i$ induces a linear transformation that amounts to a permutation of the basis elements. Since $(\mathbf{a}_j\mathbf{a}_i)(\alpha) = \mathbf{a}_j(\mathbf{a}_i(\alpha))$, the set of linear transformations thus obtained forms a representation. Denote the set of matrices representing these linear transformations by $R(G)$. $R(G)$ is called the *regular representation*, and any representation equivalent to $R(G)$ is called a regular representation.

Let $R(\mathbf{a}) = [r_{ij}(\mathbf{a})]$. Then $r_{ij}(\mathbf{a}) = 1$ if $\mathbf{aa}_j = \mathbf{a}_i$, and otherwise $r_{ij}(\mathbf{a}) = 0$. Since $\mathbf{aa}_j = \mathbf{a}_j$ if and only if $\mathbf{a} = \mathbf{e}$, we have

$$S_R(\mathbf{e}) = g, \tag{9.26}$$

$$S_R(\mathbf{a}) = 0 \quad \text{for} \quad \mathbf{a} \neq \mathbf{e}. \tag{9.27}$$

Thus,

$$\sum_{\mathbf{a}} \chi^r(\mathbf{a}^{-1}) S_R(\mathbf{a}) = \chi^r(\mathbf{e}) S_R(\mathbf{e}) = gn_r, \tag{9.28}$$

so that by (9.21) a representation equivalent to $D^r(G)$ occurs n_r times in the regular representation.

Theorem 9.10. *There are only finitely many inequivalent irreducible representations.*

PROOF. Every irreducible representation is equivalent to a component of the regular representation. The regular representation is finite dimensional and has only a finite number of components. $\square$

Furthermore,

$$\sum_{\mathbf{a}} S_R(\mathbf{a}^{-1}) S_R(\mathbf{a}) = S_R(\mathbf{e}) S_R(\mathbf{e}) = g^2 \tag{9.29}$$

so that by (9.22)

$$\sum_r n_r^2 = g. \tag{9.30}$$

Let C_i denote a conjugate class in G and h_i the number of elements in the class C_i. Let m be the number of classes, and m' the number of inequivalent representations. Since the characters are constant on conjugate classes, they are really functions of the classes, and we can define $\chi_i^r = \chi^r(\mathbf{a})$ for any $\mathbf{a} \in C_i$. With this notation formula (9.20)′ takes the form

$$\sum_{i=1}^{m} h_i \overline{\chi_i^r} \chi_i^s = g\delta_{rs}. \tag{9.31}$$

Thus, the m' m-tuples $(\sqrt{h_1}\chi_1^r, \sqrt{h_2}\chi_2^r, \ldots, \sqrt{h_m}\chi_m^r)$ are orthogonal and, hence, linearly independent. This gives $m' \leq m$.

We can introduce a multiplication in V determined by the underlying group relations. If $\alpha = \sum_{i=1}^{g} x_i \mathbf{a}_i$ and $\beta = \sum_{j=1}^{g} y_j \mathbf{a}_j$, we define

$$\alpha\beta = \left(\sum_{i=1}^{g} x_i \mathbf{a}_i \right) \left(\sum_{j=1}^{g} y_j \mathbf{a}_j \right)$$

$$= \sum_{k=1}^{g} \left(\sum_{\mathbf{a}_i \mathbf{a}_j = \mathbf{a}_k} x_i y_j \right) \mathbf{a}_k. \tag{9.32}$$

This multiplication is associative and distributive with respect to the previously detained addition. The unit element is $\mathbf{e}$. Multiplication is not commutative unless G is commutative.

Consider the elements

$$\gamma_i = \sum_{a \in C_i} a$$

$$\gamma_i b = \sum_{a \in C_i} ab = b \sum_{a \in C_i} b^{-1}ab = b\gamma_i \qquad (9.33)$$

and, hence, $\gamma_i \alpha = \alpha \gamma_i$ for all $\alpha \in V$. Similarly, any sum of the γ_i commutes with every element of V.

The importance of the γ_i is that the converse of the above statement is also true. Any element in V which commutes with every element in V is a linear combination of the γ_i. Let $\gamma = \sum_{i=1}^{g} c_i a_i$ be a vector such that $\gamma \alpha = \alpha \gamma$ for every $\alpha \in V$. Then, in particular, $\gamma b = b\gamma$, or $b^{-1}\gamma b = \gamma$ for every $b \in G$. If $b^{-1}a_i b$ is denoted by a_j, we have

$$\gamma = b^{-1}\left(\sum_{i=1}^{g} c_i a_i\right) b = \sum_{i=1}^{g} c_i b^{-1} a_i b$$

$$= \sum_{i=1}^{g} c_i a_j = \sum_{j=1}^{g} c_j a_j,$$

so that $c_i = c_j$. This means that all basis elements from a given conjugate class must have the same coefficients. Thus, in fact,

$$\gamma = \sum_{i=1}^{m} c_i \gamma_i.$$

Since $\gamma_i \gamma_j$ also commutes with every element in V, we must have

$$\gamma_i \gamma_j = \sum_{k=0}^{m} c_{jk}^{i} \gamma_k. \qquad (9.34)$$

Now, let $C_i^{r} = \sum_{a \in C_i} D^r(a)$. By exactly the same argument as before, C_i^{r} commutes with every matrix in $D^r(G)$. Thus, C_i^{r} must be a diagonal matrix of the form

$$C_i^{r} = \eta_i^{r} I^{r} \qquad (9.35)$$

where I^{r} is the identity of the rth representation. But $S(C_i^{r}) = n_r \eta_i^{r}$ and $S(C_i^{r}) = \sum_{a \in C_i} S(D^r(a)) = h_i \chi_i^{r}$. Thus,

$$\eta_i^{r} = \frac{h_i \chi_i^{r}}{n_r}, \qquad (9.36)$$

and

$$\eta_1^{r} = \frac{h_1 \chi_1^{r}}{n_r} = \frac{n_r}{n_r} = 1, \qquad (9.37)$$

where we agree that C_1 is the conjugate class containing the identity e. Also,

$$C_i^{r} C_j^{r} = \sum_{k=1}^{m} c_{jk}^{i} C_k^{r} \qquad (9.38)$$

where these c_{jk}^i are the same as those appearing in equation (9.34). This means

$$(\eta_i{}^r I^r)(\eta_j{}^r I^r) = \sum_{k=1}^{m} c_{jk}^i \eta_k{}^r I^r,$$

or

$$\eta_i{}^r \eta_j{}^r = \sum_{k=1}^{m} c_{jk}^i \eta_k{}^r. \tag{9.39}$$

In view of equation 9.36, this becomes

$$\frac{h_i \chi_i{}^r}{n_r} \frac{h_j \chi_j{}^r}{n_r} = \sum_{k=1}^{m} c_{jk}^i \frac{h_k \chi_k{}^r}{n_r},$$

or

$$h_i \chi_i{}^r h_j \chi_j{}^r = n_r \sum_{k=1}^{m} c_{jk}^i h_k \chi_k{}^r$$

$$= \chi_1{}^r \sum_{k=1}^{m} c_{jk}^i h_k \chi_k{}^r. \tag{9.40}$$

Thus,

$$\sum_{r=1}^{m'} h_i \chi_i{}^r h_j \chi_j{}^r = \sum_{k=1}^{m} c_{jk}^i h_k \sum_{r=1}^{m'} n_r \chi_k{}^r$$

$$= \sum_{k=1}^{m} c_{jk}^i h_k S_R(\mathbf{a}), \qquad \text{where } \mathbf{a} \in C_k,$$

$$= c_{j1}^i g, \tag{9.41}$$

remembering that C_1 is the conjugate class containing the identity.

• Suppose that C_k contains the inverses of the elements of C_i. Then $\chi_i{}^r = \overline{\chi_k{}^r}$. Also, observe that $\gamma_i \gamma_j$ contains the identity h_i times if C_j contains the inverses of the elements of C_i, and otherwise $\gamma_i \gamma_j$ does not contain the identity. Thus $c_{j1}^i = h_i$ if $j = k$, and $c_{j1}^i = 0$ if $j \neq k$. Thus,

$$\sum_{r=1}^{m'} \overline{\chi_k{}^r} \chi_j{}^r = \frac{g}{h_j} \delta_{jk}. \tag{9.42}$$

Theorem 9.11. *The number of inequivalent irreducible representations of a finite group* G *is equal to the number of conjugate classes in* G.

PROOF. With m the number of conjugate classes and m' the number of inequivalent irreducible representations, we have already shown that $m' \leq m$. Formula (9.42) shows that the m'-tuples $(\chi_i{}^1, \chi_i{}^2, \ldots, \chi_i{}^{m'})$ are mutually orthogonal. Thus $m \leq m'$, and $m = m'$. $\square$

So far the only numbers that can be computed directly from the group G are the c_{jk}^i. Formula (9.39) is the key to an effective method for computing all the relevant numbers. Formula (9.39) can be written as a matrix equation in the form

$$\eta_i^r \begin{bmatrix} \eta_1^r \\ \eta_2^r \\ \cdot \\ \cdot \\ \cdot \\ \eta_m^r \end{bmatrix} = [c_{jk}^i] \begin{bmatrix} \eta_1^r \\ \eta_2^r \\ \cdot \\ \cdot \\ \cdot \\ \eta_m^r \end{bmatrix} \tag{9.43}$$

where $[c_{jk}^i]$ is a matrix with i fixed, j the row index, and k the column index. Thus, η_i^r is an eigenvalue of the matrix $[c_{jk}^i]$ and the vector $(\eta_1^r, \eta_2^r, \ldots, \eta_m^r)$ is an eigenvector for this eigenvalue. This eigenvector is uniquely determined by the eigenvalue if and only if the eigenvalue is a simple solution of the characteristic equation for $[c_{jk}^i]$. For the moment, suppose this is the case. We have already noted that $\eta_1^r = 1$. Thus, normalizing the eigenvector so that $\eta_1^r = 1$ will yield an eigenvector whose components are all the eigenvalues associated with the rth representation.

The computational procedure: For a fixed i find the matrix $[c_{jk}^i]$ and compute its eigenvalues. Each of the m eigenvalues will correspond to one of the irreducible representations. For each simple eigenvalue, find the corresponding eigenvector and normalize it so that the first component is 1. From formulas (9.36) and (9.31) we have

$$\sum_{i=1}^{m} \frac{\overline{\eta_i^r} \eta_i^r}{h_i} = \sum_{i=1}^{m} \frac{h_i \overline{\chi_i^r} \chi_i^r}{n_r^2} = \frac{g}{n_r^2}. \tag{9.44}$$

This gives the dimension of each representation. Knowing this the characters can be computed by means of the formula

$$\chi_i^r = \frac{n_r \eta_i^r}{h_i}. \tag{9.36}$$

Even if all the eigenvalues of $[c_{jk}^i]$ are not simple, those that are may be used in the manner outlined. This may yield enough information to enable us to compute the remaining character values by means of orthogonality relations (9.31) and (9.42). It may be necessary to compute the matrix $[c_{jk}^i]$ for another value of i. Those eigenvectors which have already been obtained are also eigenvectors for this new matrix, and this will simplify the process of finding the eigenvalues and eigenvectors for it.

Theorem 9.12. *The dimension of an irreducible representation divides the order of the group.*

PROOF. Multiplying (9.39) by η_t^r, we obtain

$$\eta_i^r \eta_i^r \eta_j^r = \sum_{k=1}^{m} c_{jk}^i \left(\sum_{p=1}^{m} c_{kp}^t \eta_p^r \right)$$

$$= \sum_{p=1}^{m} \left(\sum_{k=1}^{m} c_{jk}^i c_{kp}^t \right) \eta_p^r. \qquad (9.45)$$

Hence, $\eta_i^r \eta_i^r$ is an eigenvalue of the matrix $[c_{jk}^i][c_{kp}^t]$. If C_t is taken to be the class containing the inverses of the elements in C_i, we have

$$\sum_{i=1}^{m} \frac{g}{h_i} \eta_t^r \eta_i^r = \sum_{i=1}^{m} \frac{g}{h_i} \overline{\eta_i^r} \eta_i^r$$

$$= \frac{1}{n_r^2} \sum_{i=1}^{m} h_i \overline{\chi_i^r} \chi_i^r = \frac{g^2}{n_r^2}. \qquad (9.46)$$

Then $\dfrac{g^2}{n_r^2}$ is an eigenvalue of the matrix $\sum_{i=1}^{m} \dfrac{g}{h_i} [c_{jk}^i][c_{kp}^t]$. All the coefficients of this matrix are integers. Hence, its characteristic polynomial has integral coefficients and leading coefficient 1. A rational solution of such an equation must be an integer. Thus, $\dfrac{g^2}{n_r^2}$ and, hence, $\dfrac{g}{n_r}$ must be an integer. $\square$

It is often convenient to summarize the information about the characters in a table of the form:

	h_1	h_2	$\cdots$	h_m
	C_1	C_2	$\cdots$	C_m
D^1	$\chi_1{}^1$	$\chi_2{}^1$	$\cdots$	$\chi_m{}^1$
D^2	$\chi_1{}^2$	$\chi_2{}^2$	$\cdots$	$\chi_m{}^2$
$\cdot$	$\cdot$	$\cdot$	$\cdots$	$\cdot$
$\cdot$	$\cdot$	$\cdot$	$\cdots$	$\cdot$
$\cdot$	$\cdot$	$\cdot$	$\cdots$	$\cdot$
D^m	$\chi_1{}^m$	$\chi_2{}^m$	$\cdots$	$\chi_m{}^m.$

(9.47)

The rows satisfy the orthogonality relation (9.31) and the columns satisfy the orthogonality relation (9.42):

$$\sum_{i=1}^{m} h_i \overline{\chi_i^r} \chi_i^s = g\delta_{rs}, \qquad (9.31)$$

$$\sum_{r=1}^{m} h_i \overline{\chi_i^r} \chi_j^r = g\delta_{ij}. \qquad (9.42)$$

If some of the characters are known these relations are very helpful in completing the table of characters.

Example. Consider the equilateral triangle shown in Fig. 8. Let (123) denote the rotation of this figure through $120°$; that is, the rotation maps P_1

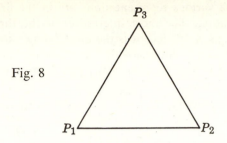

Fig. 8

onto P_2, P_2 onto P_3, and P_3 onto P_1. Similarly, (132) denotes a rotation through 240°. Let (12) denote the reflection that interchanges P_1 and P_2 and leaves P_3 fixed. Similarly, (13) interchanges P_1 and P_3 while (23) interchanges P_2 and P_3. These mappings are called *symmetries* of the geometric figure. We define multiplication of these symmetries by applying first one and then the other; that is, (123)(12) means to interchange P_1 and P_2 and then rotate through 120°. We see that (123)(12) = (13). Including the identity mapping as a symmetry, this defines a group $G = \{e, (123), (132), (12), (13), (23)\}$ of symmetries of the equilateral triangle.

The conjugate classes are $C_1 = \{e\}$, $C_2 = \{(123), (132)\}$, and $C_3 = \{(12), (13), (23)\}$. It is easy to verify that

$$\gamma_2\gamma_2 = 2\gamma_1 + \gamma_2,$$
$$\gamma_2\gamma_3 = 2\gamma_3,$$
$$\gamma_3\gamma_3 = 3\gamma_1 + 3\gamma_2.$$

Thus, we have $[c_{jk}^3]$ is

$$\begin{bmatrix} 0 & 0 & 1 \\ 0 & 0 & 2 \\ 3 & 3 & 0 \end{bmatrix}.$$

The eigenvalues are 0, 3, and -3. Taking the eigenvalue $\eta_3^1 = 3$, we get the eigenvector (1, 2, 3). From (9.44), we get $n_1 = 1$. For the eigenvalue $\eta_3^2 = -3$ we get the eigenvector (1, 2, -3) and $n_2 = 1$. For the eigenvalue $\eta_3^3 = 0$, we get the eigenvector (1, -1, 0) and $n_3 = 2$. Computing the characters by means of (9.36)′, we get the character table

	1 C_1	2 C_2	3 C_3
D^1	1	1	1
D^2	1	1	-1
D^3	2	-1	0.

The dimensions of the various representations are in the first column, the characters of the identity. The most interesting of the three possible irreducible representations is D^3 since it is the only 2-dimensional representation. Since the others are 1-dimensional, the characters are the elements of the corresponding matrices. Among many possibilities we can take

$$D^3(e) = \begin{bmatrix} 1 & 0 \\ 0 & 1 \end{bmatrix} \quad D^3((123)) = \begin{bmatrix} 0 & -1 \\ 1 & -1 \end{bmatrix} \quad D^3((132)) = \begin{bmatrix} -1 & 0 \\ -1 & 1 \end{bmatrix}$$

$$D^3((12)) = \begin{bmatrix} 0 & 1 \\ 1 & 0 \end{bmatrix} \quad D^3((13)) = \begin{bmatrix} -1 & 0 \\ -1 & 1 \end{bmatrix} \quad D^3((23)) = \begin{bmatrix} 1 & -1 \\ 0 & -1 \end{bmatrix}.$$

BIBLIOGRAPHICAL NOTES

The necessary background material in group theory is easily available in G. Birkhoff and S. MacLane, *A Survey of Modern Algebra, Third Edition*, or B. L. van der Waerden, *Modern Algebra, Vol.* 1. More information on representation theory is available in V. I. Smirnov, *Linear Algebra and Group Theory*, and B. L. van der Waerden, *Gruppen von Linearen Transformationen*. F. D. Murnaghan, *The Theory of Group Representations*, is encyclopedic.

EXERCISES

The notation of the example given above is particularly convenient for representing permutations. The symbol (123) is used to represent the permutation, "1 goes to 2, 2 goes to 3, and 3 goes to 1." Notice that the elements appearing in a sequence enclosed by parentheses are cyclically permuted. The symbol (123)(45) means that the elements of $\{1, 2, 3\}$ are permuted cyclically, and the elements of $\{4, 5\}$ are independently permuted cyclically (interchanged in this case). Elements that do not appear are left fixed.

1. Write out the full multiplication table for the group G given in the example above. Is this group commutative?

2. Verify that the set of all permutations described in Chapter III-1 form a group. Write the permutation given as illustration in the notation of this section and verify the laws of combination given. The group of all permutations of a finite set $S = \{1, 2, \ldots, n\}$ is called the *symmetric group on n objects* and is denoted by $\mathfrak{S}_n$. Show that $\mathfrak{S}_n$ is of order $n!$. A subgroup of $\mathfrak{S}_n$ is called a *group of symmetries*, or a *permutation group*.

3. Show that the subset of $\mathfrak{S}_n$ consisting of even permutations forms a subgroup of $\mathfrak{S}_n$. This subgroup is called the *alternating group* and is denoted by $\mathfrak{A}_n$. Show that $\mathfrak{A}_n$ is a normal subgroup.

4. For any group G and any $a \in G$, let $D^1(a)$ be the 1×1 unit matrix, $D^1(a) = [1]$. Show that $D^1(G) = \{D^1(a) \mid a \in G\}$ is a representation of G. This representation is called the *identity representation*.

5. $\mathfrak{S}_3$ be the symmetric group described in the example given above. We showed there that $\mathfrak{S}_3$ has three inequivalent irreducible representations. One of them is the identity representation; another is the 2×2 representation which we described. Find the third one.

6. Show that any 1-dimensional representation of a finite group is always in unitary form.

7. Give the 2×2 irreducible representation of $\mathfrak{S}_3$ in unitary form.

8. Show that a finite group G is commutative if and only if every irreducible representation is of dimension 1.

9. Show that a finite commutative group of order n has n inequivalent irreducible representations.

10. Let G be a cyclic group of order n. Find the n irreducible inequivalent representations of G.

11. There are two non-isomorphic groups of order 4. One is cyclic. The other is of the form $\mathfrak{B} = \{e, a, b, c\}$ where $a^2 = b^2 = c^2 = e$ and $ab = c$, $ac = b$, $bc = a$. $\mathfrak{B}$ is called the *four-group*. Find the four inequivalent irreducible representations for each of these groups.

12. Show that if G is a group of order p^2, where p is a prime number, then G is commutative.

13. Show that all groups of orders, 2, 3, 4, 5, 7, 9, 11, and 13 are commutative.

14. Show that there is just one commutative group for each of the orders 6, 10, 14, 15.

15. Show that a non-commutative group of order 6 must have three irreducible representations, two of dimension 1 and one of dimension 2. Show that this information and the knowledge that one of the representations must be the identity representation determines five of the nine numbers that appear in the character table. How many conjugate classes can there be? What are their orders? Show that we now know enough to determine the remaining elements of the character table. Show that this information determines the group up to an isomorphism; that is, any two non-commutative groups of order 6 must be isomorphic.

16. Show that if every element of a group is of order 1 or 2, then the group must be commutative.

17. There are five groups of order 8; that is, every group of order 8 is isomorphic to one of them. Three of them are commutative and two are non-commutative. Of the three commutative groups, one contains an element of order 8, one contains an element of order 4 and no element of higher order, and one contains elements of order 2 and no higher order. Write down full multiplication tables for these three groups. Determine the associated character tables.

18. There are two non-commutative groups of order 8. One of them is generated by the elements $\{a, b\}$ subject to the relations, a is of order 4, b is of order 2, and $ab = ba^3$. Write out the full multiplication table and determine the associated character table. An example of this group can be obtained by considering the group of symmetries of a square. If the four corners of this square are numbered,

a representation of this group as a permutation group can be obtained. The other group of order 8 is generated by $\{\mathbf{a}, \mathbf{b}, \mathbf{c}\}$ where each is of order 4, $\mathbf{ab} = \mathbf{c}$, $\mathbf{bc} = \mathbf{a}$, $\mathbf{ca} = \mathbf{b}$, and $\mathbf{a}^2 = \mathbf{b}^2 = \mathbf{c}^2$. Show that $\mathbf{ab} = \mathbf{b}^3\mathbf{a} = \mathbf{ba}^3$. Write out the full multiplication table for this group and determine the associated character table. Compare the character tables for these two non-isomorphic groups of order 8.

The above exercises have given us a reservoir of representations of groups of relatively small order. There are several techniques for using these representations to find representations of groups of higher order. The following exercises illustrate some of these techniques.

19. Let G_1 be a group which is the homomorphic image of the group G_2. Let $D(G_1)$ be a representation of G_1. Define a homomorphism of G_2 onto $D(G_1)$ and show that $D(G_1)$ is also a representation of G_2.

20. Consider the two non-commutative groups of order 8 given in Exercise 18. Show that $H = \{e, \mathbf{a}^2\}$ is a normal subgroup (using the appropriate interpretation of the symbol "$\mathbf{a}$" in each case). Show that, in either case, G/H is isomorphic to the four-group $\mathfrak{B}$. Show how we can use this infromation to obtain the four 1-dimensional representations for each of these groups. Show how the characters for the remaining representation can be obtained by use of the orthogonality relations (9.31) and (9.42).

21. In a commutative group every subgroup is a normal subgroup. In Exercise 10 we determined the character tables for a cyclic group. Using this information and the technique of Exercise 19, find the character tables for the three commutative groups of order 8.

22. Show that $\mathfrak{S}_n$ has a 1-dimensional representation in which every element of $\mathfrak{A}_n$ is mapped onto $[1]$ and every element not in $\mathfrak{A}_n$ is mapped onto $[-1]$.

23. Show that if $D^r(G)$ is a representation of G of dimension n, where $D^r(\mathbf{a}) = [a_{ij}^r(\mathbf{a})]$, and $D^s(G)$ is a representation of dimension m, where $D^s(\mathbf{a}) = [a_{kl}^s(\mathbf{a})]$, then $D^{r\times s}(G)$, where $D^{r\times s}(\mathbf{a}) = [a_{ik;jl}^{r,s}(\mathbf{a}) = a_{ij}^r(\mathbf{a})a_{kl}^s(\mathbf{a})]$, is a representation of G of dimension mn. $D^{r\times s}(G)$ is known as the *Kronecker product* of $D^r(G)$ and $D^s(G)$.

24. Let $S^r(\mathbf{a})$ be the trace of $\mathbf{a}$ for the representation $D^r(G)$, $S^s(\mathbf{a})$ the trace for $D^s(G)$, and $S^{r\times s}(\mathbf{a})$ the trace for $D^{r\times s}(G)$. Show that $S^{r\times s}(\mathbf{a}) = S^r(\mathbf{a})S^s(\mathbf{a})$.

25. The commutative group of order 8, with no element of order higher than 2, has the following three rows in the associated character table:

$$
\begin{array}{cccccccc}
1 & -1 & 1 & 1 & -1 & -1 & 1 & -1 \\
1 & 1 & -1 & 1 & -1 & 1 & -1 & -1 \\
1 & 1 & 1 & -1 & 1 & -1 & -1 & -1.
\end{array}
$$

Find the remaining five rows of the character table.

26. The commutative group of order 8 with an element of order 4 but no element of order higher than 4 has the following two rows in the associated character table:

$$
\begin{array}{cccccccc}
1 & 1 & 1 & 1 & -1 & -1 & -1 & -1 \\
1 & i & -1 & -i & 1 & i & -1 & -i.
\end{array}
$$

Find the remaining six rows of the character table.

27. Let π and σ be permutations of the set $S = \{1, 2, \ldots, n\}$, and let $\sigma' = \pi^{-1}\sigma\pi$ be conjugate to σ. Show that if $\sigma'(i) = j$, then $\sigma(\pi(i)) = \pi(j)$. Let σ' be represented in the notation of the above example in the form $\sigma' = (\cdots ij \cdots) \cdots$. Show that σ is represented in the form $\sigma = (\cdots \pi(i)\pi(j) \cdots) \cdots$. As an example, let $\sigma' = (123)(45)$ and $\pi = (1432)$. Compute $\sigma = \pi\sigma'\pi^{-1}$ directly, and also replace each element in $(123)(45)$ by its image under π.

28. Use Exercise 27 to show that two elements of $\mathfrak{S}_n$ are conjugate if and only if their cyclic representations have the same form; for example, $(123)(45)$ and $(253)(14)$ are conjugate. (*Warning:* This is not true in a permutation group, a subgroup of a symmetric group.) Show that $\mathfrak{S}_4$ has five conjugate classes.

29. Use Exercise 28, Theorems 9.11 and 9.12, and formula (9.30) to determine the dimensions of the irreducible representations of $\mathfrak{S}_4$.

30. Show that three of the conjugate classes of $\mathfrak{S}_4$ fill out $\mathfrak{A}_4$.

31. Use Exercises 22 and 30 to determine the characters for the two 1-dimensional representations of $\mathfrak{S}_4$. Use Exercise 29 to determine one column of the character table for $\mathfrak{S}_4$, the column of the conjugate class containing the identity element.

32. Show that $\mathfrak{B} = \{e, (12)(34), (13)(24), (14)(23)\}$ is a subgroup of $\mathfrak{S}_4$ isomorphic to the four-group. Show that $\mathfrak{B}$ is a normal subgroup of $\mathfrak{S}_4$. Show that each coset of $\mathfrak{B}$ contains one and only one of the elements of the set $\{e, (123), (132), (12), (13), (23)\}$. Show that the factor group $\mathfrak{S}_4/\mathfrak{B}$ is isomorphic to $\mathfrak{S}_3$.

33. Use Exercises 19 and 32 and the example preceding this set of exercises to determine a 2-dimensional representation of $\mathfrak{S}_4$. Determine the character values of this representation.

34. To fix the notation, let us now assume that we have obtained part of the character table for $\mathfrak{S}_4$ in the form:

	1	6	8	6	3
	C_1	C_2	C_3	C_4	C_5
D^1	1	1	1	1	1
D^2	1	-1	1	-1	1
D^3	2	0	-1	0	2
D^4	3				
D^5	3				

Show that if D^4 is a representation of $\mathfrak{S}_4$, then the matrices obtained by multiplying the matrices in D^4 by the matrices in D^2 is also a representation of $\mathfrak{S}_4$. Show that this new representation is also irreducible. Show that this representation must be different from D^4, unless D^4 has zero characters for C_2 and C_4.

35. Let the characters of D^4 be denoted by

$$D^4 \quad 3 \quad a \quad b \quad c \quad d.$$

Show that

$$3 + 6a + 8b + 6c + 3d = 0$$
$$3 - 6a + 8b - 6c + 3d = 0$$
$$6 \quad\quad - 8b \quad\quad + 6d = 0.$$

Determine b and d and show that $a = -c$. Show that $a^2 = 1$. Obtain the complete character table for $\mathfrak{S}_4$. Verify the orthogonality relations (9.31) and (9.42) for this table.

10 | Application of Representation Theory to Symmetric Mechanical Systems

This section depends directly on the material in the previous two sections, 8 and 9.

Consider a mechanical system which is symmetric when in an equilibrium position. For example, the·ozone molecule consisting of three oxygen atoms at the corners of an equilateral triangle is very symmetric (Fig. 9). This particular system can be moved into many new positions in which it looks and behaves the same as it did before it was moved. For example, the system can be rotated through an angle of 120° about an axis through the centroid perpendicular to the plane of the triangle. It can also be reflected in a plane containing an altitude of the triangle and perpendicular to the plane of the triangle. And it can be reflected in the plane containing the triangle. Such a motion is called a *symmetry* of the system. The system above has twelve symmetries (including the identity symmetry, which is to leave the system fixed).

⁃ Since any sequence of symmetries must result in a symmetry, the symmetries form a group G under successive application of the symmetries as the law of combination.

Let $X = (x_1, \ldots, x_n)$ be an n-tuple representing a displacement of the system. Let **a** be a symmetry of the system. The symmetry will move the

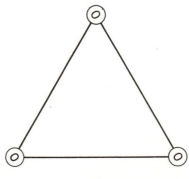

Fig. 9

system to a new configuration in which the displacement is represented by X'. The mapping of X onto X' will be represented by a matrix $D(\mathbf{a})$; that is, $D(\mathbf{a})X = X'$. If a new symmetry $\mathbf{b}$ is now applied, the system will be moved to another configuration represented by X'' where $X'' = D(\mathbf{b})X'$. But since $\mathbf{ba}$ moves the system to the configuration X'' in one step, we have $X'' = D(\mathbf{ba})X = D(\mathbf{b})D(\mathbf{a})X$. This holds for any X so we have $D(\mathbf{ba}) = D(\mathbf{b})D(\mathbf{a})$. Thus, the set $D(G)$ of matrices obtained in this way is a representation of the group of symmetries.

The idea behind the application of representation theory to the analysis of symmetric mechanical systems is that the irreducible invariant subspaces under the representation $D(G)$ are closely related to the principal axes of the system.

Suppose that a group G is represented by a group of linear transformations $\{\sigma_\mathbf{a}\}$ on a vector space V. Let f be a Hermitian form and let g be the symmetrization of f defined by

$$g(\alpha, \beta) = \frac{1}{g} \sum_\mathbf{a} f(\sigma_\mathbf{a}(\alpha), \sigma_\mathbf{a}(\beta)). \tag{10.1}$$

Let $A = \{\ldots, \alpha_1{}^r, \ldots, \alpha_{n_r}^r, \ldots\}$ be a basis for V such that $\{\alpha_1{}^r, \ldots, \alpha_{n_r}^r\}$ is a basis for the irreducible subspace on which G is represented by $D^r(G)$ in unitary form; that is,

$$\sigma_\mathbf{a}(\alpha_i^r) = \sum_{j=1}^{n_r} a_{ji}^r(\mathbf{a})\alpha_j^r \tag{10.2}$$

where $D^r(\mathbf{a}) = [a_{ji}^r(\mathbf{a})]$ is unitary. Then, by $(9.17)'$,

$$\begin{aligned}
g(\alpha_i^r, \alpha_l^s) &= \frac{1}{g} \sum_\mathbf{a} f(\sigma_\mathbf{a}(\alpha_i^r), \sigma_\mathbf{a}(\alpha_l^s)) \\
&= \frac{1}{g} \sum_\mathbf{a} f\left[\sum_{j=1}^{n_r} a_{ji}^r(\mathbf{a})\alpha_j^r, \sum_{k=1}^{n_s} a_{kl}^s(\mathbf{a})\alpha_k^s \right] \\
&= \frac{1}{g} \sum_\mathbf{a} \sum_{j=1}^{n_r} \sum_{k=1}^{n_s} \overline{a_{ji}^r(\mathbf{a})} a_{kl}^s(\mathbf{a}) f(\alpha_j^r, \alpha_k^s) \\
&= \sum_{j=1}^{n_r} \sum_{k=1}^{n_s} \left[\frac{1}{g} \sum_\mathbf{a} \overline{a_{ji}^r(\mathbf{a})} a_{kl}^s(\mathbf{a}) \right] f(\alpha_j^r, \alpha_k^s) \\
&= \sum_{j=1}^{n_r} \sum_{k=1}^{n_s} \frac{1}{n_r} \delta_{jk}\delta_{il}\delta_{rs} f(\alpha_j^r, \alpha_k^s). \tag{10.3}
\end{aligned}$$

If there is at most one invariant subspace corresponding to each irreducible representation of G, the matrix representing g with respect to the basis A would be a diagonal matrix. If a given irreducible representations occurs more than once as a component of $D(G)$, then terms off the main diagonal can occur, but their appearance depends on the values of $f(\alpha_j^r, \alpha_k^s)$. If f is left invariant under the group of transformations—that is, $f(\sigma_a(\alpha), \sigma_a)(\beta)) = f(\alpha, \beta)$ for all $\mathbf{a} \in G$—then $g = f$ and the same remarks apply to f.

By a symmetry of a mechanical system we mean a motion which preserves the mechanical properties of the system as well as the geometric properties. This means that the quadratic forms representing the potential energy and the kinetic energy must be left invariant under the group of symmetries. If a coordinate system is chosen in which the representation is unitary and decomposed into its irreducible components, considerable progress will be made toward finding the principal axes of the system. If $D(G)$ contain each irreducible representation at most once, the decomposition of the representation will yield the principal axes of the system. If the system is not very symmetric, the group of symmetries will be small, there will be few inequivalent irreducible representations, and it is likely that the reduced form will fall short of yielding the principal axes. (As an extreme case, consider the situation where the system has no symmetries except the identity.) However, in that part of the representing matrices where $r \neq s$ the terms will be zero. The problem, then, is to find effective methods for finding the basis A which reduces the representation.

The first step is to find the irreducible representations contained in $D(G)$. This is achieved by determining the trace S_D for D and using formula (9.21)′. The trace is not difficult to determine. Let $U(\mathbf{a})$ be the number of particles in the system left fixed by the symmetry $\mathbf{a}$. Only coordinates attached to fixed particles can contribute to the trace $S_D(\mathbf{a})$. If a local coordinate system is chosen at each particle so that corresponding axes are parallel, they will be parallel after the symmetry is applied. Thus, each local coordinate system (at a fixed point) undergoes the same transformation. If the local coordinate system is Euclidean, the local effect of the symmetry must be represented by a 3×3 orthogonal matrix since the symmetry is distance preserving. The trace of a matrix is the sum of its eigenvalues, since that is the case when it is in diagonal form. The eigenvalues of an orthogonal matrix are of absolute value 1. Since the matrix is real, at least one must be real and the others real or a pair of complex conjugate numbers. Thus, the local trace is

$$\pm 1 + e^{i\theta} + e^{-i\theta} = \pm 1 + 2 \cos \theta,$$

and

$$S_D(\mathbf{a}) = U(\mathbf{a})(\pm 1 + 2 \cos \theta). \tag{10.4}$$

The angle θ is the angle of rotation about some axis and it is easily determined from the geometric description of the symmetry. The $+1$ occurs if the symmetry is a local rotation, and the -1 occurs if a mirror reflection is present.

Once it is determined that the representation $D^r(G)$ is contained in $D(G)$, the problem is to find a basis for the corresponding invariant subspace.

If $\{\alpha_1^r, \ldots, \alpha_{n_r}^r\}$ is the required basis, we must have

$$\sigma_{\mathbf{a}}(\alpha_i^r) = \sum_{j=1}^{n_r} a_{ji}^r(\mathbf{a})\alpha_j^r. \tag{10.5}$$

If α_i^r is represented by X_i (unknown) in the given coordinate system, then $\sigma_{\mathbf{a}}(\alpha_i^r)$ is represented by $D(\mathbf{a})X_i$. Thus, we must solve the equations

$$D(\mathbf{a})X_i = \sum_{j=1}^{n_r} a_{ji}^r(\mathbf{a})X_j, \qquad i = 1, \ldots, n_r, \tag{10.6}$$

simultaneously for all $\mathbf{a} \in G$. The $a_{ji}^r(\mathbf{a})$ can be computed once for all and are presumed known. Since each X_i has n coordinates, there are $n \cdot n_r$ unknowns. Each matric equation of the form (10.6) involves n linear equations. Thus, there are $g \cdot n \cdot n_r$ equations. Most of the equations are redundant, but the existence or non-existence of the rth representation in $D(G)$ is what determines the solvability of the system. Even when many equations are eliminated as redundant, the system of linear equations to be solved is still very large. However, the system is linear and the solution can be worked out.

There are ways the work can be reduced considerably. Some principal axes are obvious for one reason or another. Suppose that Y is an n-tuple representing a known principal axis. Then any other principal axis represented by X must satisfy the condition

$$Y^T B X = 0 \tag{10.7}$$

since the principal axes are orthogonal with respect to the quadratic form B.

There is also the possibility of using irreducible representations in other than unitary form in the equations of (10.6). The basis obtained will not necessarily reduce A and B to diagonal form. But if the representation uses matrices with integral coefficients, the computation is sometimes easier, and the change to an orthonormal basis can be made in each invariant subspace separately.

BIBLIOGRAPHICAL NOTES

There are a number of good treatments of different aspects of the applications of group theory to physical problems. None is easy because of the degree of sophistication required for both the physics and the mathematics. Recommended are: B. Higman, *Applied Group-Theoretic and Matrix Methods;* J. S. Lomont, *Applications of Finite Groups.* T. Venkatarayudu, *Applications of Group Theory to Physical Problems;* H. Weyl, *Theory of Groups and Quantum Mechanics;* E. P. Wigner, *Group Theory and Its Application to the Quantum Mechanics of Atomic Spectra.*

EXERCISES

The following exercises all pertain to the ozone molecule described at the beginning of this section. However, to reduce the complexity of analyzing this system we make a simplification of the problem. As described at the beginning of this section, the phase space for the system is of dimension 9 and the group of symmetries is of order 12. This system has already been discussed in the exercises of Section 8. There we assumed that the displacements in a direction perpendicular to the plane of the triangle could be neglected. This has the effect of reducing the dimension of the phase space to 6 and the order of the group of symmetries to 6. This greatly simplifies the problem without discarding essential information, information about the vibrations of the system.

1. Show that if the ozone molecule is considered as embedded in a 2-dimensional space, the group of symmetries is of order 6 (instead of 12 when it is considered as embedded in a 3-dimensional space). Show that this group is isomorphic to $\mathfrak{S}_3$, the symmetric group on three objects.

2. Let $(x_1, y_1, x_2, y_2, x_3, y_3)$ denote the coordinates of the phase space as illustrated in Fig. 6 of Section 8. Let (12) denote the symmetry of the figure in which P_1 and P_2 are interchanged. Let (123) denote the symmetry of the figure which corresponds to a counterclockwise rotation through 120°. Find the matrices representing the permutations (12) and (123) in this coordinate system. Find all matrices representing the group of symmetries. Call this representation $D(G)$.

3. Find the traces of the matrices in $D(G)$ as given in Exercise 2. Determine which irreducible representations (as given in the example of Section 9) are contained in this representation.

4. Show that since $D(G)$ contains the identity representation, 1 is in eigenvalue of every matrix in $D(G)$. Show that the corresponding eigenvector is the same for every matrix in $D(G)$. Show that this eigenvector spans the irreducible invariant subspace corresponding to the identity representation. On a drawing like Fig. 7, draw the displacement corresponding to this eigenvector. Give a physical interpretation of this displacement.

5. There is one other 1-dimensional representation in $D(G)$. Show that for a 1-dimensional representation a character value is also an eigenvalue. Determine the vector spanning the irreducible invariant subspace corresponding to this representation. Draw the displacement represented by this eigenvector and give a physical interpretation of this displacement.

6. There are two 2-dimensional representations. One of them always corresponds to an invariant subspace that always appears in this type of problem. It corresponds to a translation of the molecule in the plane containing the molecule. Such a translation does not distort the molecule and does not play a role in determining the vibrations of the molecule. Find a basis for the irreducible invariant subspace corresponding to this representation.

7. In the previous exercises we have determined two 1-dimensional subspaces and one 2-dimensional subspace of the representation space for $D(G)$. There remains one more 2-dimensional representation to determine. At this stage the easiest thing to do is to use the orthogonality relations in formula (10.7) to find a basis for the remaining irreducible invariant subspace. Find this subspace.

8. Consider the displacement vectors $\{\xi_5 = (0, 1, \sqrt{3}/2, -\frac{1}{2}, -\sqrt{3}/2, -\frac{1}{2})$, $\xi_6 = (-\sqrt{3}/2, -\frac{1}{2}, 0, 1, \sqrt{3}/2, -\frac{1}{2})\}$. Show that they span an irreducible invariant subspace of the phase space under $D(G)$. Draw these displacements on a figure like Fig. 7. Similarly, draw $-(\xi_5 + \xi_6)$. Interpret these displacements in terms of distortions of the molecule and describe the type of vibration that would result if the molecule were started from rest in one of these positions.

Note: In working through the exercises given above we should have seen that one of the 1-dimensional representations corresponds to a rotation of the molecule without distortion, so that no energy is stored in the stresses of the system. Also, one 2-dimensional representation corresponds to translations which also do not distort the molecule. If we had started with the original 9-dimensional phase space, we would have found that six of the dimensions correspond to displacements that do not distort the molecule. Three dimensions correspond to translations in three independent directions, and three correspond to rotations about three independent axes. Restricting our attention to the plane resulted only in removing three of these distortionless displacements from consideration. The remaining three dimensions correspond to displacements which do distort the molecule, and hence these result in vibrations of the system. The 1-dimensional representation corresponds to a radial expansion and contraction of the system. The 2-dimensional representation corresponds to a type of distortion in which the molecule is expanded in one direction and contracted in a perpendicular direction.

A collection of matrices with integral elements and inverses with integral elements

Any numerical exercise involving matrices can be converted to an equivalent exercise with different matrices by a change of coordinates. For example, the linear problem

$$AX = B \qquad (A.1)$$

is equivalent to the linear problem

$$A'X = B' \qquad (A.2)$$

where $A' = PA$ and $B' = PB$ and P is non-singular. The problem (A.2) even has the same solution. The problem

$$A''Y = B \qquad (A.3)$$

where $A'' = AP$ has $Y = P^{-1}X$ as a solution if X is a solution of (A.1).

It should be clear enough how these modifications can be combined. Other exercises can be modified in a similar way. For this purpose it is most convenient to choose a matrix P that has integral elements (integral matrices). For (A.3), it is also desirable to require P^{-1} to have integral elements.

Let P be any non-singular matrix, and let D be any diagonal matrix of the same order. Compute $A = PDP^{-1}$. Then $D = P^{-1}AP$. A is a matrix similar to the diagonal matrix D. The eigenvalues of A are the elements in the main diagonal of D, and the eigenvectors of A are the columns of P. Thus, by choosing D and P appropriately, we can find a matrix A with prescribed eigenvalues (whatever we choose to enter in the main diagonal of D) and prescribed eigenvectors (whatever we put in the columns of P). If P is orthogonal, A will be orthogonal similar to a diagonal matrix. If P is unitary, A will be unitary similar to a diagonal matrix.

It is extremely easy to obtain an infinite number of integral matrices with integral inverses. Any product of integral elementary matrices will be integral.

319

If these elementary matrices have integral inverses, the product will have an integral inverse. Any elementary matrix of Type III is integral and has an integral inverse. If an elementary matrix of Type II is integral, its inverse will be integral. An integral elementary matrix of Type I does not have an integral inverse unless it corresponds to the elementary operation of multiplying by ± 1. These two possibilities are rather uninteresting.

Computing the product of elementary matrices is most easily carried out by starting with the unit matrix and performing the corresponding elementary operation in the right order. Thus, we avoid operations of Type I, use only integral multiples in operations of Type II, and use operations of Type III without restriction.

For convenience, a short list of integral matrices with integral inverses is given. In this list, the pair of matrices in a line are inverses of each other except that the inverse of an orthogonal, or unitary, matrix is not given.

$$P \qquad\qquad P^{-1}$$

$$\begin{bmatrix} 2 & 1 \\ 5 & 3 \end{bmatrix} \qquad \begin{bmatrix} 3 & -1 \\ -5 & 2 \end{bmatrix}$$

$$\begin{bmatrix} 5 & 8 \\ 3 & 5 \end{bmatrix} \qquad \begin{bmatrix} 5 & -8 \\ -3 & 5 \end{bmatrix}$$

$$\begin{bmatrix} 3 & -8 \\ 4 & -11 \end{bmatrix} \qquad \begin{bmatrix} -11 & 8 \\ -4 & 3 \end{bmatrix}$$

$$\begin{bmatrix} 3 & 8 \\ 4 & 11 \end{bmatrix} \qquad \begin{bmatrix} 11 & -8 \\ -4 & 3 \end{bmatrix}$$

$$\begin{bmatrix} 4 & 3 & 2 \\ 3 & 5 & 2 \\ 2 & 2 & 1 \end{bmatrix} \qquad \begin{bmatrix} -1 & -1 & 4 \\ -1 & 0 & 2 \\ 4 & 2 & -11 \end{bmatrix}$$

$$\begin{bmatrix} 2 & -5 & 5 \\ 2 & -3 & 8 \\ 3 & -8 & 7 \end{bmatrix} \qquad \begin{bmatrix} 43 & -5 & -25 \\ 10 & -1 & -6 \\ -7 & 1 & 4 \end{bmatrix}$$

$$P \qquad\qquad P^{-1}$$

$$\begin{bmatrix} 10 & -6 & 3 \\ 8 & -3 & 2 \\ 3 & -2 & 1 \end{bmatrix} \qquad \begin{bmatrix} 1 & 0 & -3 \\ -2 & 1 & 4 \\ -7 & 2 & 18 \end{bmatrix}$$

$$P \qquad\qquad\qquad P^{-1}$$

$$\begin{bmatrix} 2 & -5 & 5 \\ 2 & -3 & 8 \\ 3 & -8 & 7 \end{bmatrix} \qquad \begin{bmatrix} 43 & -5 & -25 \\ 10 & -1 & -6 \\ -7 & 1 & 4 \end{bmatrix}$$

$$\begin{bmatrix} 2 & 5 & 8 \\ 4 & 5 & 13 \\ 1 & -6 & -1 \end{bmatrix} \qquad \begin{bmatrix} -73 & 43 & -25 \\ -17 & 10 & -6 \\ 29 & -17 & 10 \end{bmatrix}$$

$$\begin{bmatrix} 1 & 2 & 3 \\ 2 & 3 & 4 \\ 3 & 4 & 6 \end{bmatrix} \qquad \begin{bmatrix} -2 & 0 & 1 \\ 0 & 3 & -2 \\ 1 & -2 & 1 \end{bmatrix}$$

$$\begin{bmatrix} 4 & 3 & 2 & 0 \\ 5 & 4 & 3 & 0 \\ -2 & -2 & -1 & 0 \\ 11 & 6 & 4 & 1 \end{bmatrix} \qquad \begin{bmatrix} 2 & -1 & 1 & 0 \\ -1 & 0 & -2 & 0 \\ -2 & 2 & 1 & 0 \\ -8 & 3 & -3 & 1 \end{bmatrix}$$

$$\begin{bmatrix} 2 & 1 & 0 \\ -1 & 0 & 1 \\ 0 & -1 & 1 \end{bmatrix} \qquad \begin{bmatrix} 1 & 1 & 1 \\ -1 & -2 & -2 \\ 1 & 2 & 1 \end{bmatrix}$$

$$\begin{bmatrix} 2 & -1 & 0 \\ -1 & 0 & 1 \\ 0 & 1 & 1 \end{bmatrix} \qquad \begin{bmatrix} 1 & 1 & 1 \\ 1 & 2 & 2 \\ 1 & 2 & 1 \end{bmatrix}$$

$$\begin{bmatrix} 4 & 3 & 2 & 1 \\ 5 & 4 & 3 & 1 \\ -2 & -2 & -1 & -1 \\ 11 & 6 & 4 & 3 \end{bmatrix} \qquad \begin{bmatrix} 2 & -1 & 1 & 0 \\ 7 & -3 & 1 & -1 \\ -10 & 5 & -2 & 1 \\ -8 & 3 & -3 & 1 \end{bmatrix}$$

Orthogonal

$$\frac{1}{5}\begin{bmatrix} 3 & -4 \\ 4 & 3 \end{bmatrix}$$

$$\frac{1}{3}\begin{bmatrix} 1 & 2 & 2 \\ 2 & -2 & 1 \\ 2 & 1 & -2 \end{bmatrix}$$

Orthogonal

$$\frac{1}{3}\begin{bmatrix} -2 & 1 & -2 \\ 1 & -2 & -2 \\ -2 & -2 & 1 \end{bmatrix}$$

$$\frac{1}{3}\begin{bmatrix} 1 & 2 & 2 \\ -2 & -1 & 2 \\ 2 & -2 & 1 \end{bmatrix}$$

$$\frac{1}{9}\begin{bmatrix} -7 & -4 & -4 \\ 4 & 1 & -8 \\ 4 & -8 & 1 \end{bmatrix}$$

$$\frac{1}{27}\begin{bmatrix} -23 & 10 & 10 \\ 10 & 25 & -2 \\ 10 & -2 & 25 \end{bmatrix}$$

$$\frac{1}{81}\begin{bmatrix} 17 & 56 & 56 \\ -56 & 49 & -32 \\ -56 & -32 & 49 \end{bmatrix}$$

$$\frac{1}{11}\begin{bmatrix} 2 & 6 & 9 \\ 6 & 7 & -6 \\ 9 & -6 & 2 \end{bmatrix}$$

$$\frac{1}{33}\begin{bmatrix} -17 & 4 & 28 \\ 4 & -32 & 7 \\ 28 & 7 & 16 \end{bmatrix}$$

Unitary

$$\frac{1}{5}\begin{bmatrix} 4 & 3i \\ 3i & 4 \end{bmatrix}$$

$$\frac{1}{10}\begin{bmatrix} 7+i & -1+7i \\ 1+7i & 7-i \end{bmatrix}$$

$$\frac{1}{26}\begin{bmatrix} 7+17i & -17+7i \\ 17+7i & 7-17i \end{bmatrix}$$

Unitary

$$\begin{bmatrix} \cos\theta & i\sin\theta \\ i\sin\theta & \cos\theta \end{bmatrix}$$

$$\frac{1}{2}\begin{bmatrix} (1+i)e^{-i\theta} & (1-i)e^{-i\theta} \\ -(1+i)e^{i\theta} & (1-i)e^{i\theta} \end{bmatrix} \quad (\theta \text{ real})$$

Answers
to selected
exercises

I–1

1–4. If f and g are (continuous, integrable, differentiable m times, satisfy the differential equation) and a is a constant, then $f + g$ and af also are continuous, etc.). With the exception of $A1$ and $B1$, any vector space axiom which is satisfied in a set is satisfied in any subset.

5. $B1$ is not satisfied if a is negative.

6. $\alpha = (a^{-1}a)\alpha = a^{-1}(a\alpha) = a^{-1}0 = 0$.

9. (a) $(-1, -2, 1, 0)$; (b) $(5, -8, -1, 2)$; (c) $(6, -15, 0, 3)$; (d) $(-5, -1, 3, -1)$.

I–2

1. $3p_1 + 2p_2 - 5p_3 + 4p_4 = 0$.

2. The set of all polynomials of degree 2 or less, and the zero polynomial.

3. Every subset of three polynomials is maximal linearly independent.

6. No polynomial is a linear combination of the preceding ones.

7. 1 cannot be expressed as a linear combination of polynomials divisible by $x - 1$.

8. (a) dependent; (b) dependent; (c) independent.

I–3

3. $\{(1, 2, -1, 1), (0, 1, 2, -1), (1, 1, 0, 0), (0, 0, 1, 1)\}$.

4. $\{(1, 2, 3, 4), (1, 0, 0, 0), (0, 1, 0, 0), (0, 0, 1, 0)\}$, for example.

I–4

1. (a), (b), and (d) are subspaces. (c) and (e) do not satisfy $A1$.

3. The equation used in (c) is homogeneous. This condition is essential if $A1$ and $B1$ are to be satisfied. Why?

5. $(2, -1, 3, 3) = -(1, 1, 0, 0) + 3(1, 0, 1, 1)$, $(0, 1, -1, -1) = (1, 1, 0, 0) -$ $(1, 0, 1, 0)$; $(1, 1, 0, 0) = \frac{1}{2}(2, -1, 3, 3) + \frac{3}{2}(0, 1, -1, -1)$, $(1, 0, 1, 1) =$ $\frac{1}{2}(2, -1, 3, 3) + \frac{1}{2}(0, 1, -1, -1)$.

6. $\{(1, 1, 1, 1, 1), (1, 0, 1, 0, 1), (0, 1, 1, 1, 0), (2, 0, 0, 1, 1)\}$ is a basis for W.

8. $W_1 \cap W_2 = \langle(-1, 2, 1, 2)\rangle$, $W_1 = \langle(-1, 2, 1, 2), (1, 2, 3, 6)\rangle$, $W_2 = \langle(-1, 2, 1, 2), (1, -1, 1, 1)\rangle$, $W_1 + W_2 = \langle(-1, 2, 1, 2), (1, 2, 3, 6), (1, -1, 1, 1)\rangle$.

9. $W_1 \cap W_2$ is the set of all polynomials divisible by $(x - 1)(x - 2)$. $W_1 + W_2 = P$.

11. If $W_1 \not\subset W_2$ and $W_2 \not\subset W_1$, there exist $\alpha_1 \in W_1 - W_2$ and $\alpha_2 \in W_2 - W_1$. $\alpha_1 + \alpha_2 \notin W_1$ since $\alpha_1 \in W_1$ and $\alpha_2 \notin W_1$. Similarly, $\alpha_1 + \alpha_2 \notin W_2$. Since $\alpha_1 + \alpha_2 \notin W_1 \cup W_2$, $W_1 \cup W_2$ is not a subspace.

12. $\{(1, 1, 1, 0), (2, 1, 0, 1)\}$ is a basis for the subspace of solutions.

16. Since $W_1 \subset W$, $W_1 + (W \cap W_2) \subset W$. Every $\alpha \in W \cap (W_1 + W_2)$ can be written in the form $\alpha = \alpha_1 + \alpha_2$ where $\alpha_1 \in W_1$ and $\alpha_2 \in W_2$. Since $W_1 \subset W$, $\alpha_1 \in W$. Thus $\alpha_2 \in W$ and $\alpha \in (W \cap W_1) + (W \cap W_2) = W_1 + (W \cap W_2)$. Thus $W = W \cap (W_1 + W_2) = W_1 + (W \cap W_2)$. Finally, it is easily seen that this last sum is direct.

II–1

2. $(\sigma_1 + \sigma_2)((x_1, x_2)) = (x_1 + x_2, -x_1 - x_2)$, $\sigma_1\sigma_2((x_1, x_2)) = (-x_2, -x_1)$, $\sigma_2\sigma_1((x_1, x_2)) = (x_2, x_1)$.

4. The kernel of σ is the set of solutions of the equations given in Exercise 12 of Section 4, Chapter I.

5. $\{(1, 0, 1), (0, 1, -2)\}$ is a basis of $\sigma(U)$. $\{(-4, -7, 5)\}$ is a basis of $K(\sigma)$.

7. See Exercise 2.

12. By Theorem 1.6, $\rho(\sigma) = \dim V' = \dim \{K(\sigma) \cap V'\} + \dim \tau(V') \leq \dim K(\tau) + \dim \tau\sigma(U) = \nu(\tau) + \rho(\tau\sigma)$.

13. By Exercise 12, $\rho(\tau\sigma) \geq \rho(\sigma) - \nu(\tau) = \rho(\sigma) - (m - \rho(\tau)) = \rho(\sigma) + \rho(\tau) - m$. The other part of the inequality is Theorem 1.12.

14. By Exercise 13, $\nu(\tau\sigma) = n - \rho(\tau\sigma) \leq n - (\rho(\sigma) + \rho(\tau) - m) = (n - \rho(\sigma)) + (m - \rho(\tau)) = \nu(\sigma) + \nu(\tau)$. $\nu(\tau\sigma) \leq n$ since $K(\tau\sigma) \subset U$. The inequality $\nu(\sigma) \leq \nu(\tau\sigma)$ follows from the fact that $K(\sigma) \subset K(\tau\sigma)$. Since $\tau\sigma(U) \subset \tau(V)$ we have $\rho(\tau\sigma) \leq \rho(\tau)$. Thus $\nu(\tau\sigma) = n - \rho(\tau\sigma) \geq n - \rho(\tau) = n - m + \nu(\tau)$.

15. By Exercise 14, $\nu(\tau\sigma) = \nu(\sigma)$.

17. By Exercise 14, $\nu(\tau\sigma) = \nu(\tau)$.

18. $\rho(\sigma_1 + \sigma_2) = \dim(\sigma_1 + \sigma_2)(U) \leq \dim\{\sigma_1(U) + \sigma_2(U)\} = \dim\sigma_1(U) + \dim\sigma_1(U) - \dim\{\sigma_1(U) \cap \sigma_2(U)\} \leq \rho(\sigma_1) + \rho(\sigma_2)$.

19. Since $\rho(\sigma_2) = \rho(-\sigma_2)$, Exercise 18 also says that $\rho(\sigma_1 - \sigma_2) \leq \rho(\sigma_1) + \rho(\sigma_2)$. Then $\rho(\sigma_1) = \rho(\sigma_2 - (\sigma_1 + \sigma_2)) \leq \rho(\sigma_2) + \rho(\sigma_1 + \sigma_2)$. By symmetry, $\rho(\sigma_2) \leq \rho(\sigma_1) + \rho(\sigma_1 + \sigma_2)$.

II–2

2. $\begin{bmatrix} 37 \\ -8 \end{bmatrix}$.

3. $AB = \begin{bmatrix} -4 & -4 & -4 & -4 \\ 4 & 4 & 4 & 4 \\ 0 & 0 & 0 & 0 \\ 0 & 0 & 0 & 0 \end{bmatrix}$, $BA = \begin{bmatrix} 4 & 6 & 5 & 5 \\ 12 & 14 & 13 & 13 \\ -4 & -6 & -5 & -5 \\ -12 & -14 & -13 & -13 \end{bmatrix}$.

Notice that AB and BA have different ranks.

4. $\begin{bmatrix} 3 & -1 \\ -1 & 2 \end{bmatrix}$.   5. $\begin{bmatrix} 3 & -1 \\ -5 & 2 \end{bmatrix}$.   6. $\begin{bmatrix} \frac{2}{5} & \frac{1}{5} \\ \frac{1}{5} & \frac{3}{5} \end{bmatrix}$.

8. (a) $\begin{bmatrix} -1 & 0 \\ 0 & -1 \end{bmatrix}$, (b) $\begin{bmatrix} 0 & -1 \\ -1 & 0 \end{bmatrix}$, $\begin{bmatrix} -1 & 0 \\ 0 & 1 \end{bmatrix}$ (c) $\begin{bmatrix} \frac{5}{2} & -\frac{1}{2} \\ -\frac{1}{2} & \frac{5}{2} \end{bmatrix}$, $\begin{bmatrix} 2 & 0 \\ 0 & 3 \end{bmatrix}$.

(d) $\begin{bmatrix} 1 & 1 \\ 0 & 1 \end{bmatrix}$. (e) $\begin{bmatrix} \frac{5}{13} & -\frac{12}{13} \\ \frac{12}{13} & \frac{5}{13} \end{bmatrix}$. (f) $\begin{bmatrix} \frac{2}{3} & \frac{1}{3} \\ \frac{2}{3} & \frac{1}{3} \end{bmatrix}$, $\begin{bmatrix} 1 & 0 \\ 0 & 0 \end{bmatrix}$.

9. (a) Reflection in the line $y = x$. See 8(b). (b) Projection onto x_1-axis parallel to x_2-axis, followed by $90°$ rotation counterclockwise. See 8(f) and 8(e). (c) Projection onto x_1-axis parallel to line $x_1 + x_2 = 0$. See 8(f). (d) Shear parallel to x_2-axis. See 8(d). (e) Stretch x_1-axis by factor b, stretch x_2-axis by factor c. See 8(c). (f) Rotation counterclockwise through acute angle $\theta = \arccos \frac{3}{5}$.

10. (a) $y = x$ (onto, fixed), (b) none, (c) $x_2 = 0$ (onto, fixed), $x_1 + x_2 = 0$ (into), (d) $x_1 = 0$ (onto, fixed), (e) $x_2 = 0$ (onto, stretched by factor b), $x_1 = 0$ (onto, stretched by factor a), (f) none.

11. $\begin{bmatrix} -1 & 1 \\ 0 & 1 \\ 1 & 1 \end{bmatrix}$.

12. If $\{d_1, d_2, \ldots, d_n\}$ is the set of elements in the diagonal of D, then multiplying A by D on the right multiplies column k by d_k and multiplying A by D on the left multiplies row k by d_k.

13. A is a diagonal matrix.

15. Let f be represented by $\begin{bmatrix} a & b \\ c & d \end{bmatrix}$.

16. The linear transformation, and thereby the function $f(x + yi)$, can be represented by the matrix $\begin{bmatrix} a & -b \\ b & a \end{bmatrix}$.

17. For example, $A = \begin{bmatrix} 1 & 0 \\ 0 & 0 \end{bmatrix}$ $B = \begin{bmatrix} 0 & 0 \\ 1 & 0 \end{bmatrix}$.

II–3

2. $A^2 = I$.

3. $(0, 0, 0)$. See Theorem 3.3.

4. $\begin{bmatrix} 2 & 1 \\ 5 & 3 \end{bmatrix}$.   5. $\begin{bmatrix} \frac{3}{5} & \frac{4}{5} \\ -\frac{4}{5} & \frac{3}{5} \end{bmatrix}$.   6. $\begin{bmatrix} 0 & -1 \\ -1 & 0 \end{bmatrix}$.

7. $\begin{bmatrix} 1 & -1 \\ 0 & 1 \end{bmatrix}$.

8. $\begin{bmatrix} 0 & 0 & 1 \\ -1 & 0 & 0 \\ 0 & 1 & 0 \end{bmatrix}, \quad \begin{bmatrix} 0 & -1 & 0 \\ 0 & 0 & 1 \\ 1 & 0 & 0 \end{bmatrix}$.

9. If σ is an automorphism and S is a subspace, $\sigma(S)$ is of dimension less than or equal to dim S. Since $K(\sigma)$ is of dimension zero in V, it is of dimension zero when confined to S. Thus dim $\sigma(S) = $ dim S.

10. For example, $A = \begin{bmatrix} 1 & 0 & 0 \\ 0 & 1 & 0 \end{bmatrix} \quad B = \begin{bmatrix} 1 & 0 \\ 0 & 1 \\ 0 & 0 \end{bmatrix}$.

II–4

1. $\begin{bmatrix} 1 & -2 & -1 \\ 1 & -1 & 1 \\ 1 & 1 & 1 \end{bmatrix}$.

3. $\begin{bmatrix} \dfrac{1}{2} & \dfrac{\sqrt{3}}{2} \\ \dfrac{-\sqrt{3}}{2} & \dfrac{1}{2} \end{bmatrix}$

5. $P = \begin{bmatrix} 0 & 1 & 1 \\ 1 & 0 & 1 \\ 1 & 1 & 0 \end{bmatrix}, \quad P^{-1} = \dfrac{1}{2} \begin{bmatrix} -1 & 1 & 1 \\ 1 & -1 & 1 \\ 1 & 1 & -1 \end{bmatrix}$.

6. PQ is the matrix of transition from A to C. The order of multiplication is reversed in the two cases.

II–5

1. (a), (c), (d).
2. (a) 3, (b) 3, (c) 3, (d) 3, (e) 4.
3. Both A and B are in Hermite normal form. Since this form is unique under possible changes in basis in R^2, there is no matrix Q such that $B = Q^{-1}A$.

II–6

1. (a) 2, (b) 2, (c) 3.
2. (a) Subtract twice row 1 from row 2. (b) Interchange row 1 and row 3. (c) Multiply row 2 by 2.
3. (From right to left.) Add row 1 to row 2, subtract row 2 from row 1, add row 1 to row 2, multiply row 1 by -1. The result interchanges row 1 and row 2.

5. (a) $\begin{bmatrix} 1 & 0 & 2 & 0 \\ 0 & 1 & -1 & 0 \\ 0 & 0 & 0 & 1 \end{bmatrix}$,

(b) $\begin{bmatrix} 1 & 0 & -1 & 0 & 1 & 1 \\ 0 & 1 & 2 & 0 & 0 & 1 \\ 0 & 0 & 0 & 1 & 3 & 1 \\ 0 & 0 & 0 & 0 & 0 & 0 \end{bmatrix}$.

6. (a) $\begin{bmatrix} 2 & 1 \\ 5 & 3 \end{bmatrix}$,

(b) $\begin{bmatrix} -2 & 0 & 1 \\ 0 & 3 & -2 \\ 1 & -2 & 1 \end{bmatrix}$.

II-7

1. For every value of x_3 and x_4, $(x_1, x_2, x_3, x_4) = x_3(1, 1, 1, 0) + x_4(2, 1, 0, 1)$ is a solution of the system of equations. Since the system of homogeneous equations is of rank 2 and these two solutions are linearly independent, they span the space of solutions.
2. $\langle(-1, 2, 1, 0)\rangle$.
3. (a) $(2, 3, 0, 0) + \langle(1, -2, 1, 0), (-1, 1, 0, 1)\rangle$, (b) no solution.
4. $(3, 5, 0) + \langle(5, 7, 1)\rangle$.
5. $(-3, 2, 0, 0, 1) + \langle(-3, 0, 1, 0, 0), (1, 2, 0, 1, 0)\rangle$.
6. If the system has m equations and n unknowns, the augmented matrix is $m \times (n + 1)$. The reduced form of the augmented matrix contains the reduced form of the coefficient matrix in the first n columns. Their ranks are equal if and only if the last column of the reduced form of the augmented matrix does not start a non-zero row.

II-8

2. $\{(1, 0, 0, 0, 1), (0, 1, 0, 0, 1), (0, 0, 1, 0, 0), (0, 0, 0, 1, -1)\}$.
3. $\{(1, 0, 2, -\frac{3}{2}), (0, 1, 0, \frac{3}{2})\}$ is a standard basis of the subspace spanned by the first set and $\{(1, 0, 1, 0), (0, 1, 0, \frac{3}{2})\}$ is a standard basis for the second. Hence these subspaces are not identical.
4. $\{(1, 0, 0, 1), (0, 1, 0, -\frac{7}{2}), (0, 0, 1, \frac{1}{2})\}$ is a standard basis of $W_1 + W_2$.
5. $\{(1, 0, -1, 0, 2), (0, 1, 2, 0, 1), (0, 0, 0, 1, 1)\}$ is a stand. basis. $x_1 - 2x_2 + x_3 = 0$, $-2x_1 - x_2 - x_4 + x_5 = 0$ is a characterizing system.
6. $W_1 = \langle(\frac{1}{2}, 0, \frac{1}{2}, 1), (-1, 1, 0, 0)\rangle$, $E_1 = \langle(1, 1, 0, -\frac{1}{2}), (0, 0, 1, -\frac{1}{2})\rangle$, $W_2 = \langle(-2, 3, 0, 1), (3, -4, 1, 0)\rangle$, $E_2 = \langle(1, 0, -3, 2), (0, 1, 4, -3)\rangle$, $E_1 + E_2 = \langle(1, 0, 0, \frac{1}{2}), (0, 1, 0, -1), (0, 0, 1, -\frac{1}{2})\rangle$, $W_1 \cap W_2 = \langle(-\frac{1}{2}, 1, \frac{1}{2}, 1)\rangle$, $W_1 = \langle(-\frac{1}{2}, 1, \frac{1}{2}, 1), (0, 0, 1, -\frac{1}{2})\rangle$, $W_2 = \langle(-\frac{1}{2}, 1, \frac{1}{2}, 1), (0, 1, 4, -3)\rangle$, $W_1 + W_2 = \langle(-\frac{1}{2}, 1, \frac{1}{2}, 1), (0, 1, 4, -3), (0, 0, 1, -\frac{1}{2})\rangle$.

III-1

2. The permutations $\begin{pmatrix} 1 & 2 & 3 \\ 2 & 3 & 1 \end{pmatrix}$, $\begin{pmatrix} 1 & 2 & 3 \\ 3 & 1 & 2 \end{pmatrix}$, and identity permutation are even.
3. Of the 24 permutations, eight leave exactly one object fixed. They are permutations of three objects and have already been determined to be even. Six leave exactly two objects fixed and they are odd. The identity permutation is even. Of the remaining nine permutations, six permute the objects cyclically and three interchange two pairs of objects. These last three are even since they involve two interchanges.
4. Of the nine permutations that leave no object fixed, six are of one parity and three others are of one parity (possibly the same parity). Of the 15 already considered, nine are known to be even and six are known to be odd. Since half the 24 permutations must be even, the six cyclic permutations must be odd and the remaining even.
5. π is odd.

III-2

1. Since for any permutation π not the identity there is at least one i such that $\pi(i) < i$, all terms of the determinant but one vanish. $|\det A| = \prod_{i=1}^{n} a_{ii}$.
2. 6.
4. (a) -32, (b) -18.

III–3

1. 145; 134.
2. -114.
4. By Theorem 3.1, $A \cdot \tilde{A} = (\det A)I$. Thus $\det A \cdot \det \tilde{A} = (\det A)^n$. If $\det A \neq 0$, then $\det \tilde{A} = (\det A)^{n-1}$. If $\det A = 0$, then $A \cdot \tilde{A} = 0$. By (3.5), $\sum_j a_{ij}A_{kj} = 0$ for each k. This means the columns of $\tilde{A}$ are linearly dependent and $\det \tilde{A} = 0 = (\det A)^{n-1}$.
5. If $\det A \neq 0$, then $\tilde{A}^{-1} = A/\det A$ and $\tilde{A}$ is non-singular. If $\det A = 0$, then $\det \tilde{A} = 0$ and $\tilde{A}$ is singular.

6.

$$\begin{vmatrix} a_{11} & \cdots & a_{1n} & x_1 \\ \cdot & \cdots & \cdot & \cdot \\ \cdot & \cdots & \cdot & \cdot \\ \cdot & \cdots & \cdot & \cdot \\ a_{n1} & \cdots & a_{nn} & x_n \\ y_1 & \cdots & y_n & 0 \end{vmatrix} = \sum_{i=1}^{n} x_i(-1)^{n+i-1} \det B_i,$$

where B_i is the matrix obtained by crossing out row i and the last column,

$$= \sum_{i=1}^{n} x_i(-1)^{n+i-1}\left(\sum_{j=1}^{n} y_j(-1)^{(n-1)+j-1} \det C_{ij}\right),$$

where C_{ij} is the matrix obtained by crossing out column j and the last row of B_i,

$$= \sum_{i=1}^{n} \sum_{j=1}^{n} x_i y_j(-1)^{2n+i+j-3}(-1)^{i+j}A_{ij}$$

$$= -\sum_{i=1}^{n} \sum_{j+1}^{n} x_i A_{ij} y_i = -Y^T \tilde{A} X.$$

III–4

2. $-x^3 + 2x^2 + 5x - 6.$
3. $-x^3 + 6x^2 - 11x + 6.$

4. $\begin{bmatrix} 0 & 0 & 0 & -6 \\ 1 & 0 & 0 & 1 \\ 0 & 1 & 0 & -2 \\ 0 & 0 & 1 & -3 \end{bmatrix}.$

5. If $A^2 + A + I = 0$, then $A(-A - I) = I$ so that $-A - I = A^{-1}$.

6. If A is a real 3×3 matrix its characteristic polynomial $f(x)$ is real of degree 3. If A satisfies $x^2 + 1 = 0$, the minimum polynomial would divide $x^2 + 1$ and could not have as an irreducible factor the real factor of degree one which $f(x)$ must have.

7. $\begin{bmatrix} 0 & -1 \\ 1 & -1 \end{bmatrix}.$
   8. $x^2 - 81$ is the minimum polynomial.

III–5

1. If ξ is an eigenvector with eigenvalue 0, then ξ is a non-zero vector in the kernel of σ. Conversely, if σ is singular, then any non-zero vector in the kernel of σ is an eigenvector with eigenvalue 0.
2. If $\sigma(\xi) = \lambda \xi$, then $\sigma^2(\xi) = \sigma(\lambda \xi) = \lambda^2 \xi$. Generally, $\sigma^n(\xi) = \lambda^n \xi$.
3. If $\sigma(\xi) = \lambda_1 \xi$ and $\tau(\xi) = \lambda_2 \xi$, then $(\sigma + \tau)(\xi) = \sigma(\xi) + \tau(\xi) = \lambda_1 \xi + \lambda_2 \xi = (\lambda_1 + \lambda_2)\xi$. Also, $(a\sigma)(\xi) = a\sigma(\xi) = a\lambda_1 \xi$.

4. Consider, for example, $\begin{bmatrix} 1 & 0 \\ 2 & 2 \end{bmatrix}$ and $\begin{bmatrix} 2 & 2 \\ 0 & 1 \end{bmatrix}$.

5. $p(\lambda)$ is an eigenvalue of $p(\sigma)$ corresponding to ξ.
6. If $\sigma(\xi) = \lambda \xi$, then $\xi = \sigma^{-1}(\lambda \xi) = \lambda \sigma^{-1}(\xi)$ so that ξ is an eigenvector of σ^{-1} corresponding to λ^{-1}.
7. Let $\{\xi_1, \ldots, \xi_n\}$ be a linearly independent set of eigenvectors with eigenvalues $\{\lambda_1, \ldots, \lambda_n\}$. Then $\sigma(\sum_i \xi_i) = \sum_i \sigma(\xi_i) = \sum_i \lambda_i \xi_i$. But since $\sum_i \xi_i$ is also an eigenvector (by assumption), we have $\sigma(\sum_i \xi_i) = \lambda \sum_i \xi_i$. Thus $\sum_i (\lambda_i - \lambda)\xi_i = 0$. Since the set is linearly independent, $\lambda_i - \lambda = 0$ for each i.
8. x^n is the characteristic polynomial and also the minimum polynomial. An eigenvector $p(x)$ must satisfy the equation $D(p(x)) = kp(x)$. The constants are the eigenvectors of D.
9. c is the corresponding eigenvalue.
11. If $\xi_1 + \xi_2$ is an eigenvector with eigenvalue λ, then $\lambda(\xi_1 + \xi_2) = \sigma(\xi_1 + \xi_2) = \lambda_1 \xi_1 + \lambda_2 \xi_2$. Since $(\lambda - \lambda_1)\xi_1 + (\lambda - \lambda_2)\xi_2 = 0$ we have $\lambda - \lambda_1 = \lambda - \lambda_2 = 0$.
12. If $\xi = \sum_{i=1}^{r} a_i \xi_i$ is an eigenvector with eigenvalue λ, then $\lambda \sum_{i=1}^{r} a_i \xi_i = \lambda \xi = \sigma(\xi) = \sum_{i=1}^{r} a_i \sigma(\xi_i) = \sum_{i=1}^{r} a_i \lambda_i \xi_i$. Then $\sum_{i=1}^{r} a_i(\lambda - \lambda_i)\xi_i = 0$ and $a_i(\lambda - \lambda_i) = 0$ for each i. Since the λ_i are distinct at most one of the $\lambda - \lambda_i$ can be zero. For the other terms we must have $a_i = 0$. Since ξ is an eigenvector, and hence non-zero, not all $a_i = 0$.

III–6

1. $-2, (1, -1); \ 7, (4, 5).$
2. $3 + 2i, (1, i); \ 3 - 2i, (1, -i).$
3. $-3, (1, -2); \ 2, (2, 1).$
4. $2, (\sqrt{2}, -1); \ 3, (1, -\sqrt{2}).$
5. $4, (1, 0, 0); \ -2, (3, -2, 0); \ 7, (24, 8, 9).$
6. $1, (-1, 0, 1); \ 2, (2, -1, 0); \ 3, (0, 1, -1).$
7. $9, (4, 1, -1); \ -9, (1, -4, 0), (1, 0, 4).$
8. $1, (i, 1, 0); \ 3, (1, i, 0), (0, 0, 1).$

III–7

1. For each matrix A, P has the components of the eigenvectors of A in its columns. Every matrix in the exercises of Section 6 can be diagonalized.
2. The minimum polynomial is $(x - 1)^2$.

3. Let σ be the corresponding linear transformation. Since $\sigma \neq 1$, there is a non-zero vector ξ_1 such that $\sigma(\xi_1) = \xi_2 \neq \xi_1$. $\xi_2 \neq 0$ since σ is non-singular. If $\{\xi_1, \xi_2\}$ is a basis, then the matrix representing σ with respect to this basis has the desired form. On the other hand, suppose that for every choice of ξ_1, $\{\xi_1, \xi_2\}$ is dependent. Then ξ_2 is a multiple of ξ_1 and every vector is an eigenvector. By Exercise 7 of Section 5 this is impossible.

4. $A^{-1}(AB)A = BA$.

5. If π_1 and π_2 are projections of the same rank k, we are asked to find a non-singular linear transformation σ such that $\sigma^{-1}\pi_1\sigma = \pi_2$. Let $\{\alpha_1, \ldots, \alpha_n\}$ be a basis such that $\pi_1(\alpha_i) = \alpha_i$ for $i \leq k$ and $\pi_1(\alpha_i) = 0$ for $i > k$. Let $\{\beta_1, \ldots, \beta_n\}$ be a basis having similar properties with respect to π_2. Define σ by the rule $\sigma(\beta_i) = \alpha_i$. Then $\sigma^{-1}\pi_1\sigma(\beta_i) = \sigma^{-1}\pi_1(\alpha_i) = \sigma^{-1}(\alpha_i) = \beta_i$ for $i \leq k$, and $\sigma^{-1}\pi_1\sigma(\beta_i) = \sigma^{-1}\pi_1(\alpha_i) = \sigma^{-1}(0) = 0$ for $i > k$. Thus $\sigma^{-1}\pi_1\sigma = \pi_2$.

6. By (7.2) of Chapter II, $\text{Tr}(A^{-1}BA) = \text{Tr}(BAA^{-1}) = \text{Tr}(B)$.

IV–1

1. (a) $[1 \quad 1 \quad 1]$; (c) $[\sqrt{2} \quad 0 \quad 0]$; (d) $[-\frac{1}{2} \quad 1 \quad 0]$. (b) and (e) are not linear functionals.

2. (a) $\{[1 \quad 0 \quad 0], [0 \quad 1 \quad 0], [0 \quad 0 \quad 1]\}$; (b) $\{[1 \quad -1 \quad 0], [0 \quad 1 \quad -1], [0 \quad 0 \quad 1]\}$; (c) $\{[\frac{1}{2} \quad \frac{1}{2} \quad -\frac{1}{2}], [-\frac{1}{2} \quad \frac{1}{2} \quad -\frac{1}{2}], [\frac{1}{2} \quad \frac{1}{2} \quad \frac{1}{2}]\}$.

5. If $\alpha = \sum_{i=1}^{n} x_i\alpha_i$, then $\phi_j(\alpha) = \sum_{i=1}^{n} x_i\phi_j(\alpha_i) = x_j$.

6. Let $A = \{\alpha_1, \ldots, \alpha_n\}$ be a basis such that $\alpha_1 = \alpha$ and $\alpha_2 = \beta$. Let $\hat{A} = \{\phi_1, \ldots, \phi_n\}$ be the dual basis. Then $\phi_1(\alpha) = 1$ and $\phi_1(\beta) = 0$.

7. Let $p(x) = x$. Then $\sigma_a(x) \neq \sigma_b(x)$.

8. The space of linear functionals obtained in this way is of dimension 1, and $\hat{P}_n$ is of dimension $n > 1$.

9. $f'(x) = \sum_{k=1}^{n} \left\{ \prod_{\substack{j=1 \\ j \neq k}}^{n} (x - a_j) \right\} = \sum_{k=1}^{n} h_k(x)$. $f'(a_j) = h_j(a_j)$.

10. $\sigma_j(\sum_{k=1}^{n} b_k h_k(x)) = \sum_{k=1}^{n} b_k \sigma_j(h_k(x)) = \sum_{k=1}^{n} b_k \frac{1}{f'(a_j)} \sigma_{a_j}(h_k(x)) = b_j$. If $\sum_{k=1}^{n} b_k h_k(x) = 0$, then $\sigma_j(0) = b_j = 0$. Thus $\{h_1(x), \ldots, h_n(x)\}$ is linearly independent. Since $\sigma_i(h_j(x)) = \delta_{ij}$, the set $\{\sigma_1, \ldots, \sigma_n\}$ is a basis in the dual space.

11. By Exercise 5, $p(x) = \sum_{k=1}^{n} \sigma_k(p(x))h_k(x) = \sum_{k=1}^{n} \frac{p(a_k)}{f'(a_k)} h_k(x)$.

12. Let $\{\alpha_1, \ldots, \alpha_r\}$ be a basis of W. Since $\alpha_0 \notin W$, $\{\alpha_1, \ldots, \alpha_r, \alpha_0\}$ is linearly independent. Extend this set to a basis $\{\alpha_1, \ldots, \alpha_n\}$ where $\alpha_{r+1} = \alpha_0$. Let $\{\phi_1, \ldots, \phi_n\}$ be the dual basis. Then ϕ_{r+1} has the desired property.

13. Let $A = \{\alpha_1, \ldots, \alpha_n\}$ be a basis of V such that $\{\alpha_1, \ldots, \alpha_r\}$ is a basis of W. Let $\hat{A} = \{\phi_1, \ldots, \phi_n\}$ be the basis dual to A. Let $\psi_0 = \sum_{j=1}^{r} \psi(\alpha_j)\phi_j$. Then for each $\alpha_i \in W$, $\psi_0(\alpha_i) = \psi(\alpha_i)$. Thus ψ and ψ_0 coincide on all of W.

14. The argument given for Exercise 12 works for $W = \{0\}$. Since $\alpha \notin W$, there is a ϕ such that $\phi(\alpha) \neq 0$.

15. Let $W = \langle \beta \rangle$. If $\alpha \notin W$, by Exercise 12 there is a ϕ such that $\phi(\alpha) = 1$ and $\phi(\beta) = 0$.

IV–2

1. This is the dual of Exercise 5 of Section 1.
2. Dual of Exercise 6 of Section 1.
3. Dual of Exercise 12 of Section 1.
4. Dual of Exercise 14 of Section 1.
5. Dual of Exercise 15 of Section 1.

IV–3

1. $P = (P^{-1})^T = (P^T)^{-1}$.

2. $P = \begin{bmatrix} 1 & 1 & 0 \\ 1 & 0 & 1 \\ 1 & 1 & -1 \end{bmatrix}, \quad (P^{-1})^T = \begin{bmatrix} -1 & 2 & 1 \\ 1 & -1 & 0 \\ 1 & -1 & -1 \end{bmatrix}.$

   Thus $\hat{A}' = \{[-1 \quad 1 \quad 1], [2 \quad -1 \quad -1], [1 \quad 0 \quad -1]\}$.

3. $\{[1 \quad -1 \quad 0], [0 \quad 1 \quad -1], [0 \quad 0 \quad 1]\}$.

4. $\{[\frac{1}{2}, \quad \frac{1}{2} \quad -\frac{1}{2}], [-\frac{1}{2} \quad \frac{1}{2} \quad -\frac{1}{2}], [\frac{1}{2} \quad \frac{1}{2} \quad \frac{1}{2}]\}$.

5. $BX = B(PX') = (BP)X' = B'X'$.

IV–4

1. (a) $\{[1 \quad 1 \quad 1]\}$ (b) $\{[-1 \quad -1 \quad 0 \quad 1 \quad 1]\}$.

2. $[1 \quad -1 \quad 1]$.

3. Let W be the space spanned by $\{\alpha\}$. Since dim $W = 1$, dim $W^\perp = $ dim $V - 1 < $ dim V. Thus there is a $\phi \notin W^\perp$.

4. If $\phi \in T^\perp$, then $\phi\alpha = 0$ for all $\alpha \in T$. But since $S \subset T$, $\phi\alpha = 0$ for all $\alpha \in S$ also. Thus $T^\perp \subset S^\perp$.

5. Since $S^{\perp\perp}$ is a subspace containing S, $\langle S \rangle \subset S^{\perp\perp}$. By Exercise 4, $S \subset \langle S \rangle$, $S^\perp \supset \langle S \rangle^\perp$, $S^{\perp\perp} \subset \langle S \rangle^{\perp\perp} = \langle S \rangle$.

6. Since $S \subset S + T$, $S^\perp \supset (S + T)^\perp$. Similarly, $T^\perp \supset (S + T)^\perp$. Thus $(S + T)^\perp \subset S^\perp \cap T^\perp$. Since $S \cap T \subset S$, $(S \cap T)^\perp \supset S^\perp$. Similarly, $(S \cap T)^\perp \supset T^\perp$. Since $(S \cap T)^\perp$ is a subspace, $(S \cap T)^\perp \supset S^\perp + T^\perp$.

7. If S and T are subspaces, $S^{\perp\perp} = S$ and $T^{\perp\perp} = T$. Thus $S \cap T = S^{\perp\perp} \cap T^{\perp\perp} \supset (S^\perp + T^\perp)^\perp$ and hence $(S \cap T)^\perp \subset (S^\perp + T^\perp)^{\perp\perp} = (S^\perp + T^\perp)$. Similarly, $S + T = S^{\perp\perp} + T^{\perp\perp} \subset (S^\perp \cap T^\perp)^\perp$ and hence $(S + T)^\perp \supset (S^\perp \cap T^\perp)^{\perp\perp} = S^\perp \cap T^\perp$.

8. $S^\perp + T^\perp = (S \cap T)^\perp = \{0\}^\perp = \hat{V}$.

9. $S^\perp \cap T^\perp = (S + T)^\perp = V^\perp = \{0\}$.

10. By Exercises 9 and 10, $S^\perp + T^\perp = V$, and the sum is direct. For each $\psi \in \hat{V}$ define $\psi_1 \in \hat{S}$ by the rule: $\psi_1\alpha = \psi\alpha$ for all $\alpha \in S$. The mapping of $\psi \in \hat{V}$ onto $\psi_1 \in \hat{S}$ is linear and the kernel is $S^\perp$. By Exercise 13 of Section 1 every functional on S can be obtained in this way. Since $\hat{V} = S^\perp \oplus T^\perp$, $\hat{S}$ is isomorphic to $T^\perp$.

11. $f(t) = \phi(t\alpha + (1 - t)\beta)$ is a continuous function of t. Since $\phi(t\alpha + (1 - t)\beta) = t\phi(\alpha) + (1 - t)\phi(\beta)$, $f(t) > 0$ if α, $\beta \in S^+$ and $0 < t < 1$. Thus $\overline{\alpha\beta} \subset S^+$. If $\alpha \in S^+$ and $\beta \in S^-$, then $f(0) < 0$ and $f(1) > 0$. Since f is continuous, there is a $t, 0 < t < 1$, such that $f(t) = 0$.

IV–5

1. Let τ be a mapping of U into V and σ a mapping of V into W. Then, if $\phi \in \hat{W}$, we have for all $\xi \in U$, $(\widehat{\sigma\tau}(\phi))(\xi) = \phi[\sigma\tau(\xi)] = \phi\sigma[\tau(\xi)] = \hat{\sigma}(\phi)[\tau(\xi)] = \{\hat{\tau}[\hat{\sigma}(\phi)]\}(\xi)$.

2. $\{[-6 \quad 2 \quad 1]\}$.

3. $[1 \quad -2 \quad 1]$.

IV–8

1. $\begin{bmatrix} 1 & 2 & 6 \\ -1 & -1 & 0 \end{bmatrix}$.

2. $\begin{bmatrix} 1 & 3 & 5 \\ 3 & 5 & 7 \\ 5 & 7 & 9 \end{bmatrix} + \begin{bmatrix} 0 & -1 & -2 \\ 1 & 0 & -1 \\ 2 & 1 & 0 \end{bmatrix}$.

5. $\det A^T = \det A$ and $\det A^T = \det(-A) = (-1)^n \det A$. Thus $\det A = -\det A$.

7. $\sigma_f(\alpha) = 0$ if and only if $\sigma_f(\alpha)(\beta) = f(\alpha, \beta) = 0$ for all $\beta \in V$.

9. Let $\dim U = m$; $\dim V = n$. $f(\alpha, \beta) = 0$ for all $\beta \in V$ means $\alpha \in [\tau_f(V)]^\perp$ or, equivalently, $\alpha \in \sigma_f^{-1}(0)$. Thus $\rho(\tau_f) = \dim \tau_f(V) = m - \dim [\tau_f(V)]^\perp = m - \dim \sigma_f^{-1}(0) = m - \nu(\sigma_f) = \rho(\sigma_f)$.

10. If $m \neq n$, then either $\rho(\sigma_f) < m$ or $\rho(\tau_f) < n$.

11. U_0 is the kernel of σ_f and V_0 is the kernel of τ_f.

12. $m - \dim U_0 = m - \nu(\sigma_f) = \rho(\sigma_f) = \rho(\tau_f) = n - \nu(\tau_f) = n - \dim V_0$.

13. $0 = f(\alpha + \beta, \alpha + \beta) = f(\alpha, \alpha) + f(\alpha, \beta) + f(\beta, \alpha) + f(\beta, \beta) = f(\alpha, \beta) + f(\beta, \alpha)$.

14. If $AB = BA$, then $(AB)^T = (BA)^T = A^T B^T = AB$.

15. B is skew-symmetric.

16. (a) $(A^2)^T = A^T A^T = (-A)(-A) = A^2$;
(b) $(AB - BA)^T = (AB)^T - (BA)^T = B(-A) - (-A)B = AB - BA$;
(c) $(AB)^T = (BA)^T = A^T B^T = (-A)B = -AB$. If $(AB)^T = -AB$, then $AB = -(AB)^T = -B^T A^T = -B(-A) = BA$.

IV–9

1. (a) $\begin{bmatrix} 2 & \frac{3}{2} \\ \frac{3}{2} & 6 \end{bmatrix}$,

(c) $\begin{bmatrix} 1 & 1 & 2 \\ 1 & 3 & \frac{1}{2} \\ 2 & \frac{1}{2} & 7 \end{bmatrix}$,

(e) $\begin{bmatrix} 1 & 2 & 1 \\ 2 & 4 & 2 \\ 1 & 2 & 1 \end{bmatrix}$.

2. (a) $2x_1x_2 + \frac{3}{2}x_1y_2 + \frac{3}{2}y_1x_2 + 6y_1y_2$ (if (x_1, y_1) and (x_2, y_2) are the coordinates of the two points), (c) $x_1x_2 + x_1y_2 + x_2y_1 + 2x_1z_2 + 2x_2z_1 + 3y_1y_2 + \frac{1}{2}y_1z_2 + \frac{1}{2}y_2z_1 + 7z_1z_2$, (e) $x_1x_2 + 2x_1y_2 + 2x_2y_1 + 4y_1y_2 + x_1z_2 + x_2z_1 + z_1z_2 + 2y_1z_2 + 2y_2z_1$.

IV–10

1. (In this and the following exercises the matrix of transition P, the order of the elements in the main diagonal of $P^T BP$, and their values, which may be multiplied by perfect squares, are not unique. The following answers can only be thought of as representative possibilities.) (a) $P = \begin{bmatrix} 1 & -2 & 2 \\ 0 & 1 & -2 \\ 0 & 0 & 1 \end{bmatrix}$.

The diagonal of $P^T BP$ is $\{1, -3, 9\}$; (b) $\begin{bmatrix} 1 & 2 & 2 \\ 0 & -1 & -7 \\ 0 & 0 & 4 \end{bmatrix}$, $\{1, -4, 68\}$;

(c) $\begin{bmatrix} 0 & 1 & 0 & -1 \\ 1 & 0 & 0 & 2 \\ 0 & 0 & 1 & 2 \\ 0 & 1 & 0 & 1 \end{bmatrix}$, $\{1, 4, -1, -4\}$.

2. (a) $P = \begin{bmatrix} 1 & -3 \\ 0 & 4 \end{bmatrix}$, $\{2, 78\}$ (c) $\begin{bmatrix} 1 & 1 & -11 \\ 0 & -1 & 3 \\ 0 & 0 & 4 \end{bmatrix}$, $\{1, 2, 30\}$;

(e) $\begin{bmatrix} 1 & -2 & -1 \\ 0 & 1 & 0 \\ 0 & 0 & 1 \end{bmatrix}$, $\{1, 0, 0\}$.

IV–11

1. (a) $r = 2$, $S = 2$; (b) 2, 0; (c) 3, 3; (d) 2, 0; (e) 1, 1; (f) 2, 0; (g) 3, 1.

2. If $P = \begin{bmatrix} 1 & -b/2 \\ 0 & a \end{bmatrix}$, then $P^T BP = \begin{bmatrix} a & 0 \\ 0 & (a/4)(-b^2 + 4ac) \end{bmatrix}$.

3. There is a non-singular Q such that $Q^T AQ = I$. Take $P = Q^{-1}$.

4. Let $P = A^{-1}$.

5. There is a non-singular Q such that $Q^T AQ = B$ has r 1's along the main diagonal. Take $R = BQ^{-1}$.

6. For real $Y = (y_1, \ldots, y_n)$, $Y^T Y = \sum_{i=1}^{n} y_i^2 \geq 0$. Thus $X^T A^T AX = (AX)^T(AX) = Y^T Y \geq 0$ for all real $X = (x_1, \ldots, x_n)$.

7. If $Y = (y_1, \ldots, y_n) \neq 0$, then $Y^T Y > 0$. If $A \neq 0$, there is an $X = (x_1, \ldots, x_n)$ such that $AX = Y \neq 0$ (why?). Then we would have $0 = X^T A^T AX = Y^T Y > 0$.

9. If $A_i X \neq 0$ for any i, then $0 = X^T(\sum_{i=1}^{r} A_i^2)X = \sum_{i=1}^{r} X^T A_i^T A_i X > 0$. Thus $A_i X = 0$ for all X and $A_i = 0$.

IV–12

1. (a) $P = \begin{bmatrix} 1 & i \\ 0 & 1 \end{bmatrix}$, diagonal $= \{1, 0\}$ (b) $\begin{bmatrix} 1 & -1 + i \\ 0 & 1 \end{bmatrix}$, $\{1, -1\}$.

3–9. Proofs are similar to those for Exercises 3–9 of Section 11.

10. Similar to Exercise 14 of Section 8.

V–1

1. 6.   2. $2i$.

6. $(\alpha - \beta, \alpha + \beta) = (\alpha, \alpha) - (\beta, \alpha) + (\alpha, \beta) - (\beta, \beta) = \|\alpha\|^2 - \|\beta\|^2 = 0$.

7. $\|\alpha + \beta\|^2 = \|\alpha\|^2 + 2(\alpha, \beta) + \|\beta\|^2$.

9. $\left\{ \dfrac{1}{\sqrt{3}}(1, -1, 1), \dfrac{1}{\sqrt{2}}(1, 1, 0), \dfrac{1}{\sqrt{6}}(-1, 1, 2) \right\}$.

11. $\left\{ \dfrac{1}{\sqrt{2}}(0, 1, 1, 0), \tfrac{1}{3}(0, 2, -2, -1), \tfrac{1}{9}(-3, -2, 2, -8) \right\}$.

12. $x^2 - \frac{1}{3}$, $x^3 - 3x/5$.

13. (a) If $\sum_{j=1}^{m} a_j \xi_j = 0$, then $\sum_{j=1}^{m} a_j(\xi_i, \xi_j) = (\xi_i, \sum_{j=1}^{m} a_j \xi_j) = (\xi_i, 0) = 0$ for each i. Thus $\sum_{j=1}^{m} g_{ij} a_j = 0$ and the columns of G are dependent. (b) If $\sum_{j=1}^{m} g_{ij} a_j = 0$ for each i, then $0 = \sum_{j=1}^{m} a_j(\xi_i, \xi_j) = (\xi_i, \sum_{j=1}^{m} a_j \xi_j)$ for each i. Hence, $\sum_{i=1}^{m} \bar{a}_i(\xi_i, \sum_{j=1}^{m} a_j \xi) = (\sum_{i=1}^{m} a_i \xi_i, \sum_{j=1}^{m} a_j \xi_j) = 0$. Thus $\sum_{j=1}^{m} a_j \xi_j = 0$. (c) Let $A = \{\alpha_1, \ldots, \alpha_n\}$ be orthonormal, $\xi_j = \sum_{i=1}^{n} a_{ij} \alpha_j$. Then $g_{ij} = (\xi_i, \xi_j) = \sum_{k=1}^{n} \bar{a}_{ki} a_{kj}$. Thus $G = A^*A$ where $A = [a_{ij}]$.

V–2

1. If $\alpha = \sum_{j=1}^{n} a_j \xi_j$, then $(\xi_i, \alpha) = \sum_{j=1}^{n} a_j(\xi_i, \xi_j) = a_i$.

2. X is linearly independent. Let $\alpha \in V$ and consider $\beta = \alpha - \sum_{i=1}^{n} \frac{(\xi_i, \alpha)\xi_i}{\|\xi_i\|^2}$. Since $(\xi_j, \beta) = 0$, $\beta = 0$.

V–4

1. $((\sigma\tau)^*(\alpha), \beta) = (\alpha, \sigma\tau(\beta)) = (\sigma^*(\alpha), \tau(\beta)) = (\tau^*\sigma^*(\alpha), \beta)$.

2. $(\sigma(\alpha), \sigma(\alpha)) = (\alpha, \sigma^*\sigma(\alpha)) = 0$ for all α.

3. $(\sigma^*(\alpha), \beta) = (\alpha, \sigma(\beta)) = f(\alpha, \beta) = -f(\beta, \alpha) = -(\beta, \sigma(\alpha)) = -(\sigma(\alpha), \beta) = (-\sigma(\alpha), \beta)$.

5. Let ξ be an eigenvector corresponding to λ. Then $\lambda(\xi, \xi) = (\xi, \sigma(\xi)) = (\sigma^*(\xi), \xi) = (-\lambda\xi, \xi) = -\bar{\lambda}(\xi, \xi)$. Thus $(\lambda + \bar{\lambda}) = 0$.

6. σ is skew-symmetric.   7. σ is skew-symmetric.

8. Let $\xi \in W^{\perp}$. Then for all $\eta \in W$, $(\sigma^*(\xi), \eta) = (\xi, \sigma(\eta)) = 0$.

10. Since $(\pi^*)^2 = (\pi^2)^* = \pi^*$, π^* is a projection. $\xi \in K(\pi^*)$ if and only if $(\pi^*(\xi), \eta) = (\xi, \pi(\eta)) = 0$ for all η; that is, if and only if $\xi \in S^{\perp}$. Finally, $(\pi^*(\xi), \eta) = (\xi, \pi(\eta))$ vanishes for all ξ if and only if $\pi(\eta) = 0$; that is, if and only if $\eta \in T$. Then $\pi^*(V) \in T^{\perp}$. Since $\pi^*(V)$ and $T^{\perp}$ have the same dimension, $\pi^*(V) = T^{\perp}$.

11. $(\xi, \sigma(\eta)) = (\sigma^*(\xi), \eta) = 0$ for all η if and only if $\sigma^*(\xi) = 0$, or $\xi \in W^{\perp}$.

13. By Theorem 4.3, $V = W \oplus W^{\perp}$. By Exercise 11, $\sigma^*(V) = \sigma^*(W)$.

14. $\sigma^*(V) = \sigma^*(W) = \sigma^*\sigma(V)$. $\sigma(V) = \sigma\sigma^*(V)$ is the dual statement.

15. $\sigma^*(V) = \sigma^*\sigma(V) = \sigma\sigma^*(V) = \sigma(V)$.

16. By Exercises 15 and 11, $W^{\perp}$ is the kernel of σ^* and σ.

21. By Exercise 15, $\sigma(V) = \sigma^*(V)$. Then $\sigma^2(V) = \sigma\sigma^*(V) = \sigma(V)$ by Exercise 14.

V–5

1. Let ξ be the corresponding eigenvector. Then $(\xi, \xi) = (\sigma(\xi), \sigma(\xi)) = (\lambda\xi, \lambda\xi) = \lambda\bar{\lambda}(\xi, \xi)$.

3. It also maps ξ_2 onto $\pm \dfrac{\xi_1 - \xi_2}{\sqrt{2}}$.

4. For example, ξ_2 onto $\frac{1}{3}(2\xi_1 - 2\xi_2 + \xi_3)$ and ξ_3 onto $\frac{1}{3}(2\xi_1 + \xi_2 - 2\xi_3)$.

V–6

1. (a) and (c) are orthogonal. 2. (a).
5. (a) Reflection in a plane, (x_1, x_2-plane). (b) $180°$ rotation about an axis (x_3-axis). (c) Inversion with respect to the origin. (d) Rotation through θ about an axis (x_3-axis). (e) Rotation through θ about an axis (x_3-axis) and reflection in the perpendicular plane (x_1, x_2-plane). The characteristic equation of a third-order orthogonal matrix either has three real roots (the identity and (a), (b), and (c) represent all possibilities) or two complex roots and one real root ((d) and (e) represent these possibilities).

V–7

1. Change basis in V as in obtaining the Hermite normal form. Apply the Gram-Schmidt process to this basis.
2. If $\sigma(\eta_j) = \sum_{i=1}^{j} a_{ij}\eta_i$, then $\sigma^*(\eta_k) = \sum_{j=1}^{n} (\eta_j, \sigma^*))\eta_j = \sum_{j=1}^{n} (\sigma(\eta_j), \eta_k)\eta_i = \sum_{j=1}^{n} (\sum_{i=1}^{j}, a_{ij}\eta_i, \eta_k)\eta_i = \sum_{j=k}^{n} \bar{a}_{ki}\eta_j$.
3. Choose an orthogonal basis such that the matrix representing σ^* is in super-diagonal form.

V–8

1. (a) normal; (b) normal; (c) normal; (d) symmetric, orthogonal; (e) orthogonal, skew-symmetric; (f) Hermitian; (g) orthogonal; (h) symmetric, orthogonal; (i) skew-symmetric; normal; (j) non-normal; (k) skew-symmetric normal.
2. All but (c) and (j). 3. $A^T A = (-A)A = -A^2 = AA^T$.

5. $\begin{bmatrix} 0 & -1 & -i \\ 1 & 0 & -1 \\ i & 1 & 0 \end{bmatrix}$. 6. Exercise 1(c).

V–9

4. $(\sigma^*(\alpha), \beta) = (\alpha, \sigma(\beta)) = f(\alpha, \beta) = \overline{f(\beta, \beta)} = \overline{(\beta, \sigma(\alpha))} = (\sigma(\alpha), \beta)$.
5. $f(\alpha, \beta) = (\alpha, \sigma(\beta)) = (\sum_{i=1}^{n} a_i\xi_i, \sum_{j=1}^{n} b_j\sigma(\xi_j)) = (\sum_{i=1}^{n} a_i\xi_i, \sum_{j=1}^{n} b_j\lambda_j\xi_j) = \sum_{i=1}^{n} \bar{a}_i\beta_i\lambda_i$.
6. $q(\alpha) = f(\alpha, \alpha) = \sum_{i=1}^{n} |a_i|^2 \lambda_i$. Since $\sum_{i=1} |a_i|^2 = 1$, $\min \{\lambda_i\} \le q(\alpha) \le \max \{\lambda_i\}$ for $\alpha \in S$, and both equalities occur. If $\alpha \ne 0$, there is a real positive scalar a such that $a\alpha \in S$. Then $q(\alpha) = \dfrac{1}{a^2} q(a\alpha) \ge \min \{\lambda_i\} > 0$, if all eigenvalues are > 0.

V–10

1. (a) unitary, diagonal is $\{1, i\}$. (b) Hermitian, $\{2, 0\}$. (c) orthogonal, $\{\cos \theta + i \sin \theta, \cos \theta - i \sin \theta\}$, where $\theta = \arccos 0.6$. (d) Hermitian, $\{1, 4\}$. (e) Hermitian, $\{1, 1 + \sqrt{2}, 1 - \sqrt{2}\}$.

V–11

1. Diagonal is $\{15, -5\}$. (d) $\{9, -9, -9\}$. (e) $\{18, 9, 9\}$. (f) $\{-9, 3, 6\}$. (g) $\{-9, 0, 0\}$. (h) $\{1, 2, 0\}$. (i) $\{1, -1, -\frac{61}{11}\}$. (j) $\{3, 3, -3\}$. (k) $\{-3, 6, 6\}$.

2. (d), (h).

3. Since $P^T B P = B'$ is symmetric, there is an orthogonal matrix R such that $R^T B' R = B''$ is diagonal matrix. Let $Q = PR$. Then $(PR)^T A (PR) = R^T P^T A P R = R^T R = 1$ and $(PR)^T B (PR) = R^T P^T B P R = R^T B' R = B''$ is diagonal.

4. $A^T A$ is symmetric. Thus there is an orthogonal matrix Q such that $Q^T (A^T A) Q = D$ is diagonal. Let $B = Q^T A Q$.

5. Let P be given as in Exercise 3. Then $\det(P^T B P - xI) = \det(P^T(B - xI)P) = \det P^2 \cdot \det(B - xA)$. Since $P^T B P$ is symmetric, the solutions of $\det(B - xA) = 0$ are also real.

7. Let $A = \begin{bmatrix} a & b \\ c & d \end{bmatrix}$. Then A is normal if and only if $b^2 = c^2$ and $ab + cd = ac + bd$. If b or c is zero, then both are zero and A is symmetric. If $a = 0$, then $b^2 = c^2$ and $cd = bd$. If $d \neq 0$, $c = d$ and A is symmetric. If $d = 0$, either $b = c$ and A is symmetric, or $b = -c$ and A is skew-symmetric.

8. If $b = c$, A is symmetric. If $b = -c$, then $a = d$.

9. The first part is the same as Exercise 5 of Section 3. Since the eigenvalues of a linear transformation must be in the field of scalars, σ has only real eigenvalues.

10. $\sigma^2 = -\sigma^*\sigma$ is symmetric. Hence the solutions of $|A^2 - xI| = 0$ are all real. Let λ be an eigenvalue of σ^2 corresponding to ξ. Then $(\sigma(\xi), \sigma(\xi)) = (\xi, \sigma^*\sigma(\xi)) = (\xi, \sigma^2(\xi)) = -\lambda(\xi, \xi)$. Thus $\lambda \leq 0$. Let $\lambda = -\mu^2$.
$$\sigma(\eta) = \frac{1}{\mu}\sigma^2(\xi) = \frac{1}{\mu}(-\mu^2\xi) = -\mu\xi. \qquad \sigma^2(\eta) = -\mu\sigma(\xi) = -\mu^2\eta. \qquad (\xi, \eta) =$$
$$\left(\xi, \frac{1}{\mu}\sigma(\xi)\right) = \frac{1}{\mu}(\sigma^*(\xi), \xi) = \frac{1}{\mu}(-\sigma(\xi), \xi) = -(\xi, \eta).$$

11. $\sigma(\xi) = \mu\eta$, $\sigma(\eta) = -\mu\xi$.

12. The eigenvalues of an isometry are of absolute value 1. If $\sigma(\xi) = \lambda\xi$ with λ real, then $\sigma^*(\xi) = \lambda\xi$, so that $(\sigma + \sigma^*)(\xi) = 2\lambda\xi$.

13. If $(\sigma + \sigma^*)(\xi) = 2\mu\xi$ and $\sigma(\xi) = \lambda\xi$, then $\lambda = \pm 1$ and $2\mu = 2\lambda = \pm 2$. Since $(\xi, \sigma(\xi)) = (\sigma^*(\xi), \xi) = (\xi, \sigma^*(\xi))$. $2\mu(\xi, \xi) = (\xi, (\sigma + \sigma^*)(\xi)) = 2(\xi, \sigma(\xi))$. Thus $|\mu| \cdot \|\xi\|^2 = |(\xi, \sigma(\xi))| \leq \|\xi\| \cdot \|\sigma(\xi)\| = \|\xi\|^2$, and hence $|\mu| \leq 1$. If $|\mu| = 1$, equality holds in Schwarz's inequality and this can occur if and only if $\sigma(\xi)$ is a multiple of ξ. Since ξ is not an eigenvector, this is not possible.

14. $(\xi, \eta) = \dfrac{1}{\sqrt{1 - \mu^2}} \{(\xi, \sigma(\xi)) - \mu(\xi, \xi)\} = 0$. Since $\sigma(\xi) + \sigma^*(\xi) = 2\mu\xi$,

$\sigma^2(\xi) + \xi = 2\mu\sigma(\xi)$. Thus
$$\sigma(\eta) = \frac{\sigma^2(\xi) - \mu\sigma(\xi)}{\sqrt{1 - \mu^2}} = \frac{\mu\sigma(\xi) - \xi}{\sqrt{1 - \mu^2}} = \frac{\mu^2\xi + \mu\sqrt{1 - \mu^2}\eta - \xi}{\sqrt{1 - \mu^2}}$$
$$= -\sqrt{1 - \mu^2} + \mu\eta.$$

15. Let ξ_1, η_1 be associated with μ_1, and ξ_2, η_2 be associated with μ_2, where $\mu_2 \neq \mu_2$. Then $(\xi_1, (\sigma + \sigma^*)(\xi_2)) = (\xi_1, 2\mu_2\xi_2) = ((\sigma + \sigma^*)(\xi_1), \xi_2) = (2\mu_1\xi_1, \xi_2)$. Thus $(\xi_1, \xi_2) = 0$.

V–12

2. $A + A^* = \begin{bmatrix} \frac{2}{3} & -\frac{4}{3} & 0 \\ -\frac{4}{3} & \frac{2}{3} & 0 \\ 0 & 0 & -\frac{2}{3} \end{bmatrix}$. The eigenvalues of $A + A^*$ are $\{-\frac{2}{3}, -\frac{2}{3}, 2\}$.

Thus $\mu = -\frac{1}{3}$ and 1. To $\mu = 1$ corresponds an eigenvector of A which is $\frac{1}{\sqrt{2}}(1, -1, 0)$. An eigenvector of $A + A^*$ corresponding to $-\frac{2}{3}$ is $(0, 0, 1)$. If this represents ξ, the triple representing η is $\frac{1}{\sqrt{2}}(1, 1, 0)$. The matrix representing σ with respect to the basis $\{(0, 0, 1), \frac{1}{\sqrt{2}}(1, 1, 0), \frac{1}{\sqrt{2}}(1, -1, 0)\}$ is

$$\begin{bmatrix} -\frac{1}{3} & -\frac{2\sqrt{2}}{3} & 0 \\ \frac{2\sqrt{2}}{3} & -\frac{1}{3} & 0 \\ 0 & 0 & 1 \end{bmatrix}.$$

VI–1

1. (1) $(1, 0, 1) + t_1(1, 1, 1) + t_2(2, 1, 0)$; $\begin{bmatrix} 1 & -2 & 1 \end{bmatrix}(x_1, x_2, x_3) = 2$.
   (2) $(1, 2, 2) + t_1(2, 1, -2) + t_2(2, -2, 1)$; $\begin{bmatrix} 1 & 2 & 2 \end{bmatrix}(x_1, x_2, x_3) = 9$.
   (3) $(1, 1, 1, 2) + t_1(0, 1, 0, -1) + t_2(2, 1, -2, 3)$; $\begin{bmatrix} 1 & 0 & 1 & 0 \end{bmatrix}(x_1, x_2, x_3, x_4) = 2$, $\begin{bmatrix} -2 & 1 & 0 & 1 \end{bmatrix}(x_1, x_2, x_3, x_4) = 1$.
2. $(6, 2, -1) + t(-6, -1, 4)$; $\begin{bmatrix} 1 & -2 & 1 \end{bmatrix}(x_1, x_2, x_3) = 2$, $\begin{bmatrix} 1 & 2 & 2 \end{bmatrix}(x_1, x_2, x_3) = 9$.
3. $L = (2, 1, 2) + \langle (0, 1, -1), (-3, 0, \frac{3}{2}) \rangle$.
4. $\frac{1}{25}(1, 1) + \frac{11}{25}(-6, 7) + \frac{13}{25}(5, -6) = (0, 0)$.
6. Let L_1 and L_2 be linear manifolds. If $L_1 \cap L_2 \neq \emptyset$, let $\alpha_0 \in L_1 \cap L_2$. Then $L_1 = \alpha_0 + S_1$ and $L_2 = \alpha_0 + S_2$, where S_1 and S_2 are subspaces. Then $L_1 \cap L_2 = \alpha_0 + (S_1 \cap S_2)$.
7. Clearly, $\alpha_1 + S_1 \subset \alpha_1 + \langle \alpha_2 - \alpha_1 \rangle + S_1 + S_2$ and $\alpha_2 + S_2 = \alpha_1 + (\alpha_2 - \alpha_1) + S_2 \subset \alpha_1 + \langle \alpha_2 - \alpha_1 \rangle + S_1 + S_2$. On the other hand, let $\alpha_1 + S$ be the join of L_1 and L_2. Then $L_1 = \alpha_1 + S_1 \subset \alpha_1 + S$ implies $S_1 \subset S$, and $L_2 = \alpha_2 + S_2 \subset \alpha_1 + S$ implies $\alpha_2 - \alpha_1 + S_2 \subset S$. Since S is a subspace, $\langle \alpha_2 - \alpha_1 \rangle + S_1 + S_2 \subset S$. Since, $\alpha_1 + S$ is the smallest linear manifold containing L_1 and L_2, $\alpha_1 + S = \alpha_1 + \langle \alpha_2 - \alpha_1 \rangle + S_1 + S_2$.
8. If $\alpha_0 \in L_1 \cap L_2$, then $L_1 = \alpha_0 + S_1$ and $L_2 = \alpha_0 + S_2$. Thus $L_1 J L_2 = \alpha_0 + \langle \alpha_0 - \alpha_0 \rangle + S_1 + S_2 = \alpha_0 + S_1 + S_2$. Since $\alpha_1 \in L_1 J L_2$, $L_1 J L_2 = \alpha_1 + S_1 + S_2$.
9. If $\alpha_2 - \alpha_1 \in S_1 + S_2$, then $\alpha_2 - \alpha_1 = \beta_1 + \beta_2$ where $\beta_1 \in S_1$ and $\beta_2 \in S_2$. Hence $\alpha_2 - \beta_2 = \alpha_1 + \beta_1$. Since $\alpha_1 + \beta_1 \in \alpha_1 + S_1 = L_1$ and $\alpha_2 - \beta_2 \in \alpha_2 + S_2 = L_2$, $L_1 \cap L_2 \neq \emptyset$.
10. If $L_1 \cap L_2 \neq \emptyset$, then $L_1 J L_2 = \alpha_1 + S_1 + S_2$. Thus $\dim L_1 J L_2 = \dim (S_1 + S_2)$. If $L_1 \cap L_2 = \emptyset$, then $L_1 J L_2 = \alpha_1 + \langle \alpha_2 - \alpha_1 \rangle + S_1 + S_2$ and $L_1 J L_2 \neq \alpha_1 + S_2 + S_2$. Thus $\dim L_1 J L_2 = \dim (S_1 + S_2) + 1$.

VI–2

1. If $Y = [y_1 \quad y_2 \quad y_3]$, then Y must satisfy the conditions $Y(1, 1, 0) \geq 0$, $Y(1, 0, -1) \geq 0$, $Y(0, -1, 1) \geq 0$
2. $\{[1 \quad -1 \quad 1], [1 \quad -1 \quad -1], [1 \quad 1 \quad 1]\}$.
3. $\{(1, 1, 0), (1, 0, -1), (0, -1, 1), (0, 1, 1), (1, -1, 0), (1, 1, 1)\}$.
4. $\{(1, 0, -1), (0, -1, 1), (0, 1, 1)\}$. Express the omitted generators in terms of the elements of this set.
5. $\{[-1 \quad -1 \quad 2], [1 \quad 1 \quad -1], [1 \quad -1 \quad -1]\}$.
6. $\{(1, 0, 1), (3, 1, 2), (1, -1, 0)\}$.
7. Let $Y = [-1 \quad -1 \quad 2]$. Since $YA \geq 0$ and $YB = -2 < 0$, $(1, 1, 0) \notin C_2$.
8. Let $Y = [-2 \quad -2 \quad 1]$.   9. Let $Y = [1 \quad -2]$.
10. This is the dual of Theorem 2.14.   11. Let $Y = [2 \quad 2 \quad 1]$.
12. Let $\hat{A} = \{\phi_1, \ldots, \phi_n\}$ be the dual basis to A. Let $\phi_0 = \sum_{i=1}^{n} \phi_i$. Then ξ is semi-positive if and only if $\xi \geq 0$ and $\phi_0 \xi > 0$. In Theorem 2.11, take $\beta = 0$ and $g = 1$. Then $\psi\beta = 0 < g = 1$ for all $\psi \in \hat{V}$ and the last condition in (2) of Theorem 2.11 need not be stated. Then the stated theorem follows immediately from Theorem 2.11.
14. Using the notation of Exercise 13, either (1) there is a semi-positive ξ such that $\sigma(\xi) = 0$, that is, $\xi \in W$, or (2) there is a $\psi \in \hat{V}$ such that $\hat{\sigma}(\psi) > 0$. Let $\phi = \hat{\sigma}(\psi)$. For $\xi \in W$, $\phi\xi = \hat{\sigma}(\psi)\xi = \psi\sigma(\xi) = 0$. Thus $\phi \in W^{\perp}$.
15. Take $\beta = 0$, $g = 1$, and $\phi = \sum_{i=1}^{n} \phi_i$, where $\{\phi_1 \ldots, \phi_n\}$ is the basis of $P^{\perp}$.

VI–3

1. Given A, B, C, the primal problem is to find $X \geq 0$ which maximizes CX subject to $AX \leq B$. The dual problem is to find $Y \geq 0$ which minimizes YB subject to $YA \leq C$.
2. Given A, B, C, the primal problem is to find $X \geq 0$ which maximizes CX subject to $AX = B$. The dual problem is to find Y which minimizes YB subject to $YA \geq C$.
6. The pivot operation uses only the arithmetic operations permitted by the field axioms. Thus no tableau can contain any numbers not in any field containing the numbers in the original tableau.
7. Examining Equation (3.7) we see that $\phi\xi'$ will be smaller than $\phi\xi$ if $c_k - d_k < 0$. This requires a change in the first selection rule. The second selection rule is imposed so that the new ξ' will be feasible, so this rule should not be changed. The remaining steps constitute the pivot operation and merely carry out the decisions made in the first and second steps.
8. Start with the equations

$$2x_1 + x_2 + x_3 \qquad\qquad = 6$$
$$4x_1 + x_2 \qquad + x_4 \qquad = 10$$
$$-x_1 + x_2 \qquad\qquad + x_5 = 3.$$

The first feasible solution is $(0, 0, 6, 10, 3)$. The optimal solution is $(2, 2, 0, 0, 3)$. The numbers in the indicator row of the last tableau are $(0, 0, -\frac{3}{2}, -\frac{1}{2}, 0)$.
9. The last three elements of the indicator row of the previous exercise give $y_1 = \frac{3}{2}, y_2 = \frac{1}{2}, y_3 = 0$.

10. The problem is to minimize $6y_1 + 10y_2 + 3y_3 + My_4 + My_5$, where M is very large, subject to

$$
\begin{aligned}
y_1 + y_2 + y_3 + y_4 \quad\quad - y_6 \quad\quad &= 2 \\
2y_1 + 4y_2 - y_3 \quad\quad + y_5 \quad\quad - y_7 &= 5.
\end{aligned}
$$

When the last tableau is obtained, the row of $\{d_i\}$ will be [6 10 0 2 2 -2 -2]. The fourth and fifth elements correspond to the unit matrix in the original tableau and give the solution $x_1 = 2$, $x_2 = 2$ to Exercise 8.

11. Maximum $= 12$ at $x_1 = 2$, $x_2 = 5$.

12. $(0, 0)$, $(0, 2)$, $(1, 4)$, $(2, 5)$.

15. X and Y meet the test for optimality given in Exercise 14, and both are optimal.

16. $AX = B$ has a non-negative solution if and only if min $FZ = 0$.

17. $(\frac{79}{164},\ 0,\ \frac{3}{164},\ \frac{13}{164}\ 0,\ \frac{69}{164})$.

VI-6

1. (a) $A = (-1)\begin{bmatrix} \frac{1}{2} & -\frac{1}{2} \\ -\frac{1}{2} & \frac{1}{2} \end{bmatrix} + 3\begin{bmatrix} \frac{1}{2} & \frac{1}{2} \\ \frac{1}{2} & \frac{1}{2} \end{bmatrix}.$

(b) $A = 2\begin{bmatrix} \frac{4}{5} & \frac{2}{5} \\ \frac{2}{5} & \frac{1}{5} \end{bmatrix} + (-3)\begin{bmatrix} \frac{1}{5} & -\frac{2}{5} \\ -\frac{2}{5} & \frac{4}{5} \end{bmatrix},$

(c) $A = 5\begin{bmatrix} \frac{1}{5} & \frac{2}{5} \\ \frac{2}{5} & \frac{4}{5} \end{bmatrix},$ $A = 2\begin{bmatrix} \frac{9}{10} & \frac{3}{10} \\ \frac{3}{10} & \frac{1}{10} \end{bmatrix} + (-8)\begin{bmatrix} \frac{1}{10} & -\frac{3}{10} \\ -\frac{3}{10} & \frac{9}{10} \end{bmatrix},$

(e) $A = 3\begin{bmatrix} \frac{4}{5} & \frac{2}{5} \\ \frac{2}{5} & \frac{1}{5} \end{bmatrix} + (-7)\begin{bmatrix} \frac{1}{5} & -\frac{2}{5} \\ -\frac{2}{5} & \frac{4}{5} \end{bmatrix}.$

2. $A = \begin{bmatrix} 0 & -\frac{11}{2} & -11 \\ 0 & 7 & 14 \\ 0 & -3 & -6 \end{bmatrix} + 2\begin{bmatrix} 0 & 18 & 42 \\ 0 & -6 & -14 \\ 0 & 3 & 7 \end{bmatrix} + 3\begin{bmatrix} 1 & -\frac{25}{2} & -31 \\ 0 & 0 & 0 \\ 0 & 0 & 0 \end{bmatrix}.$

4. $A = AE_1 + AE_2$ where $E_1 = \begin{bmatrix} 3 & 3 & 0 \\ -2 & -2 & 0 \\ -2 & -3 & 1 \end{bmatrix}$, $E_2 = \begin{bmatrix} -2 & -3 & 0 \\ 2 & 3 & 0 \\ 2 & 3 & 0 \end{bmatrix},$

and $AE_1 = 2E_1 + N_1$, where $N_1 = \begin{bmatrix} -3 & -6 & 3 \\ 2 & 4 & -2 \\ 1 & 2 & -1 \end{bmatrix}.$

6. $e^A = e^2\begin{bmatrix} 0 & -3 & 3 \\ 0 & 2 & -2 \\ -1 & -1 & 0 \end{bmatrix} + e\begin{bmatrix} -2 & -3 & 0 \\ 2 & 3 & 0 \\ 2 & 3 & 0 \end{bmatrix}.$

VI-8

1. $V = \dfrac{k}{2}\left\{(x_2 - x_1)^2 + \left[\tfrac{1}{2}(x_2 - x_3) - \dfrac{\sqrt{3}}{2}(y_2 - y_3)\right]^2 + \left[\tfrac{1}{2}(x_3 - x_1) + \dfrac{\sqrt{3}}{2}(y_3 - y_1)\right]^2\right\}.$

2. $\dfrac{M}{2} I.$

4. These displacements represent translations of the molecule in the plane containing it. They do not distort the molecule, do not store potential energy, and do not lead to vibrations of the system.

VI-9

2. $\pi = (124)$, $\sigma = (234)$, $\sigma\pi = (134)$, $\rho = \sigma\pi^{-1} = (12)(34)$.

3. Since the subgroup is always one of its cosets, the alternating group has only two cosets in the full symmetric group, itself and the remaining elements. Since this is true for both right and left cosets, its right and left cosets are equal.

5. $(e) = D((123)) = D((132)) = [1]$, $D((12)), = D((13)) = D((23)) = [-1]$.

7. The matrix appearing in (9.7) is $H = 4 \begin{bmatrix} 2 & -1 \\ -1 & 2 \end{bmatrix}$. The matrix of transition is then $P = \dfrac{1}{2\sqrt{6}} \begin{bmatrix} 3 & 1 \\ 0 & 2 \end{bmatrix}$.

8. G is commutative if and only if every element is conjugate only to itself. By Theorem 9.11 and Equation 9.30, each $n_r = 1$.

10. Let $\zeta = e^{2\pi i/n}$ be a primitive nth root of unity. If a is a generator of the cyclic group, let $D^k(a) = [\zeta^k]$, $k = 0, \ldots, n - 1$.

11.

C_4	e	a	a_2	a^3		v	e	a	b	c
D^1	1	1	1	1		D^1	1	1	1	1
D^2	1	i	-1	$-i$		D^2	1	1	-1	-1
D^3	1	-1	1	-1		D^3	1	-1	1	-1
D^4	1	$-i$	-1	i		D^4	1	-1	-1	1

12. By Theorem 9.12 each $n_r \mid p^2$. But $n_r = p$ or p^2 is impossible because of (9.30) and the fact that there is at least one representation of dimension 1. Thus each $n_r = 1$, and the group is commutative.

16. Since ab must be of order 1 or 2, we have $(ab)^2 = e$, or $ab = b^{-1}a^{-1}$. Since a and b are of order 1 or 2, $a^{-1} = a$ and $b^{-1} = b$.

17. If G is cyclic, let a be a generator of G, let $\zeta = e^{\pi i/4}$, and define $D^k(a) = [\zeta^k]$. If G contains an element a of order 4 and no element of higher order, then G contains an element b which is not a power of a. b is of order 2 or 4. If b is of order 4, then b^2 is of order 2. If b^2 is a power of a, then $b^2 = a^2$. Then $c = ab$ is of order 2 and not a power of a. In any event there is an element c of order 2 which is not a power of a. Then G is generated by a and c. If G contains elements of order 2 and no higher, let a, b, c be three distinct elements of order 2. They generate the group. Hints for obtaining the character tables for these last two groups are given in Exercises 21, 25, and 26.

18. The character tables for these two non-isomorphic groups are identical.

29. $1^2 + 1^2 + 2^2 + 3^2 + 3^2$.

30. $\mathfrak{U}_4$ contains C_1 (the conjugate class containing only the identity), C_3 (the class containing the eight 3-cycles), and C_5 (the class containing the three pairs of interchanges).

VI–10

2. The permutation (123) is represented by

$$
\begin{bmatrix}
0 & 0 & 0 & 0 & -\tfrac{1}{2} & -\sqrt{3}/2 \\
0 & 0 & 0 & 0 & \sqrt{3}/2 & -\tfrac{1}{2} \\
-\tfrac{1}{2} & -\sqrt{3}/2 & 0 & 0 & 0 & 0 \\
\sqrt{3}/2 & -\tfrac{1}{2} & 0 & 0 & 0 & 0 \\
0 & 0 & -\tfrac{1}{2} & \sqrt{3}/2 & 0 & 0 \\
0 & 0 & \sqrt{3}/2 & -\tfrac{1}{2} & 0 & 0
\end{bmatrix}.
$$

The representation of (12) is

$$
\begin{bmatrix}
0 & 0 & -1 & 0 & 0 & 0 \\
0 & 0 & 0 & 1 & 0 & 0 \\
-1 & 0 & 0 & 0 & 0 & 0 \\
0 & 1 & 0 & 0 & 0 & 0 \\
0 & 0 & 0 & 0 & -1 & 0 \\
0 & 0 & 0 & 0 & 0 & 1
\end{bmatrix}.
$$

3. $c_1 = 1,\ c_2 = 1,\ c_3 = 2.$

4. $\xi_1 = (-\sqrt{3}/2, -\tfrac{1}{2}, \sqrt{3}/2, -\tfrac{1}{2}, 0, 1).$

The displacement is a uniform expansion of the molecule.

5. $\xi_2 = (-\tfrac{1}{2}, \tfrac{3}{2}, -\tfrac{1}{2}, -\tfrac{3}{2}, 1, 0).$

This displacement is a rotation of the molecule without storing potential energy.

6. $\{\xi_3 = (1, 0, 1, 0, 1, 0),\ \xi_4 = (0, 1, 0, 1, 0, 1)\}$. This subspace consists of translations without distortion in the plane containing the molecule.

7. This subspace is spanned by the vectors ξ_5 and ξ_6 given in Exercise 8.

ξ_5 $\qquad\qquad\qquad$ ξ_6

Notation

345

Index